Praise for *A Theory of Everyone*

'*A Theory of Everyone* flavorfully mixes a stunning breadth of scholarship with an impressive knowledge of pop-culture and current issues, boldly going where most social scientists fear to tread. Lucidly discussing ideas surrounding IQ, race, sex differences, inheritance taxes, religion, Microsoft and even monogamy, readers are treated to a fascinating intellectual flight that thoughtfully offers many new perspectives on old issues. Buckle up!'

Joseph Henrich, Professor of Human Evolutionary Biology, Harvard University and author of *The WEIRDest People in the World* and *The Secret of Our Success*

'The best book I've read in a decade. A sprightly page-turner that entertains with specifics, astonishes with universals, and reframes the big issues facing humanity'   Robert Klitgaard, author of *Controlling Corruption* and *Tropical Gangsters*

'Buzzing with ideas, *A Theory of Everyone* encourages us to rethink what it is to be human. A compelling and essential read for anyone interested in building a better, more sustainable future'

David Bodanis, author of *The Art of Fairness*

'There is a truly wonderful idea at the heart of this book: that by exchanging things and thoughts, human beings became capable of doing and knowing far more than their meagre brains would have otherwise made possible. It is not an entirely new idea, but Michael Muthukrishna explores its extraordinary and hopeful implications with rich and thrilling energy'

Matt Ridley, author of *The Evolution of Everything*

'Do you know your own species? You might think so, but Muthukrishna will make you think again. With clarity, humor, and energy, he opens new vistas on how genes and cultures shaped who we are and how we can improve our lives together. *A Theory of Everyone* is for everyone'

Walter Sinnott-Armstrong, Chauncey Stillman Professor of Practical Ethics, Duke University, author of *Think Again: How to Reason and Argue*

'*A Theory of Everyone* uses the latest social science research to answer the critical question of how all human communities can be made to work better. Magisterial in scope and practical in application, this book should be required reading for CEOs, community organisers, head teachers, and Presidents'

Jamie Heywood, CEO of zolar and former
head of Uber, Northern & Eastern Europe

'*A Theory of Everyone* is your guide to some of the most important advances in the social sciences, written by a foremost researcher, beautifully illustrated, and positively overflowing with fascinating facts and ideas'

Erez Yoeli, Director, Applied Cooperation
Initiative, MIT, co-author of *Hidden Games*

'Michael brings the reader up to date on this powerful theoretical framework – including much of his own innovative work on corruption, cooperation, and collective intelligence – and thoughtfully discusses how this framework can be applied to address pressing societal issues, ranging from diversity to taxation to free speech'

Moshe Hoffman, Visiting Lecturer on Economics,
Harvard University, co-author of *Hidden Games*

'This book, which I read with great fascination, shows how we can move beyond neoclassical economics with a firmer foundation in the natural sciences and energy. This is extremely important as the world soon, and Europe now, increasingly faces critical energy shortages. I hope this book helps more people understand the critical importance of energy in generating our current affluence, and its diminution as a probable root cause of future inflation. A failure to understand these relations is likely to cause our societies to become impossible to govern'

Charles Hall, ESF Foundation Distinguished Professor
at State University of New York, inventor of the EROI
metric & author of *Energy and the Wealth of Nations*

A Theory of Everyone

The New Science of Who We Are,
How We Got Here, and Where We're Going

MICHAEL MUTHUKRISHNA

The MIT Press
Cambridge, Massachusetts
London, England

Illustrations on pages 19, 24, 42, 57, 73, 83, 84, 115, 116, 165, 251, and 293 created by Veronika Plant in collaboration with Michael Muthukrishna.

The MIT Press would like to thank the anonymous peer reviewers who provided comments on drafts of this book. The generous work of academic experts is essential for establishing the authority and quality of our publications. We acknowledge with gratitude the contributions of these otherwise uncredited readers.

This book was set in Janson Text LT by Palimpsest Book Production Ltd., Falkirk, Stirlingshire, UK. Printed and bound in the United States of America.

Library of Congress Cataloging-in-Publication Data

Names: Muthukrishna, Michael, author.
Title: A theory of everyone : the new science of who we are, how we
got here, and where we're going / Michael Muthukrishna.
Description: Cambridge, Massachusetts : The MIT Press, [2023] |
Includes bibliographical references and index.
Identifiers: LCCN 2023007538 (print) | LCCN 2023007539 (ebook) |
ISBN 9780262048378 (hardcover) | ISBN 9780262375795 (epub) |
ISBN 9780262375788 (pdf)
Subjects: LCSH: Cooperation. | Social change. |
Evolutionary psychology. | Human behavior.
Classification: LCC HD2963 .M89 2023 (print) | LCC HD2963 (ebook) |
DDC 334—dc23/eng/20230509
LC record available at https://lccn.loc.gov/2023007538
LC ebook record available at https://lccn.loc.gov/2023007539

10 9 8 7 6 5 4 3 2 1

publication supported by a grant from
The Community Foundation for Greater New Haven
as part of the **Urban Haven Project**

To
Robert Mathes,
Alexandra Lucille,
and Gabriele Elizabeth,
may this book guide you in leaving the
world a better place than you found it

Contents

Introduction

There are these two young fish swimming along and they happen to meet an older fish swimming the other way, who nods at them and says 'Morning, boys. How's the water?' And the two young fish swim on for a bit, and then eventually one of them looks over at the other and goes, 'What the hell is water?'

The novelist David Foster Wallace told this story in his 2005 commencement address at Kenyon College. Wallace's point was that there are some things that are so familiar, so integral to one's perception of the world, that we don't notice them any more. They melt into our experience, permeate our senses, and become part of what we might call the background conditions of life.

This book is about the species called *Homo sapiens*, who are in precisely this position. From ancient bacteria-like life forms, humans have evolved through various laws that we shall explore in this book. But the forces that shape our thinking, our economies, and our societies have become invisible to us. And this leaves us with a deep, potentially existential problem. If we do not know who we are and how we got here, we cannot choose where we go next. If we cannot perceive the forces that shape us, we are impotent to shape these forces.

Fish can't live without water. It's part of their background, cheerfully ignored in favor of what biologists call the four Fs of life: feeding, fighting, fleeing, and mating. But if the water suddenly changes then the fish will notice.

We can feel in our bones that the world is breaking – that something is wrong. America – until recently widely regarded as one of the most successful modern democracies – is teetering. Civil conversations in which we agree to disagree have given way to enraged moralizing aimed at those who hold beliefs different to our own. Polarization is on the rise almost everywhere. We are in the midst of yet another economic crisis, and for the first time in a long time the lives of our children will be less abundant, lower in opportunity, and just more difficult than our own. A war in the West – civil or international – once more feels plausible, if not inevitable. One crisis seems to melt into another.

A number of books have been published predicting a looming cataclysm, and even if they seem alarmist, they speak to a sense that we face trouble ahead. Many thinkers, myself included, believe that this century could be the most important century in human history.

In this book we will start with the single most important quantity in the universe: energy. Energy is an abstract notion even in physics but it is central to life; in fact to everything. We cannot move without energy, cannot reproduce, cannot do anything at all. And yet since the Industrial Revolution, which unlocked unfathomable quantities of unexploited energy in the form of fossil fuels, we have stopped thinking about it.

To simplify only a little, we have come to take energy for granted. We flip a switch and lights go on. We fill our cars with fuel and go where we need to go. We microwave leftovers without ever worrying about where the food came from or how easy it is to feed ourselves.

We thought that energy was the gift that would keep on giving. Our models of the economy have what we earn and what we buy continually cycling between companies and people like a perpetual motion machine with no inputs. Our models of economic growth hide the limits on technology and the ultimate constraints on labor and capital. We imagined that we could keep doing more and more, growing our economies, becoming more prosperous, and developing ever more fantastical technologies, without realizing that we were exhausting the cheap fuel that made it all possible.

Energy is to the human species what water is to the fish in Wallace's metaphor.

But this is not a book solely about energy. It's not just about fossil fuels versus renewables or electric versus gas cars. It's about the way in which energy breakthroughs across the grand timescale of our species have led to periods of abundance that have in turn led to increases in the number of people and the scale at which they work together, which in turn have led to scarcity and conflict. This dance of energy and evolution eventually turns abundance to scarcity, but along the way it offers opportunities for critical social and technological breakthroughs. When these breakthroughs raise the energy ceiling then we reach a new threshold – a new period of abundance begins. The details of how this happens are the key to ensuring an abundant future.

Energy may be the key to understanding our current predicament, but to grasp how to use it more effectively and to harness new sources, we also have to understand the fundamental dynamics of human behavior. Why do we sometimes go to war and at other times work in harmony? Why are we both cruel and kind? And what determines which of these instincts win out?

These are just some of the questions we will tackle in this book. By understanding the constraints imposed by energy, we will transform the way we think about economics, politics, and conservation. But by deepening our understanding of human behavior, we will develop original insights about how to more effectively exploit energy in ways that help increase our prosperity and reduce the risks of conflict, both within and between societies. By the end we will have a theory that encompasses both; a unified theory of human affairs. A theory of everyone.

But why, you may ask, am I writing this book? What's my story?

Let's go back to 1997. I am a young boy crouching in my bedroom furtively sneaking peeks through my window at the angry men armed with M16s screeching by in military trucks. They are on their way to Papua New Guinea's Parliament House, an arrow-shaped edifice adorned with carvings and artwork reflecting traditional architecture and the hundreds of tribes without a common language now forged into a nation. Our house, set within the confines of a barbed-wire walled compound, is just 500 yards to the south of Parliament. I try to calm the cries of my eight-year-old sister as gunfire, looting, and explosions turn PNG's capital, Port Moresby, from an everyday level of deadly threat – on an ordinary day, armed robbery and rape are so common that they are rarely reported – to a violent coup that later became known as the Sandline Affair.

Sandline referred to the British mercenary corporation, Sandline International. Prime Minister Julius Chan, the Australian-educated son of a Chinese trader and a native from PNG's New Ireland province, had lost control of the military and the Bougainville region. His solution: bypass the army by hiring mercenaries. Violent protests and a military coup followed. Chan was replaced by Bill Skate, a well-known gang leader who was caught on tape boasting, 'If I tell my gang members to kill, they kill . . . I'm the godfather.' In many countries Skate would be a wanted criminal; in PNG he was the new prime minister.

Papua New Guinea, like it's pidgin English creole official language, is a chimera. Australians had brought a British parliamentary system to the most linguistically diverse country on earth. The 5.5 million people who lived in Papua New Guinea were split by over 840 distinct languages. Australia and Papua New Guinea are both rich in natural resources and share the same governmental institutions. But unlike Australia, Papua New Guinea was and is poor, violent, and unstable. As I grew up, I needed to understand why.

During my time in PNG I had a front-row ticket to a terrifying clash of Western institutions and tribal politics. But it wasn't my only formative experience, or even my first.

In Sri Lanka, where I was born, I learned how two peoples who looked so similar to outsiders – Tamils and Sinhalese – could come to hate each other. I learned how ordinary everyday existence can be, even during a civil war. Wallace was right about fish ignoring the water until it changes. Oppression, military checkpoints, or even the ever-present danger of explosions and sudden chaos can all fade into the background until punctured by the reality of violence. My grandmother worked across the road from the Central Bank when it was rammed by a Tamil Tiger truck loaded with 440 pounds of explosives. That was the first time I saw my father cry. First from the uncertainty and then from the relief when he brought her home, still wearing clothes soaked with blood from exploded shards of glass; shaken but alive.

I spent most of my childhood in Botswana, South Africa's northern neighbor. My memories are filled with the dusty streets of Gaborone, camping deep in the Kalahari Desert under the unobstructed majesty of the Milky Way, and the splendor of South Africa during the nineties. I loved the beautiful plateaus of Table Mountain, framing the sprawl of Cape Town as it met the sea; the smells of fusion foods – biltong jerky, braai BBQ, potjiekos stew, bunny chow curry – devoured on Durban's expansive white sandy beaches; the bustle of Johannesburg; the excitement of Sun City. I also remember the exhilaration and trepidation as South Africa transitioned from apartheid. Splashed across every television and newspaper was the powerful image of the last white Afrikaner president, a somber F. W. de Klerk, his face set with a faint smile next to the beaming new President Nelson Mandela, their arms raised and hands clasped together as they ushered in a new era filled with hope and uncertainty.

The waters of the world may be very different, but they are all part of the same ocean.

I was in London when bombs exploded on three busy Underground trains and the top deck of an iconic red double-decker bus. It was a coordinated attack designed to terrorize ordinary British people on an ordinary Thursday on their ordinary commute to work. But what struck me most was the identity of the bombers: ordinary British citizens. Unlike 9/11 four years prior, this was not an act committed by outsiders. Three of the terrorists, Hasib Hussain, Mohammad Sidique Khan, and Shehzad Tanweer, were born in Britain. The fourth, Germaine Lindsay, had moved to the UK when he was five. I couldn't shake the sound of Khan's thick Yorkshire accent as he explained in perfect English, 'Until we feel secure, you will be our targets.' The 'you' he refers to in his grainy video are his fellow Brits; the 'we' are a people who live thousands of miles away in countries that he had only briefly visited yet to whom he feels a greater connection.

These were second-generation migrants, roughly my age and who looked a lot like me. Yet somehow Khan and others like him felt like outsiders in their own country. What had gone wrong? What could be done better?

These were formative memories set against my otherwise unremarkable, if peripatetic life, living in these countries and also Australia, Canada, America, and most recently Britain. When you live in so many places, you see how we differ and how we are connected. We swim in different shoals but we are fish in the same body of water.

For the last two decades I've been obsessed with understanding these differences and these connections. Why was Botswana less corrupt and on many metrics more successful than South Africa? Why was Papua New Guinea so much poorer and less peaceful than Australia? What are the differences between the multicultural and immigration policies of Australia, Canada, the United States, and the countries of Europe?

When I graduated from high school, I was on a quest to figure this all out. I enrolled in an engineering degree, which seemed like a secure, well-paid career. Unlike law and medicine, it also had international accreditation, a great fit for someone with itchy feet.

Engineering was fun and I was good at it – but engineering alone didn't seem like it could answer the questions that possessed me, so I enrolled in a dual degree. In parallel with courses on calculus, discrete math, and machine learning, I took courses in economics, political science, biology, philosophy, and psychology. In each discipline I found solutions to a piece of the puzzle.

I ended up majoring in psychology in my second degree. Psychology was asking the most relevant questions about human behavior. But it seemed to flout what I was learning about the scientific method in engineering and philosophy of science. There was little attempt to falsify predictions and the idea of selecting between theories – model selection – was difficult without a theory of human behavior. Evolutionary biology was a good candidate to develop that theory of human behavior.

When evolutionary biological theories were applied to humans, they could make good predictions about human behavior, but evolutionary theories developed in psychology independent of biology relied on too many imprecise assumptions about ancestral conditions and for some reason didn't use the powerful biological mathematical toolkit. Could we build better models of human behavior?

I eventually gave up trying to answer these questions – it just seemed too difficult. I focused instead on a dissertation about smart home technologies. But the questions kept bugging me, bubbling away at the back of my mind.

Around 2007 I saw Al Gore's climate documentary, *An Inconvenient Truth*. Gore argued that we urgently needed to reduce carbon emissions. As our planet heated, so too would politics, and as places became too dry, too hot, or under water, millions of people would need to flee as their homes and livelihoods disappeared. The more I read, the more convinced I became that Gore was right about the problem but too optimistic about the solution. Would we really slow the economy to save the planet? This wasn't like the successful ban on chlorofluorocarbons (CFCs) from deodorants and refrigerators that was helping to close the hole in the ozone layer. In that case, alternatives were available – the market had a solution. But for climate change, Gore was asking us to cut back our production, wealth, and lifestyles in a world where every country was trying to outcompete every other country, every company was trying to outcompete every other company, and every person wanted a better lifestyle than their neighbors'. It seemed to me that in the absence of a global government and credible enforcement, no amount of documentaries or finger wagging would work.

It made sense that we should still try to reduce our carbon footprint, but it made even more sense to also start preparing for a climate-changed world. And neither Greenpeace nor Captain Planet were asking us to pay attention to the latter. In the meantime, reports from the Pentagon and the Intergovernmental Panel on Climate Change (IPCC) were predicting climate fluctuations and the mass movement of displaced people from places like the Middle East, Bangladesh, and the South Pacific. I could see research on climate engineering to deal with carbon capture and wild weather, but not enough adaptation research on how to deal with mass-scale refugee resettlement or ensuing conflicts over scarce resources. We desperately needed a science of culture. One mature enough to be trusted and that could be used to develop social technologies. It was in engineering that I finally saw a breakthrough that might help us get there. And it came from the design of smart homes.

Smart homes require what are called control systems. As the name suggests, these systems control functions such as temperature and lighting. A thermostat is a simple control system that manages heating and cooling in many homes. It measures the temperature and turns on heaters or air conditioners to keep a house at the right temperature.

Control systems rely on a body of math called control theory – the math of feedback loops. My insight was that perhaps control theory could be applied to model the feedback loops of people trying to influence one another to develop a science of norms. And from a science of norms we might begin to develop a science of culture and institutions. I needed to find someone who studied the psychological foundations of culture. What do you do when you want to find someone who studies the psychological foundations of culture? You google *the psychological foundations of culture*.

This led me to a book with that very title edited by an evolutionary psychologist called Mark Schaller, from the University of British Columbia. I emailed Mark describing my background and goals and asking if we could meet. Mark suggested I also meet his colleagues, cultural psychologist Steve Heine, social psychologist of religion Ara Norenzayan, and in particular, former aerospace engineer turned anthropologist then appointed in economics and psychology, Joe Henrich.

Joe was working in an area called dual inheritance theory and cultural evolution, mathematical frameworks for modeling the co-evolution of human genetics and culture (our dual inheritance) and the evolution of culture and institutions. He was applying these models to psychology and economics. After a short conversation I knew that between Joe, Mark, Steve, Ara, and their colleagues, I would have an ideal team to help me tackle the questions I so desperately wanted to answer.

After completing my dissertation at the University of British Columbia a year early, having cross-trained in evolutionary biology, statistics and data science, economics, and psychology, I moved to Harvard's Department of Human Evolutionary Biology and then to the London School of Economics, where I am currently professor of economic psychology and affiliate in developmental economics and data science.

Working across multiple disciplines has allowed me to take a non-disciplinary – or perhaps 'undisciplined' – approach, pulling on strands deep within psychology, economics, biology, anthropology, and elsewhere, tying them together into a tapestry that reveals who we are, how we got here, and where we're going.

Once you see the links between energy, innovation, cooperation, and evolution, you can't unsee them. These are underlying laws of life that apply to bacteria and businesses, cells and societies. Remember the parable of the blind men who encounter an elephant and try to describe it? One feels its trunk, others its tusks, body, or tail. From their individual vantage points each describes the elephant as a snake, spear, wall, or rope. By necessity, different disciplines focus on different parts of the system, but when you put the pieces together you can't ignore the elephant in the room: energy, the innovations that lead to more efficient use of energy, our capacity to cooperate for mutual benefit in the quest for greater energy, and the forces of evolution that shape all three.

But this book is not about coal, it's not about oil, it's not even about renewables or nuclear. It is about the future of humanity; about how each of our actions contributes to a collective brain. It's about how *Homo sapiens* can reach the next level of abundance that leads to a better life for everyone and perhaps one day a civilization that spans the galaxy. And it's about the things that stand in the way of getting where we need to be and what we can do to overcome them. Because today we stand on the shore of a sea of possibilities. We must be careful in how we address the coming waves ahead of us; waves that threaten our now precarious fossil-fueled civilizations.

In the first part of this book I zoom in on the details of the human animal and the theory of everyone. We'll discuss how one goes about building a science of us; how energy, innovation, cooperation, and evolution have shaped all of life and all human activity; how we learn from one another, what shapes our intelligence, how we can become more creative and increase our capacity for innovation, how we work together and build institutions, and how the laws of life have shaped every aspect of us and our societies. That is, we will see how an unremarkable African ape ended up able to make Zoom calls across the planet.

In Part II we will zoom out to explain why the world is changing, what we can do about it, and why the twenty-first century may be *the most important* in human history. It is imperative that we reach a new level of energy abundance. But there are barriers standing in our way. Polarization and corruption threaten to tear us apart. Inequality can (though not necessarily) lead to inefficient allocation of our energy budget. These in turn lead to an inefficient allocation of talent and opportunity, stifling the next creative explosion that we so desperately need. There are many diagnoses for the problems we face, but fewer solutions. Yet solutions do exist. These solutions include how we can design better immigration policies or target taxes on unproductive money. Other solutions are more radical but worth pursuing, such as start-up cities and programmable politics. In essence, we will discover how this comprehensive theory of everyone can lead to practical policy applications – things you and I can advocate for to ensure that our children and all *Homo sapiens* who follow – have a future.

PART I

Who We Are and How We Got Here

Who are we? How did we get here? These most profound of questions have been pondered by generations of philosophers, theologians, and college roommates. Scientists have also been studying these questions, developing better theories and more convincing evidence for how our species evolved, the secret to our creativity and intelligence, how we work together to create corporations, governments, and other structures within our societies, and how these elements interact with one another. This 'science of us' is studied in different ways by different disciplines within the human and social sciences. But up until very recently, it was most accurate to describe both the human and social sciences as 'young' sciences.

A young science behaves like a child. It spends most of its time observing the world and coming up with explanations for what it sees, some wilder and less credible than others. It gets into everything, plays with switches, knobs, and runs whatever experiments it comes up with. But it doesn't yet know how to properly make sense of what it sees, how things connect with one another, or how to confidently act in the world.

A young physics was Galileo thinking comets were atmospheric disturbances akin to aurora borealis, the northern lights. A young chemistry was when the wise sages across Eurasia tried to turn lead into gold. A young biology was Lamarck assuming that giraffes grew their necks through generations of stretching to reach leaves high on trees.

Like those of a child, the claims of a young science aren't always trustworthy. Did young Alex really see a fox or was it the neighbor's dog? Are supermarket Santas really Santa or one of Santa's helpers? Will a watermelon grow in my stomach if I swallow a seed?

Nutritional science, for example, is still a young science. Even its most carefully planned studies can't be trusted. Different investigations uncover different findings, and the young science seems to be perpetually changing its mind.

Coffee is good for you.

Actually it's bad.

But not as bad as red wine, which is actually quite good for your health.

Unlike bacon, which will cause cancer.

Or maybe not.

For a science to become a mature adult science, it first needs to go through puberty. Like human puberty, this is an exciting, embarrassing, and often awkward affair, and requires some major changes. Chief among these is the discovery of an overarching theoretical framework that can sift sense from nonsense, make trustworthy and useful predictions, and offer pathways from discoveries to technologies. Some scientists and philosophers of science would argue it is only really after the discovery of this mature theoretical framework that a science can even call itself a science. As the French polymath Henri Poincaré put it, 'Science is built up of facts, as a house is built of stones; but an accumulation of facts is no more a science than a heap of stones is a house.'

The house in Poincaré's analogy is the theoretical framework, the architecture that tells us what to expect, what not to expect, why and how things work, and how to intervene. The theories that can be derived from a mature theoretical framework are like underground subway maps, road maps, and topographic maps, reducing the reality of the world in different ways to highlight and hide different information so as to get us to where we need to go.

Sometimes theory comes before the data, the data distinguishing between competing theories. This is what happened when Einsteinian physics revealed the limits of Newtonian physics at the turn of the twentieth century. Newtonian physics works very well for calculating how fast a tennis ball will fall based on the angle, speed, and spin with which it's hit given the acceleration caused by the pull of the Earth's gravity. Einsteinian physics taught us that the Earth's gravity isn't pulling the tennis ball at all; it's warping the fabric of space-time.

According to Einstein's theory, a large mass like the Sun 'bends' space-time. Newton's theory makes no such prediction. This warping of space-time leads to phenomena such as 'gravitational lensing' where the light of distant stars appears to be in different locations when they pass by a large mass like the Sun. We don't normally see this lensing because stars aren't visible during the day when the Sun is out, but a solar eclipse in 1919 allowed scientists to observe what the Sun's gravity was doing to the light from distant stars. The stars around the Sun appeared to have moved from their normal positions in the night sky. The shift was much larger than Newton's theory predicted, but exactly in the positions predicted by Einstein's theory. Sorry Newton!

Sometimes data seems disconnected and theory helps make sense of it. The discovery of elements in the periodic table turned alchemy into chemistry. The discovery of Darwinian natural selection turned butterfly collecting into modern biology. When we get the theory right, it completely revolutionizes our understanding of what was previously confusing, inconsistent, and seemingly unrelated.

The human and social sciences are going through puberty. Its curves are showing; its muscles are growing. We are in the midst of a scientific revolution on the scale of Newtonian and Einsteinian physics, the periodic table, and Darwinian evolution. This scientific revolution is a theory of human behavior that, when combined with theories of social evolution, is close to being a theory of everyone. This theory of everyone is as profound as the revolutions in these other now adult sciences. It is a revolution that is bringing order to chaos and laying the path from science to technology – in this case, policy applications. For the first time, it is enabling us to see the causes of the problems we face and what we need to do to overcome them. The human and social sciences are moving from alchemy to chemistry.

Once upon a time the physical world seemed chaotic. It was a world of apples falling to the ground to be closer to the Earth and capricious gods creating the weather. Then folks like Newton, Maxwell, and Einstein brought order to this chaos. It's astonishing that at a time of muskets, whale oil, and horse-drawn carriages, Maxwell was able to write down equations for electromagnetism. This was before Edison and Tesla developed technology that allowed us to control electricity. A popular meme of Maxwell's equations says:

And God said:

$$\nabla \cdot D = \rho$$

$$\nabla \cdot B = 0$$

$$\nabla \times E = -\frac{\partial D}{\partial t}$$

$$\nabla \times H = J + \frac{\partial D}{\partial t}$$

And then there was light.

Today, physics is arguably the oldest of the grown-up sciences. The weather is still difficult to predict, but at least we now understand how it and other aspects of the physical world work. This understanding allows us to predict a clear day and the motions of celestial bodies precisely enough to launch a rocket and land a spacecraft on Mars. Thanks to the laws of physics, we can go beyond intuitions based on life experience or purely the results of past experiments to make predictions and distinguish what is unusual and interesting – a particle decay producing more of one particle than another – from what is unusual and probably wrong – neutrinos travelling faster than the speed of light.

In 2011 neutrinos travelling faster than the speed of light was precisely what was found by the CERN OPERA experiment when it fired particles through a tunnel from Switzerland to Italy. The Swiss particles arrived earlier than the Italians expected, indeed faster than the speed of light. Nothing is supposed to travel faster than the speed of light and so physicists knew something was up. The experiment received a large amount of scrutiny not because it violated intuitions nor because physicists don't like being wrong, but because the speed of light, rather than being a purely experimental discovery or an isolated theory, is at the heart of a rich and mature theoretical framework that explains so much of our world. If neutrinos were traveling faster than the speed of light then the science of physics would be shattered across so many subdisciplines. It turned out to be a measurement error.

We see the same pattern in the history of chemistry. Once upon a time the chemical world seemed chaotic. Metals mixed with liquid to create gases. Sulfur, carbon, and saltpeter could be combined to create gunpowder. But no matter how hard even Newton himself tried, we couldn't turn lead into gold. Then folks like Lavoisier, Mendeleev, and Meyer brought order to this chaos. The periodic table and an understanding of elements and chemical reactions arrived. Alchemy grew up and became chemistry.

More complex compounds, such as proteins, are still difficult to predict, but at least we now know how they work. We know why lithium fizzes and sodium explodes in water, and why lead cannot be turned into gold. This understanding allows us to develop material sciences, a world of plastics, and to turn crude oil into fuel, medicines, and Vaseline. Thanks to the periodic table, we can go beyond intuitions based on life experience or purely the results of past experiments to make predictions and distinguish what is unusual and interesting – an AI predicting a protein's shape from its amino acid sequence – from what is unusual and probably wrong – paraffin dissolving in water.

And we see the same pattern in the history of biology. Once upon a time the biological world seemed chaotic. There seemed to be no rhyme or reason for why some animals laid eggs and others had live births or why the peacock had a giant elaborate tail while the peahen was a drab brown. Then folks like Darwin, Fisher, Wright, and Hamilton brought order to this chaos. Biology grew out of merely counting, classifying, and measuring and became a mature science.

Species are still difficult to predict and ecologies are still chaotic, but at least we now know how they work. This understanding allows us to develop gene editing and mRNA vaccines. Thanks to the theory of evolution, we can go beyond intuitions based on life experience or purely the results of past experiments to make predictions and distinguish what is unusual and interesting – a new human species – from what is unusual and probably wrong – a mammal fossil found in the Precambrian geological record.

We now see the same pattern in the human and social sciences.

A scientific revolution is starting to bring order to the chaotic world of human affairs. Everything is starting to make more sense. Sapiens are still difficult to predict, but at least we now know the rules by which we work. We know the rules that govern how people decide whom to trust and learn from, how organizations and societies discover new innovations in norms and technologies, and the rules that shape our actions in helping or harming others and determining who is 'us' and who is 'them'. We can use these rules to improve ourselves, our technologies, our governments, companies, schools, and societies; to develop strategies, policies, and interventions – social technologies – to chart a better future. We can go beyond intuitions based on life experience or purely the results of past experiments to make predictions and distinguish what is unusual and interesting from what is unusual and probably wrong.

The theoretical framework – the periodic table – of human behavior and social change is studied under different labels that describe its different elements.

Dual inheritance theory refers to the two lines of inheritance humans have – genes and culture. Our ability to transcend instincts and become cleverer than our short lifetimes should allow is a result of acquiring accumulated cultural information from our societies – beliefs, values, technologies, institutions, know-how. Culture makes us a new kind of animal.

Culture–gene co-evolution refers to the way genes have adapted to our cultures and cultures to our genes.

The *extended evolutionary synthesis* refers to the extension of the biological theoretical framework beyond genes into socially transmitted information and environments.

And *cultural evolution* refers to the way in which companies, countries, and other aspects of our societies change, adapt, and evolve.

Physicists refer to a unifying theory that connects diverse effective theoretical frameworks, such as general relativity – the physics of the very vast – and quantum mechanics – the physics of the extremely tiny – as a *theory of everything*. I shall refer to this revolutionary body of work that links genes, culture, learning, and the environment as a *theory of everyone*.

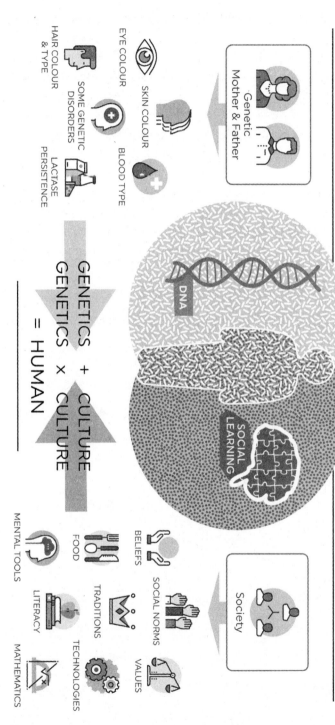

DUAL INHERITANCE THEORY

GENETIC Inheritance

Genetic Mother & Father

- EYE COLOUR
- HAIR COLOUR & TYPE
- SKIN COLOUR
- SOME GENETIC DISORDERS
- BLOOD TYPE
- LACTASE PERSISTENCE

DNA

GENETICS + CULTURE
GENETICS × CULTURE
= HUMAN

SOCIAL LEARNING

CULTURAL Inheritance

Society

- BELIEFS
- SOCIAL NORMS
- VALUES
- FOOD
- TRADITIONS
- TECHNOLOGIES
- MENTAL TOOLS
- LITERACY
- MATHEMATICS

At the heart of this theory of everyone is a quest for the capture and control of energy. All organisms, including humans, harness the energy around them – from the rays of the sun to the movement of the wind and water – to evolve. Humans have evolved an entirely new way of capturing and controlling energy through cultural evolution. But ultimately energy is at the heart of all that we do and all we can do. And when we see energy in this light, we are like the fish finally seeing the water around us. Suddenly our experiences and potential futures come into clearer vision.

All life has been on this quest for energy since its beginning. This quest is so central to all that happens that the way in which energy is captured and controlled, through genetic mutations, new technologies, cooperative norms and institutions, and evolutionary dynamics, is best described as the *laws of life*.

I

Laws of Life

One of my favorite literary genres is what I like to refer to as The One Thing That Explains Everything (TOTTEE). Surrounded by strangers on the overcrowded London Underground, I am transported by the awe-inspiring grandeur of *Guns, Germs, and Steel* expounding on how the east–west geographic orientation of Eurasia led to it outcompeting north–south-oriented America and Africa; or the masterful epoch-spanning epic storytelling of *Sapiens* explaining how our capacity for imagination turned us into gods. I feel informed and empowered, enraptured and entertained. But when I turn the last page or hear the final end credits about how Audible hopes I've enjoyed this audiobook, I am left unsatisfied.

You and I know – and so too do the authors of these books – that the world is complicated. Arrows of causality showing what causes what split, rejoin, point in multiple directions, and even feed back on each other. No *one thing* explains everything.

The power of a good TOTTEE book comes from highlighting a fundamental force that shapes our world. But in the real world, especially when we move from explanation to application, there are many forces that must be understood in their relationship to one another.

Geography, for example, is no doubt important, but the thin strip of land splitting Korea into North and South cannot explain the sudden disjuncture between wealthy South Korea, its brightly lit urban infrastructure visible even from space, and poor North Korea, a dark patch on the map separating South Korea from China. To explain that disjuncture, you might need to understand government institutions.

Institutions are important, but they can't explain why different ethnic groups have different outcomes in the same country. For that you might need to understand culture and intergroup competition.

Culture is important, but it can't explain how multilingual, multi-cultural, multi-religious Singaporeans became the second richest people on the planet after Luxembourg. For that you might need to understand history. But history is complicated and, unlike science, doesn't offer clear causal explanation and application by itself.

This book is TOTTEE adjacent. Rather than offering a single 'one thing' to explain everything, it offers a framework that unifies the many forces that shape all of life. These laws of life govern multiple scales, from single-celled bacteria competing for a patch of nutrients to societies of businesses competing over market share. Of course, bacteria and businesses differ in many details, and those details matter for how we should intervene in the world. The applied goal is to identify where, when, and how we should intervene.

Is the lever we need to pull a political matter, a market challenge, a technological gap, a cultural mismatch, a psychological barrier, or some combination? All are shaped by the laws of life. To effectively intervene, we need a periodic table for *Homo sapiens* and we need to be able to see the big picture and then zoom in and out of different parts. What the laws of life offer is a systems-level, ultimate view.

Systems-level ultimate explanations

Systems-level thinking is *essential* to the creation of permanent change. One of many cautionary tales about what happens when these inter-connections are ignored is the story of cane toads in Australia.

In the early twentieth century the new nation of Australia had a burgeoning sugar-cane industry. The cane crops flourished in the fertile soil, plentiful sunshine, and tropical climate of Queensland, Australia's Sunshine State. The only problem was the native Australian cane beetle, which was so fond of sugar cane, it bore its name. Cane beetle larvae feast on sugar-cane roots, stunting or even killing the plant. Something had to be done. The scientists at the Bureau of Sugar Experiment Stations saw an obvious problem and an obvious solution. Kill the cane beetle.

But how could they do this without hurting the plants?

In 1935, 101 assassins made their way from Hawai'i to Australia.

Not Dalmatians, but cane toads. The toads liked their new home. So much so that their numbers have grown to hundreds of millions if not over a billion. But they have found more than the cane beetle to their liking, and have thereby wrought havoc on the isolated Australian ecosystem. The cane toad is poisonous from egg to tadpole to toadlet to toad, and is thus dangerous to both those they eat and those that eat them. Australian native species, isolated on the island for so long, have no defenses against such a successful predator. Crocodile versus cane toad? Cane toads kill the crocodile even after they're dead.

Today, cane toads are everywhere in Queensland. They are a constant reminder of how a single-minded solution that ignores the broader system can wreak havoc and create new problems requiring even more solutions. Thinking at a systems level is difficult but necessary for successful solutions. One approach is to separate ultimate from proximate explanations.

The ultimate–proximate distinction is an important concept in evolutionary biology. Similar concepts are found in most sciences. In the business world, it's similar to root cause analysis and the five whys. The classic example used to explain the ultimate–proximate distinction in biology is the question of why animals enjoy sex. Here's a proximate explanation: sex is pleasurable and people prefer pleasurable things. This is a kind of explanation but it's tautological. Here's a better explanation: sex releases a chemical cocktail of motivation and loving desire – dopamine, endorphins, oxytocin – all associated with pleasure, love, and trust. Together, they reinforce the behavior. This is a better explanation that invokes neuroscience. But it's still proximate. All we've really done is given more details about the mechanism of 'sex is pleasurable'. What it doesn't tell us is why people prefer sex to, say, banging their heads against a wall. Understanding the full range of alternatives provides an ultimate explanation.

Here's a systems-level, ultimate explanation for sex: sex is associated with procreation and preferences are transmitted between generations. Imagine a world where some animals associate sex with pleasure and others associate wall-head-banging with pleasure. Animals that associated sex with pleasure had more offspring and so left more descendants who themselves enjoyed sex and had more offspring. And the animals that associated banging their heads against a wall with

EXPLAINING BEHAVIORAL PHENOMENA

Enjoy SEX? Love CHOCOLATE? Like CHILI in your food?

PROXIMATE explanation

It FEELS GOOD

CHEMICALS released — ENDORPHINS, OXYTOCINS, SEROTONIN, DOPAMINE

WHY?

It TASTES delicious

CHEMICALS released — DOPAMINE, SEROTONIN

ULTIMATE explanation

But WHY does THAT HAPPEN?

A long time ago......

"He's interested in many things, but having sex is rarely one of them."

BEHAVIOR not SUSTAINABLE

"The kids are asleep... How about we have some alone time?"

continued BEHAVIOR

But WHY does THAT HAPPEN?

UNripe FRUIT — less vitamins, less nutrients

I like this FRUIT — less healthy offspring

RIPE FRUIT — more vitamins, more nutrients

I like this FRUIT — HEALTHIER offspring

YES, I think CHILIES are GREAT

NO, I think CHILIES are TOO HOT!

MOST LIKELY
☑ grew up with spicy food
☑ lives somewhere with culturally spicy food
☑ embraces higher levels of spice

MOST LIKELY
☑ didn't grow up with spicy food
☑ cultural cuisine isn't spicy
☑ avoids any levels of spiciness

But WHY use CHILI?

IT BURNS!
IT BURNS!
IT BURNS!
IT DOESN'T BURN

IT BURNS!
IT BURNS! but I LIKE IT!

☑ lower disease
☑ less food spoilage

CHILI use ZONES

pleasure and preferred it to sex? Well, they didn't have children and are no longer with us. Nor are their preferences.

Ultimate explanations tell us why almost all of us like chocolate and only some people like chili peppers. When applied to the social world, taking a systems-level view of connected ultimate causes tells us where the ultimate source of a problem lies.

For example, people often blame politicians for inflation and rising energy prices. But if inflation and rising energy prices are occurring around the world then leaders of any specific country are unlikely to be at fault. An underlying ultimate explanation is more likely.

The first law of life is the law of energy.

The ultimate ceiling on the biomass and complexity of all life forms is the availability of energy. Energy is what makes matter come alive. Energy is what matter uses to scurry, fight, and make more of itself. Energy is why you and I can enjoy a life that would be the envy of the richest monarch a few centuries ago. It is thanks to our unprecedented control of energy that you can sit in your climate-controlled home; you can walk, drive, or fly to other places; you can eat a sandwich; you can reproduce; you can read this book. It is the density of stored energy in different sources, the availability and abundance of these sources, the power – energy transferred per second – and the efficiency with which we can find and use these sources that constrain what we are capable of doing. But while energy is in theory abundant – the light falling on Earth, the heat from its core escaping through geothermal vents, the rushing of rivers, or flammable fossil fuels – we need innovations that allow us to use these sources to do work. Indeed, when energy is abundant, as it was after the discovery of fossil fuels, it fades into the background as a law. It becomes like the fish's water. Instead, we begin to focus on innovation alone, particularly innovations in efficiency, to do more with the same amount of energy.

The second law of life is the law of innovation.

Life innovates new ways to efficiently capture and control available energy in competition with other life. These innovations include biological changes such as photosynthesis or the ability to digest meat or milk, technologies such as farming or the engine, and social organizations such as corporations and countries. These innovations, whether biological, technological, or social, increase the amount of energy available by discovering ways to use more of it or use it more efficiently.

Take the history of lighting as an example. Candles turn less than 0.04% of the chemical energy in wax into light. The rest is lost as heat. Edison's first light bulbs were still less than 1% efficient – more than 99% of the energy was lost to heat. Modern incandescent lights are at most around 10% efficient. Fluorescent lights around 15%. LED lights continue to become more efficient and can in theory approach 100% efficiency. LEDs are brighter, last longer, and consume less energy – so much so that the admonition to 'turn off the lights' is now one of the least effective energy-saving behaviors. All these technological innovations allowed us to do more with less. But each innovation, both biological and technological, required organisms to work together.

The third law of life is the law of cooperation.

When there is sufficient energy to exploit and more that is reachable with the help of just a few more helpers, we can make the leap and work together to capture it. Cells can bind together into complex organisms; regions can bind into nation-states; corporations can sign deals, mergers, and acquisitions. Innovations that unlock more energy through new sources or greater efficiency increase the *space of the possible*.

The more energy unlocked, the larger this space. The larger this space, the larger the possible scale of cooperation. A larger space allows for larger animals and larger states.

Think of it this way. Imagine you are starting a business. If you could run the company all by yourself and keep all the profits to yourself, you should. But by working with others – hiring employees, signing agreements with vendors, bringing investors on board – you can increase your chances of success and size of profits, even if you need to share them with others. The optimal level of cooperation is the level where you have a high probability of winning the spoils *and* your share of the spoils is larger than the share you could have got in a smaller group or larger group.

We typically don't calculate this consciously. Instead, we get there through trial and error, partial causal models, and selection. Companies with unnecessary employees fail or make less profits. Companies with too few also fail or make less profits. In other words, the level of cooperation is reached through an evolutionary process.

The fourth law of life is the law of evolution.

The exploitation of energy, the way in which we innovate, and the mechanisms of cooperation are typically not intelligently designed solutions but rather the product of millions of attempts, with successes outcompeting failures.

Energy, innovation, cooperation, and evolution are four laws; four interconnected ways to carve up the world and explain how geography, institutions, culture, and history have played out. We will revisit them in more detail at the end of this chapter. For now, let me show you how these laws manifest in each of our lives and then in the history of all life.

Energy, innovation, cooperation, and evolution in the everyday

We all face a trade-off in how much time to allocate to work, to our families, to our friends, and to ourselves. In tackling this trade-off, I personally am obsessed with efficiency. I've spent years figuring out how to maximize my use of the twenty-four hours I have each day, the fifty-two weeks I have each year, and the eighty or so years the average Western male gets for a lifetime. This obsession includes how to efficiently distribute my cognition in a way that prevents my to-do lists and project prioritization tools from getting in the way of focused deep work; how to hack my psychological limitations by doing things like leaving work unfinished at the end of a day to make it easier to restart the next (an application of the Zeigarnik effect); accepting that while it is inevitable that I will procrastinate, it is not inevitable what I will procrastinate on – I can procrastinate by working on low priority things that do actually need to get done – productive procrastination; and even how much time to spend on optimization itself and how much free time I need to ensure there's space for spontaneity.

My obsession even extends to how to efficiently be a better parent to my three children, efficiently be a better partner to my spouse, and how to efficiently relax.

To quickly relax my mind, I find sensory-deprivation tanks a cheat code to meditation. I float with earplugs in a pitch-black tank filled with body temperature Epsom-salted water, like a personal perfectly warm Dead Sea. After a few minutes, my mind wanders and then starts to self-organize. My anxiety and stress dissolve in the water.

To quickly relax my body, I stress it. Apart from lifting weights, one of my favorite ways to stress my body is by profusely sweating in a German Aufguss sauna ceremony. For ten to twelve long minutes, gloriously scented, steamy air heated to at least 85 °C (185 °F) is whirled and beaten around the room and at participants by a skilled Aufguss sauna master. Blood rushes to your brain and body. Stress-free participants stagger out of this communal ritual to a short warm shower followed by an icy cold 4 to 5 °C (40 °F) dip.

But here's the rub. No matter what weird psychology or ceremonies I use, there is a limit to my efficiency. At the end of the day, I still have only twenty-four hours, of which continued efficiency requires eight dedicated to sleep – efficient sleep of course, optimized for letting ideas ruminate. Imagine how much more you or I could do if we had more than twenty-four hours.

There are ways to get more than twenty-four hours. One way is to supplement what we do with machines. We multiply our time by harnessing energy to do work for us.

Polymath Buckminster Fuller had a thought experiment in which he asked us to imagine all the work of the machines that surrounded us being performed by animal or human labor. My car can probably manage around 200 horsepower. Imagine those were actual horses? But those 200 horses wouldn't last long without large amounts of food, water, and rest. A full tank represents the work of around fifty strong men working for a month. Energy-powered transport has shrunk our world. I can travel across town in minutes not hours to meet a friend. I can cross the globe in hours not months for my next collaboration.

Energy is required for everything. Even the food we eat. Long before I cook pasta on a stove or warm up leftovers in a microwave in minutes, the wheat in my pasta was fertilized by ammonia synthe-sized by combining the nitrogen in the air and hydrogen from natural gas in the Haber-Bosch process, pests were killed with crude-oil-derived pesticides, the ground was plowed by fossil-fueled tractors; and the pasta was delivered to the supermarket by refrigerated trucks, ships, trains, and airplanes.

As Vaclav Smil points out, half the planet – nearly 4 billion people – would not have been alive without synthetic ammonia fertilizer that led to the Green Revolution in agriculture, a second agricultural revolution that rivals the first agricultural revolution 12,000 years ago.

Energy is everywhere.

Our civilization's control of energy is a product of the laws of energy and innovation creating the space of the possible in which we all now live. Efficient, energy-powered technologies have shrunk the globe and effectively extended our time. But there's one more way that I can extend my twenty-four hours. I can also cooperate with other people.

I can build a better company, write better books and papers, and engineer better products by not doing everything myself. When I work with others, the synergies of our different expertise further extend the effective time we all have. I can pass on to my collaborators and employees tasks that they can do faster or better than I can, allowing me to focus on my comparative advantage. Indeed, we also need to cooperate just to harness energy – I can't mine, process, and convert coal to electricity all on my own.

I'm telling you all this because I want you to see that the decisions, trade-offs, and competition we face as individuals are part of a broader system. The challenges we face in our everyday lives and the challenges we face as a society are not new. They are as ancient as life on Earth. They are governed by the same underlying laws.

Energy, innovation, and cooperation are shaped by technologies and ways of working that are themselves shaped by genetic and cultural evolutionary forces. These four factors – energy, innovation, cooperation, and evolution – also affect my everyday life, which is a microcosm of the way in which they affect our society and the evolution of life itself.

These laws weave our personal stories into the larger story of life on Earth. All life forms had to solve energy crises and overcome sudden shocks, just as we do today. To see these laws at play, let's go back to the very beginning. Bear with me, it won't take long, but then you will see how fundamental these laws really are.

A brief history of everything

The universe is about 14 billion years old. Earth is about a third of that, at around 4.5 billion years old. Not long after its formation a planet-sized object, around the size of Mars, smashed into our young planet. This violent collision ejected enough debris to create the Moon. That was a happy accident, because it was the Moon and the Earth that together created life.

Compared to the moons of other planets, our moon is relatively large, over a quarter of the size of Earth. Indeed, some scientists have suggested that the Earth and Moon should be considered a binary planet – two planets orbiting each other. One definition of a binary planet is that the center of mass of the two bodies – the point they both orbit – lies beyond both. Currently, this is not true of the Earth–Moon orbit. The center of mass is around 1,000 miles under the surface of Earth. But the Moon is slowly drifting away. The drift isn't fast enough for the Earth–Luna system to reach this binary planet definition in any reasonable time. But if you run the tape backwards, it means that the Moon used to be a lot closer to Earth. Tides are created by the Moon's gravity pulling on the oceans and so the early Earth had massive tides, stirring the primordial chemical soup, moving warmth and sloshing the oceans back and forth over the land. This created tidal pools that brought ocean life to land and tidal-pool life to the oceans. It was thanks to this gravitational energy that life could begin.

Life begins

Abiogenesis is the process by which non-life became life, and there is still no consensus on exactly how that happened. We don't know how it was that the first self-replicating chemical compounds became the first self-replicating simple single cells, but by around 500 million to 1 billion years later (3.5 to 4 billion years ago) we see the beginning of life.

Energy gave motion to life. Indeed, that is what life is doing – trying to harness and control as much energy as it can to manipulate resources to make more of itself. More energy means more motion to access more resources.

Evolution describes the process by which life tries different strategies in the competition over resources and energy. It applies to all life, including ourselves, which is where this story culminates.

But before we get to societies of humans, let's talk about societies of cells to see how fundamental these laws are.

Cells are a bit like mini-bodies with internal structure. We have organs; cells have organelles. More complex early single-celled life resembled modern prokaryotic cells. All that really means is that these were cells without either a nucleus in which DNA is normally stored, or separate organelles that would normally perform specific functions. In prokaryotic cells, everything is kind of floating around, the cells themselves just kind of floating around in the ocean. This early life also lacked one particularly useful organelle – mitochondria.

Power plants

Among other seemingly useless pieces of information drilled into you in school, you may remember that 'mitochondria is the power plant of the cell'. What this really means is that mitochondria manufacture little chemical sugar batteries called ATP, which are the batteries that fit into and can power all the parts of a plant or animal. Muscles need more energy? Send ATP.

In animals, mitochondria create ATP from the food we eat. In plants, chloroplasts convert the sun's energy, which is then turned into ATP. ATP allows cells to store, manage, and move energy. But life was around for a long time before mitochondria. Without power plants manufacturing batteries, early life faced an energy problem. In fact, it's the same problem we face today with solar panels and wind turbines.

Solar and wind can generate electricity, but you need a way to store it so that it can be used when there is no sunshine or the wind isn't blowing. Before the ATP battery revolution to easily store energy, cells were dependent on the available amount of sunlight for warmth. Some lucky cells might have enjoyed the warmth of geothermal vents. Our moon continued to gently stir the warmed waters, gravitational energy moving the thermal energy and organisms around. Then came a critical mutation.

About 3.5 billion years ago there was a mutation that allowed these simple single-cell life forms to store the sun's energy for later use: photosynthesis. Like many new inventions, this proto-photosynthesis wasn't yet efficient, so the next step for evolution was to keep innovating on efficiency. Just like what's happening with battery technologies today.

Modern rechargeable lithium batteries have two to three times as much stored energy per kilogram than even the best rechargeable nickel batteries they replaced, and are an almost unrecognizable distant relative of the first batteries from the 1800s. It was the invention of rechargeable lithium batteries in the 1990s that led to viable modern electric vehicles. The likely next step, hydrogen fuel cells, may be another ten to twenty times as energy dense. It's the law of innovation. Once something is invented, the next step is to do it better – increase efficiency, robustness, power.

Around 3 billion years ago another innovative mutation adds water to the photosynthesis reaction. This innovation improves the efficiency of photosynthesis, but there's a cost. This new photosynthesis starts polluting the world.

With oxygen.

The first climate crisis: not enough greenhouse gases

We're so used to thinking about oxygen as a good thing. The air we breathe is 21% oxygen and it's so critical to animal life that we forget how corrosive it is. Oxidation is what allows things to burn in a fire, it's what turns apples and bananas brown, and it's what turns iron to rust. Just as humans pump out copious carbon dioxide today, these simple life forms start pumping out copious oxygen like there is no tomorrow. And because of this, for many of their descendants, there *was* no tomorrow.

Around 2.5 billion years ago a disaster hit. It was the kind of disaster that prokaryotic climate-change activists would have had fancy conferences to do almost nothing about. They might have warned the other prokaryotic cells that oxygen was being pumped out at such a high rate that the young Earth was heading for mass extinction: the Great Oxygenation Event.

Oxygen was poison for most life on Earth at this point. It also combined with the methane in the air to produce carbon dioxide.

Carbon dioxide (CO_2) is a modern climate villain, but compared to methane (CH_4), it's more of a sidekick than a super-villain. It's not as effective at warming the Earth as the methane it replaced (that's why climate activists keep complaining about beef – methane is released from cow burps and farts). And so because of oxygen and carbon dioxide replacing methane, Earth actually cools down and goes through a long ice age. Too little heat and too much oxygen create a hostile environment that kills almost all life. But a changed environment is also an opportunity for evolution. The Great Oxygenation Event enabled our earliest ancestors to evolve.

All those sugary photosynthetic prokaryotes were little bundles of energy waiting to be exploited. They represented a new niche for evolution to occupy. And so evolution favored new kinds of organisms. Instead of specializing in directly using solar energy through photosynthesis – a long and arduous process that offers only enough energy to grow and reproduce at plant pace – these new organisms specialized in eating other organisms. Like raiders exploiting hoarders, these new organisms skipped the step of creating energy for themselves and instead just learned how to eat stored solar energy.

The same logic applies not only to hoarders and raiders and various kinds of bacteria but also to the many wars of exploitation in human history fought with the same underlying logic – larger groups with new technologies, greater cooperation, or larger energy budgets often through industrialization exploiting the resources of groups that are smaller, have less powerful technology, lower cooperation, or smaller energy budgets.

A cell eat cell world

Life began to rely on other life for energy. Indeed, this is the process that led to mitochondria. At some point in this cell-eat-cell world, an exceedingly improbable event happened: one prokaryote ate another and rather than digest it, allowed it to live inside and keep creating energy for the host – the evolution of mitochondria.

You and I evolved from this earliest cooperative relationship, which created new life forms called eukaryotes. You could say that both competition and cooperation are in our genes. Indeed, we still do something similar by allowing billions of bacteria – your microbiome – to live within us and help us digest food. You do it deliberately if you take probiotics or eat fermented foods like yogurt and sauerkraut. These organisms in our microbiome aren't just helpful, they're essential – we would die without them.

Around 500 million years later or 2 billion years ago, this ice age ends and life begins to recover. Eukaryotes have even more energy than prokaryotes. That means a more tempting target!

Evolution leads to new innovations that allow one eukaryote to eat another. It was an arms race leading to bigger, more complex cells. The cooperative trick of allowing smaller cells to do work within larger cells is also a new niche for evolution to explore. This leads to ever more complex and efficient cells with lots of little workers – the organelles I mentioned earlier. These organelles are to cells what your lungs, liver, heart, and other organs are to you, or what grocery stores, hospitals, hairdressers, and accountants are to our society. And these organelles have membranes that separate them from other members of the cell. But it's the same game. Now these complex eukaryotes are themselves ever larger sources of energy. Can you see where this is going? Yes! The new innovations eat one another. This time, though, *cooperation* is the new secret weapon.

By working together not just within cells but between cells, groups of cells can outcompete and eat other cells that are going it alone. Evolution is exploring the laws of innovation and cooperation, because the space of the possible is still large enough to do more. The ceiling constraining this ecosystem is the availability of solar energy converted to chemical energy by plants and eaten further down the food chain. But there's plenty of sun, water, and resources for photosynthesis. And so life finds ever more complicated ways to cooperate within and between cells. Things really take off from here.

The joy of sex

At this point life only has one source of innovation with which to create diversity – mutation – mistakes during cloning. But around 1.2 billion years ago cells discover the joy of sex.

Sex, even today, is a new kind of cooperation between two individuals with different genes. Mixing genetic traits means swapping the best genetic tricks creating diversity through recombination. Just as Bernard Sadow used the insight of seeing a family struggling with heavy luggage while a porter wheeled a luggage rack with ease to create wheeled luggage in 1972, so the recombination of diverse genomes leads to a combinatorial explosion, increasing diversity and accelerating evolution.

It's unclear when exactly multicellular life emerged, but groups of cells soon learn they can better exploit other groups of cells by working together. Multicellular life consisting of colonies of the same cells seems to have evolved many times, but *complex* multicellular life consisting of different kinds of cells cooperating in more complex ways as a single animal – like today's modern multicultural societies with diverse occupational groups – evolves just once in each lineage around 600 to 800 million years ago. And it is through this cooperation and competition for energy that the diversity of life explodes.

Around 540 million years ago Darwin's 'endless forms most beautiful and most wonderful' erupt in what we call the Cambrian Explosion – the beginning of a diverse array of animals in the fossil record.

From bags to tubes

The earliest animals were like bags with a single orifice. Nutrients went in that orifice, which served as a mouth, and waste came out of that same orifice, which also served as an anus. I think we can all agree that evolution separating our mouth from our anus was a step up. This turned us from bags to tubes, an architecture we still use today.

We are still tubes. Food goes in one end and waste comes out the other. Don't get me wrong, our tube bodies have become fancier in the struggle for survival, in the competitive mating market, and in the competition to eat one another. They've absorbed entire other organisms as part of a microbiome – you are more like the Amazon rainforest, an entire ecology rather than just a single organism. Tubes like ourselves have added appendages to help move around and interact with the world – arms and legs, fins and tentacles. And they've added senses to interpret specific features of the world that allow them to find mates, eat, and avoid being eaten. For example, we can 'see' a narrow band of the electromagnetic spectrum that we call visible light, but not the electromagnetic range of your Wi-Fi's radio waves. We can 'hear' the vibrations in the air within a narrow frequency range, but not the sound of a bat's echolocation. We can 'taste' and 'smell' the rotten-egg-scented sulfur but not odorless monoxide. We can 'feel' the roughness of sandpaper and the smoothness of glass to some extent, but not weak electrical fields, like a shark can, or magnetic fields, like a homing pigeon can. Some of these senses might be useful to you – smelling carbon monoxide might save your life and detecting strong Wi-Fi signals is just handy, but these skills are either not essential or too difficult for evolution to get you to from your current physiology.

Some tubes even started storing and processing what they'd learned from their appendages and senses – the beginning of brains and cognition. And so it continued.

Dinosaurs

The first dinosaurs appear around 240 million years ago and the first mammals appear not long after – well, a few million years later, but not long on these timescales. Dinosaurs were around for a *long* time, with many going extinct before others even evolved. Many of your childhood favorites, like Stegosaurus and T-Rex, may have battled on your playroom floor, but unfortunately never met in real life.

Then 65 million years ago there was another mass extinction when a giant asteroid hit Earth, throwing up a cloud of dust that blocked the Sun. (Remember that at this point the energy that gives chemicals motion and turns them into life is ultimately limited by the energy of the Sun. With less sunlight, plants die. Then the herbivores that eat the plants die. Then the carnivores that eat the herbivores die. The energy ceiling comes crashing down and there's no room for such large animals.) At this point the dinosaurs died. This event offered an advantage to our tiny warm-blooded ancestors. The age of mammals began, eventually leading to *Homo sapiens*.

The rise of humans

Hominins and chimpanzees have a last common ancestor at around 4 to 7 million years ago. There were many hominin species. Some were our ancestors; others were our cousins. We are the last of the hominins.

Our closest hominin cousins are the Neanderthals and Denisovans, who evolved around 350,000 years ago. The ancestor to all hominins first evolved in Africa, but groups left Africa many times. For example, one group that settled in Eurasia eventually evolved into the Neanderthals. Another that settled from Siberia to South East Asia evolved into the Denisovans. Groups met and mated with the extra-African groups that left before. You could say that hominins have more of a family web than a family tree. We modern humans were no exception. We evolved in Africa around 150,000 to 250,000 years ago and had colonized much of the Earth by around 60,000 to 70,000 years ago. But it wasn't easy.

We fought and mated with the other humans we met. We faced volcanic eruptions, dangerous animals, disease, and bad weather. We kept getting killed and our genetic diversity shrank. The first group to cross from Eurasia to the Americas may have been as few as seventy in number. Many family lines died out. A handful of humans are ancestors to us all. As a result, there is greater human genetic diversity in our homeland, Africa, compared to anywhere else in the world. Each wave of migration took only a subset of the full range of human genes. Today, there is so little genetic diversity outside Africa that there is more genetic distance between two bands of chimpanzees plucked from the Congo than two randomly chosen humans plucked from Berlin and Beijing. Neanderthals and Denisovans went extinct around 40,000 years ago, leaving Sapiens as the last of the hominins.

Last of the hominins

Up until around 12,000 years ago our ancestors all lived in small hunter-gatherer groups scattered around the world. This period was called the Neolithic – literally the New Stone Age – the last of a cultural explosion of Stone Age tools. Around 10,000 years ago the size of many groups expanded significantly with the advent of agriculture. Around 6,000 years ago the Neolithic came to an end as the first cities began to be established. By 2,500 years ago Athens had a democracy. Around 260 years ago the Industrial Revolution began in Britain with the spinning jenny (1764). Around 55 years ago (1969) the first message was sent over ARPANET, the proto-Internet. Around 35 years ago (1989) Tim Berners-Lee gave us the World Wide Web. Around 25 years ago (1998) two Stanford PhD students, Larry Page and Sergey Brin, created Google. Around 15 years ago we decided to disregard our parents' advice not to go into a stranger's house or get into a stranger's car and started using home sharing (Airbnb; 2008) and ride sharing (Uber; 2009) platforms. And around 30 minutes ago you probably started reading this chapter.

Laws of life govern all

The ultimate constraint on this entire system – the ceiling on the space of the possible – was once the sun and the ability of plants to convert solar energy into something we can use. For most of history the energy ceiling was low for all life, regardless of genetic or technological innovations. There was no point in cooperating at large scales as we do today – what we could gain had to be taken from the other poor folks around us. No amount of innovation or cooperation could pierce the ceiling of the law of energy.

If you look at graphs over the last several millennia of wealth, energy capture, total population size, size of countries and polities, child survival rates, or just about any other indication of progress then you'll notice something odd. There are wiggles and bumps, but everything is pretty flat from the beginning of history to the mid 1700s. And then everything just explodes.

All those major world events – the fall of the Roman Empire, the violent conquests of Genghis Khan, the devastation of the Black Death, the innovations of the Renaissance, the discoveries of the Scientific Revolution – and much more of what you covered in high-school history are mere blips. They are completely dwarfed by the

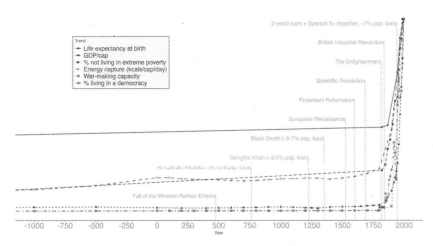

Human progress. Based on graph and data compiled by Luke
Muehlhauser, https://lukemuehlhauser.com/industrial-revolution/

enormous progress that has taken place since the Industrial Revolution.
The laws of life tell us why.

Till the 1700s the energy ceiling remained low. The amount of
work our ancestors could do was a function of their own manual labor
and the labor of their work animals, such as oxen and horses, made
more efficient by a few sparse mechanical innovations such as levers,
pulleys, and windmills. Once the technologically supplemented human
and animal labor paid for its own energy costs in terms of the energy
needed to produce the food and build and maintain the technologies,
there wasn't a lot of excess energy. With so little excess energy, no
matter what we did, no matter how clever we were, no matter how
hard we worked, there was a limit to what we could achieve. Then
we discovered a large store of densely packed solar energy that changed
everything. It was the sacrifice of past life.

Plant matter, algae, and other ancient organisms had stored solar
energy in chemical form. Time and pressure had compressed all that
chemical energy into dense coal, oil, and natural gas. We found fossil
fuels and learned to use them. Those chemical batteries took millions
of years to charge. We have been draining them in a matter of centuries, using that awesome power to raise the energy ceiling to
unimaginable levels and allow our civilizations to soar to the heights

we've reached. Underneath that almost vertical line you previously saw is a fossil-fueled fire.

Fossil fuels led to new innovations in the efficient use of the energy they contained. Industrial societies wielding this new power source were able to use it to colonize and dominate pre-industrial societies. The unlocked energy and all that we could do with it incentivized working together, leading to larger, more complex, relatively internally peaceful new civilizations, which would eventually be made up of people from all over the world. These large groups could then begin flirting with alternative sources of energy, from nuclear to solar.

Each energy source allowed us to access the next higher energy density source with greater efficiency. But energy became so abundant, the ceiling so high, that over generations we forgot that there was any limit at all. Our engineers and economists largely stared at the floor, focusing on technological innovations in efficiency. For those familiar with economic growth models, the focus was on the A technology term that multiplies our labor (L) and capital (K). Thus far, doomsayers from Thomas Malthus to M. King Hubbert seem to have been proven wrong time and again by technological advancements. Technology seems to have saved us from the Malthusian trap and delayed Hubbert's peak oil decline. But those technological advancements have been in the efficiency floor, not the energy ceiling. And in the end, it doesn't matter how fancy or efficient your gadgets are, if you can't charge them, you can't use them. Technology that unlocks more energy is fundamentally different to technologies that make life more efficient or enjoyable. We assumed that we could ignore energy because we would always have enough, but as financial advisors warn: past performance is no guarantee of future results.

The energy ceiling is falling.

Fossil fuels are becoming costlier to mine, process, and use. Once cheap and abundant, they are now expensive and scarce. Innovations in efficiency have also meant that these resources can be captured and controlled by fewer people. Continued progress, peace, and the civilization they create require us to get to the next energy level.

It is through the lens of the laws of life that all the pieces suddenly come together to make sense of who we are and how we got here.

Let's start with the ceiling constraint – the law of energy.

The law of energy

Let's say you're mining. You could dig up coal or you could dig up uranium. Uranium is much more energy dense than coal. That's really an understatement – uranium has at least 16,000 times as much energy per kilogram, 2 million times as much if enriched. That's great! But what also matters is how much energy it takes to access and use that energy source. Let's think about some of those costs.

The first step is digging some ore. You could dig the ore from the ground with your bare hands. That would take a long time and lead to dirty, bloody fingers. One innovation in efficiency would be a shovel and pickaxe. That technology would be slightly more efficient. But you are still doing all the digging. You are still limited by the energy provided by the food you consume. That in turn is limited by your agricultural technology, the energy you use to power it, and genetic limitations as an under-muscled, underpowered human ape. We are much weaker than the other great apes and many other animals.

So, a further improvement would be to use another animal with a better food-to-muscle ratio, such as an ox. An ox not only has a better energy return on energy invested in food but can also generate more power – energy released per second – than you can. You can use that power to drive a bigger plow more forcefully to prepare more unyielding rocky ground. But to have an ox, you need enough excess food to feed and domesticate the animal. Again, this is a function of your agricultural technology: can you grow enough food to feed you, your family, your community, and the oxen? Will the food grown with the help of the oxen be enough to feed your community of humans and animals?

If you already have access to industrial energy sources such as fossil fuels, you could use an engine with a more straightforward return based on the quality of the ore and the ease of finding, processing, storing, and transporting it, and the technology of your engine. You might also require the help of other people. But again, the question is the same: is the return from those fossil fuels enough to justify feeding all the people and animals and fueling all the machines involved in getting that energy?

This concept is what Charles Hall calls energy return on investment

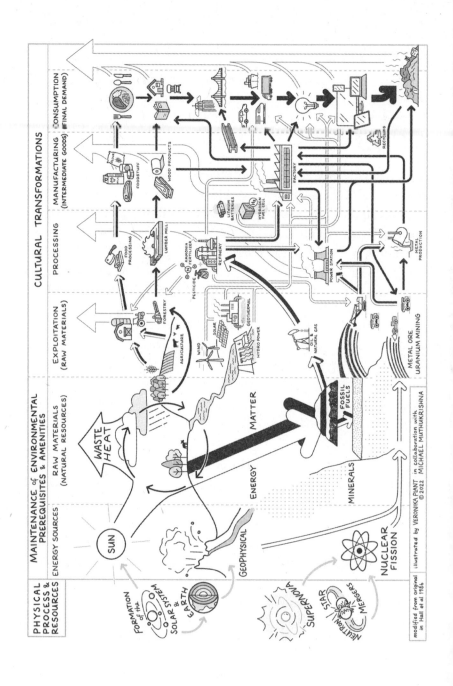

PHYSICAL PROCESS & RESOURCES

MAINTENANCE of ENVIRONMENTAL PREREQUISITES & AMENITIES

CULTURAL TRANSFORMATIONS

ENERGY SOURCES

RAW MATERIALS (NATURAL RESOURCES)

EXPLOITATION (RAW MATERIALS)

PROCESSING

MANUFACTURING (INTERMEDIATE GOODS)

CONSUMPTION (FINAL DEMAND)

SUN

FORMATION of the SOLAR SYSTEM & EARTH

GEOPHYSICAL

SUPERNOVA

NEUTRON STAR MERGERS

NUCLEAR FISSION

WASTE HEAT

ENERGY

MATTER

MINERALS

FOSSIL FUELS

AGRICULTURE

WIND

SOLAR

GEOTHERMAL

HYDRO POWER

OIL & NATURAL GAS

METAL ORE URANIUM MINING

FORESTRY

PESTICIDE

FOOD PROCESSING

LUMBER MILL

AMMONIA FERTILIZER

REFINERY

POWER STATION

METAL PRODUCTION

LITHIUM BATTERIES

HYDROGEN FUEL

FOODSTUFF

WOOD PRODUCTS

FACTORY

RECYCLING

modified from original in Hall et al 1986   illustrated by VERONIKA PLANT in collaboration with MICHAEL MUTHUKRISHNA © 2022

(EROI), sometimes also referred to as energy return on energy invested (EROEI). EROI is normally calculated as a ratio by dividing the output energy by the input energy:

$$EROI = \frac{\text{Energy output}}{\text{Energy input}}$$

It's a ratio of how much energy you expend in getting some energy back; a sense of how much excess energy you have. How many calories do you burn relative to the size of the animal you catch, cook, and consume? A hunter who expends more calories hunting than the calories they get from their kill is going to starve.

The EROI of different energy sources varies dramatically and also changes over time as an energy source becomes more difficult to access. For example, coal that's close to the ground has a higher EROI than coal that has to be laboriously dug up from deep inside a mountain. As economic historian Tony Wrigley convincingly argues in *Energy and the English Industrial Revolution*, easily accessible and abundant coal goes a long way to explaining why England was able to reach industrialization first.

EROI values also vary depending on what researchers choose to include in the energy input. If you invade a country and 'liberate' some of its oil alongside its people then should you include the energy used by all those missiles, drones, and soldiers? These challenges notwithstanding, we can compare EROI calculated using similar approaches to get a sense for the challenge in using various energy sources.

The EROI of coal ranges from around 10 to 80. Think of it like this: 1 lump of coal can get you another 10 to 80 lumps. The EROI of oil and natural gas are much more variable because of the variety of sources, but the EROI has been falling for a century. Consider oil discovery:

In 1919 1 barrel of oil found you at least another 1,000.
In 1950 1 barrel of oil found you another 100.
By 2010 1 barrel of oil found you another 5.

And the number has continued to fall as we move from abundantly available sweet crude to hard-to-refine sources such as tar sands and fracking. This is part of the reason energy prices (and consequently inflation) have risen; but it is not the only reason.

The effective EROI has also been artificially constrained by co-operation between oil-exporting nations such as Saudi Arabia and Kuwait. The Organization of the Petroleum Exporting Countries, more commonly referred to as OPEC, has artificially constrained supply to keep prices high. Their actual EROI and oil availability remain state secrets. We'll discuss this in more detail in Chapter 9.

With the availability and EROI of fossil fuels falling, one suggested path out is to transition to renewable sources of energy, but there's a problem. Indeed many.

The EROI of renewables is much lower than that of fossil fuels. Photovoltaic solar panels are currently in the single digits, typically no higher than 2 to 4, and the higher values are really only when you add a battery. So it costs you 1 watt of electricity to get 2 to 4 watts back. This value may be higher when panels exceed their planned lifespan, but it still requires a large upfront cost.

We are at the early technological stage in solar panels and so prices are falling rapidly, which is reducing that upfront cost. There are, however, fundamental limits on solar efficiency and cost sensitivity to the resources required to build them. Solar panel prices will stabilize and perhaps increase at some point based on the availability and cost of materials such as copper. And we are a long way from building enough panels to replace even our current electricity needs.

To meet the United States' current electricity consumption would require more solar panels than all the space used by roads in America. That's just the panels and doesn't include batteries, wiring, and other infrastructure. And that's just electricity usage *at the moment*.

The transition away from directly using fossil fuels such as gasoline in cars or natural gas in homes toward a fully electrified grid, such as by transitioning to electric cars and electric heating, represents a more flexible and efficient energy future. But some estimates suggest that to power this future with solar panels would require solar panels that take up more space than the entire state of California.

Wind too has similar issues. The sun doesn't shine at night, but the wind blows. Yet wind is far more erratic than sunshine and so the EROI of wind is highly variable, from around 4 to 16. Wind, and particularly solar, can play a part in our immediate energy future, and almost certainly must in the distant energy future, but alone, both lack the EROI to transition us to the next level of energy abundance.

In contrast, hydroelectric power, generated by large, fast-flowing rivers spinning a turbine, has an EROI of typically 50 to over 250 depending on the size and gradient of the river. There's very little downside to using hydropower, but it is limited to those countries lucky enough to have large, fast-flowing rivers, such as Canada, where electricity is commonly, and confusingly to non-Canadians, called 'hydro'. Canada generates 60% of its electricity through its rivers. But EROI isn't the only factor that matters. Other important factors include availability, density of energy, power, and start-up cost. The challenges of transitioning to renewables is made stark when they're contrasted with fossil fuels.

Fossil fuels are nature's batteries of densely stored solar energy. Coal is *millions of years* of densely stored sunlight in the form of plant material (peat) turned to black rock, about twice as energy dense as wood, which is merely a tree's energy surplus over its lifetime. Oil and natural gas are millions of years' worth of densely stored sunlight in the form of algae and zooplankton pressure-cooked into both oil and natural gas. Heat and pressure have made them dense, transportable batteries.

One active area of research is how to store solar and wind energy in some kind of artificial battery or fuel. There are many potential solutions, from a chemical battery of some kind, similar to what powers your electronic devices and electric car, to pumping water uphill and storing the energy as gravitational potential energy, ready to be drained by letting the water flow back downhill. One promising idea is the use of solar energy to generate hydrogen by splitting water – H_2O. This approach requires sunlight and water but has geopolitical advantages over chemical batteries. Any country – even those without the rare metals needed to make chemical batteries – can produce hydrogen in this manner.

We are currently like early life with little ability to store and distribute the power of the giant fusion reactor in the sky – the Sun. We are waiting for the modern ATP battery revolution. Remember, early life took millions of years to charge those dense fossil-fuel batteries. We have been burning through those batteries in a matter of centuries. The low density of artificial batteries is a major challenge for renewables.

We require more energy-dense batteries to be able to transport large amounts of energy and release them with sufficient power (energy per second) needed for many tasks. Electric cars are more viable than electric planes because cars can carry heavier, less energy-dense batteries than planes, which need to lift those batteries off the ground No amount of horses could power a plane nor could current batteries power anything more than a light aircraft or drone. Even though the energy transfer from batteries is more efficient than burning gasoline, jet fuel is much more energy dense and able to release sufficient power for an A380 to carry almost 1,000 people across the Atlantic.

The final major challenge with transitioning from fossil fuels to renewables is the start-up cost. A lot of energy and rare resources are required to build those wind turbines and in particular those solar

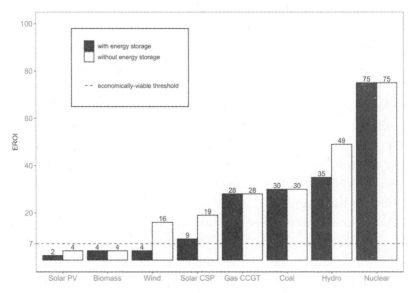

Energy return on investment for various electricity-generating power plants. Based on data and graph from: D. Weißbach, G. Ruprecht, A. Huke, K. Czerski, S. Gottlieb, and A. Hussein, (2013), 'Energy Intensities, EROIs (Energy Returned on Invested), and Energy Payback Times of Electricity Generating Power Plants', *Energy*, 52, 210–21.

panels. The availability and already low EROI of renewables mean that that initial investment in energy takes a long time to pay off. It's the same trade-off we face in using solar panels in our homes.

My mother, Shanthi, who lives in Queensland, Australia, is very fond of the solar panels on her roof, which now generate enough electricity for her needs and sometimes generate power back to the grid, turning her a small profit. But even in sunny Queensland, with 300 days of bright sunshine every year, it took years before her subsidized initial investment was paid off. Some households can afford to do this, but at a national level an initial investment in subsidies or solar power for a whole population requires a massive upfront investment. That effectively means less energy for other productive parts of the economy, which can lead to inflation and a rise in prices and cost of living at first.

Quality of life and actual wealth in terms of what we can do beyond survival is a function of excess energy, which is in turn a function of EROI and energy source availability. And so an upfront payment for renewables from our quickly falling excess energy budget will require tightening our collective belts in the meantime. In other words, it means a reduction in current productivity in other sectors – less energy for farms, hospitals, and heating homes, not to mention holidays and other leisure activities. Ultimately that means your energy bills go up while we make the transition. This debt reduces overall national wealth and drives up inflation. In the second half of this book we'll discuss in more detail the relationship between energy, money, and upfront subsidies, and how reductions in overall wealth lead to dissatisfaction and lower cooperation, both of which make people harder to govern.

These are not unsolvable problems, but they do need to be solved. We cannot ignore either the fundamental physics of energy sources or the fundamental social challenge of governance and continued cooperation. These are fundamental laws of life that have applied long before the arrival of Sapiens.

The total energy available to humans is the availability of different energy sources multiplied by their EROI. For low EROI sources such as solar, this means large tracts of land dedicated to harnessing the energy of our sun, which comes with both a material cost to build those solar panels and at an ecological cost to the land used. Renewables won't substantially increase our energy budgets. Apart from an initial energy and resource cost, renewables are also going to require a lot of innovation. But innovation too is a function of available energy.

The law of innovation

In the brief history of everything earlier in this chapter, we saw that the first major energy revolution was the photosynthetic innovation. As abundant sunlight fell on Earth, photosynthesis offered a way to store the sun's energy in chemical form for later use. Remember that energy is what makes matter move as life, and so if you can store energy then you can do bigger things. It's like saving money to buy something more expensive. Storage also requires a surplus of energy. That surplus can then be focused to harness a more energy-dense and abundant source with a higher EROI.

For humans, the earliest unlocking of energy was fire. Fire allowed us to burn one organism (dead trees) to unlock the energy of another (the chemical energy in the calories of food). Trees are dense, stored solar energy that can be unlocked through oxidation in fire. The dense trees of the past also had more energy than current less-dense, fast-growing varieties. Humans used this wood energy to stay warm, ward off predators, and cook.

Cooking is a process of pre-digestion whereby external energy is used to break down the molecules in meat and vegetables, pre-digesting them to save the mechanical movement of our teeth, the synthesizing of stomach acids, and the hosting of microbiome bacteria that would otherwise take more energy and more time to do the hard work, lowering our EROI from food. Our species needed the extra energy that cooking unlocked because it had one very energy-expensive organ: our brain.

Our brains are incredibly energy expensive. At rest, the brain uses as much energy as all your muscles do. Indeed, 1 gram of brain tissue uses twenty times more energy than 1 gram of muscle tissue.

Cooking saved us from sitting there like gorillas chewing plants all day or needing four stomachs like a cow munching on grass. We reduced the size of our gut and lost a lot of muscle, saving us a lot of energy. We used that extra energy to fuel a larger brain. What did we do with that larger brain? We learned more useful stuff, including figuring out how to hunt larger, higher EROI animals. Rather than just hunting hare or scrounging for grubs, we could focus on large animals like stag, bison, or mammoth.

Meat is the original superfood, offering a denser energy source than plants. When eaten from nose to tail, organs and all, not just the muscle meat typically eaten in the West, it also provides a dense source of all the nutrients you need, literally stealing all the things an animal needs by taking it from another animal's whole body. We figured out better techniques for making hunting and gathering more efficient, offering us more time to continue innovating. And eventually, we figured out a better way to get food: farming.

Agriculture was the next major unlocking of energy after fire. We switched from hunting and gathering to harvesting and grinding. Hunting, gathering, and cooking food with fire had a better EROI than eating raw food and digesting it inside our bodies, but we still had to expend energy wandering the plains trying to find plants that wouldn't poison us and hunting animals that didn't want to be eaten. As anyone who hunts or fishes knows, returns are uncertain. Agriculture, in contrast, meant efficiently turning areas of earth into food production. It was a solar technology, efficiently exploiting the energy of the sun to multiply our growing efforts. It was still laborious, but it was also a more reliable and higher EROI source of calories. We continued innovating and made the process so efficient that we created a bit more excess energy.

What did we do with that excess energy? We funneled it into creating more people and domesticating a higher EROI food source: other animals. Instead of just farming plants, we farmed animals. Those animals could not only be eaten but, with enough excess plant energy, could be put to work to help plow and create more food for us. Horses, in particular, also allowed us to travel faster and further, shrinking the world and helping us share ideas to further innovation. Agriculture and domestication co-evolved.

Just as the cultural innovation of fire unlocked more energy, the cultural innovations of agriculture and domestication gave us a reliable, if less diverse, food supply. Relative to hunter-gatherers, agriculturalists were – individually speaking – unhealthier and shorter in both stature and lifespan. Living in higher-density settlements alongside animals and eating nothing but grains isn't great for you. But agriculturalists' larger, more reliable food supply allowed them to expand their populations at an unprecedented rate.

Just as photosynthetic prokaryotes were exploited by other prokaryotes who were exploited by eukaryotes who were exploited by complex eukaryotes who were exploited by cooperative groups of eukaryotes, hunter-gatherers were outcompeted by the agriculturalists around them, pushing them to ecological niches less suitable for agriculture – deserts or thick forests – where even today the few remaining hunter-gatherers still live. Soon agriculture became the dominant form of subsistence. But as these agricultural groups grew in number and size relative to their agricultural output, abundance turned to scarcity. As the number of people rose, the amount of food per person decreased.

This was doubly challenging because innovations meant that fewer people were needed for farming as animals and implements made farming more efficient. So agriculturalists began to compete among their own groups and with other agriculturalists, stealing land, crops, animals, and even people. People taken from other groups could be made slaves, paid the minimum in lifestyle and food, leading to higher EROI for the enslavers (though not for the population as a whole). This continued till the next energy revolution, which would finally break us out of this Malthusian world of scarcity, violence, and conflict: the Industrial Revolution.

The Industrial Revolution of the eighteenth century made us even more efficient. Instead of burning wood like agriculturalists, we began to burn up millions of years of fossil fuels in factories that produced at a pace that would have required thousands or millions of people without them. From Henry Ford to the Toyota Way to Tesla's robots, we continued to make those factories more efficient. We also used fossil fuels to innovate efficiencies in food production.

In the mid twentieth century we launched a second agricultural revolution on the back of fossil fuels: the Green Revolution. We not only mechanized farming through fossil-fueled farming machinery but also learned how to turn that ancient fossil life into new life by turning it into fertilizers and pesticides, making farms far more productive. This led to an abundance of food, reduction in poverty and famine, increased income, higher child survival rates, and a doubling of the human population. Half the human population owe their lives and all of us owe our current standard of living to this second agricultural revolution. In 1970 the father of the Green Revolution, Norman Borlaug, won the Nobel Peace Prize.

Each of these revolutions led to new social organizations. Some hunter-gatherer groups blessed with abundant resources developed property rights and hierarchical societies, but it was really with the abundance of agriculture that these features of society became commonplace.

Excess resources meant storage. Storage meant having something that needed ownership. Owning something meant protecting it from people who would like to take it from you. Remember the prokaryotes and eukaryotes? It's not uniquely human greed, it's inevitable. Ownership and property rights increased productivity as people competed with one another to own more. And differences in ownership meant increased inequality and higher rates of violence over these stored resources. This in turn led to hierarchies and governments and greater divisions of information and labor. And in turn to innovations through *intellectual arbitrage*, the cultural equivalent of genetic recombination. Just as we combined genes to create new people, we started combining ideas to create new innovations. As Matt Ridley puts it, 'ideas had sex'.

Cultural evolution explored different norms, institutions, and political systems; new ways of cooperating but also new ways of competing and exploiting each other. But nothing in life is done alone.

Our greatest achievements and our worst atrocities are all cooperative acts. As energy becomes efficiently accessed, captured, and controlled, new niches are opened for larger units of organisms and organizations to come together to exploit energy. To do that, we needed to work together. We needed to cooperate.

The law of cooperation

Cooperation increases as organisms and societies discover new energy sources and learn to efficiently access and use them, which requires working together. More abundant, high-density, high-powered energy sources with higher EROI lead to a corresponding increase in the complexity and scale of cooperation. This increasing scale happens because the potential pay-offs are higher and it may be worth working together with others to access that reward. The reward is large enough that even if it's shared, it's still worth working with others. With energy up for grabs, smaller units enmesh as larger wholes with aligned incentives.

Organelles comprise cells, cells comprise organisms, organisms comprise colonies, colonies comprise complex societies. People come together in tribes and raiding parties. Regions come together in countries. Countries come together in unions. But for incentives to align there must be a reward for all parties.

We work together at a level that allows us to reliably access energy rewards that when divided up by the number of workers (or cells) are greater than what could be accessed by working alone, in a smaller group, or in a larger group. When a new energy source is unlocked it leads to new innovations and new abundance. But as populations grow thanks to these resources, abundance turns to scarcity.

Energy access often requires a certain size and complexity: a certain number of people to mine the ore, to build the pipelines, to work the oilfields, to ship the oil, to protect and to provide infrastructure around the entire ecosystem-like enterprise. But a positive EROI means excess energy beyond these energy costs, increasing what biologists call a carrying capacity – the maximum number of people that can be sustained. The human carrying capacity vastly increased with the Green Revolution, and our population began to catch up.

When I was born, in the late 1980s, there were only 5 billion people on the planet. We have since added another 3 billion people. More people can mean more innovation and progress, but only if those people have opportunities to express their full potential; only if they can join and participate in our collective brains and proportionally expand the space of the possible. If they do not and the space stays the same, then energy per person decreases. And when this happens a new scarcity and new conflict are inevitable: poverty, violence, and war.

The law of innovation leads to greater efficiency in energy use, which means fewer people are needed for the same energy return. Innovations can make people redundant for the energy economy. Today, with mechanized, fossil-fueled farming, fewer farmers are needed than in the past, but this was also true with past innovations and will be true of future innovations. This means that even with fewer people we can return the same or more energy per person. For this reason, the energy available per person can go up after a war or plague.

In the fourteenth century the Black Death killed a third of Europe. Rather than devastating the continent, resources per person went up for the survivors. In turn, feudalism was weakened, wages increased, and a new middle class and new social order emerged. After the Black Death the land still needed to be worked and the fewer workers alive had more bargaining power, but with innovations in efficiency fewer people were required to work the land. This in turn led to higher wages, higher life expectancy, greater equality, and a weakened aristocracy. In turn, this may have directly led to the Renaissance. Indeed, it may have contributed to Europe's Scientific Revolution, Enlightenment, and eventually the Industrial Revolution.

Similar patterns have occurred throughout history. The dearth of men during the two world wars of the twentieth century may have led to greater gender equality as women entered the workforce. Right now, efficiencies in artificial intelligence and automation are once again increasing efficiency and creating inequality as far fewer people are needed for the same productivity. AI workers are entering the economy. Those workers require vast amounts of energy for the computer servers needed to train them, but, once trained, they can be replicated and work for less energy – and money – than a human performing the same task. These AI workers are owned by a smaller group of people, which means that the fruits of production at the same or even a higher level can be controlled by a smaller number of people. These few individuals increase their energy rewards because other people are now redundant in the energy economy. But falling energy availability and EROI shrinks the space of the possible.

Clearly there's a limit to this redundancy. We still need a minimum number of people to extract energy and resources and to create the economy that surrounds and enables its efficient functioning. Take power plants for example. By one estimate, there is an almost perfect correlation ($r = 0.98$) between the worker years (how many workers and the time it takes) needed for construction of a plant and power-plant capacity (how much electricity is generated when the plant is running on a full load). But that's just the minimum for the energy portion of the economy. It is the excess energy that leads to true human wealth and a high quality of life.

In an ideal world, only a tiny fraction of our economy is dedicated to energy, because the energy returns mean that there is enough excess energy to support everything else in life. It's counterintuitive, but a growing energy sector is an indication of falling EROI, falling excess energy availability, and falling quality of life.

The laws of energy, innovation, and cooperation create an inexorable cyclical pattern. First, a new energy source increases the carrying capacity of our population. The energy ceiling is raised. This in turn creates *positive-sum conditions* that incentivize people to cooperate at a high enough scale to access that new source of energy. The higher cooperation leads to greater innovative capacity, which in turn leads to new efficiencies to do more with less energy. Those innovations in efficiency can lead to fewer people being required for the energy sector. The rest can simply enjoy the excess energy returns and use it to make life better.

But since these efficiencies mean that fewer people are required to access the same amount of energy, smaller scales of cooperation – feudalism, aristocracies, oligarchies, or corrupt cabals – emerge trying to control that wealth. This in turn leads to fewer opportunities for many people, reducing human potential, reducing quality of life, and reducing innovative and cooperative capacity.

As populations grow but the space of the possible does not proportionally grow or even shrinks because of lower innovation or no transition to the next energy level, so abundance turns to scarcity. What we call *zero-sum conditions* dominate.

Zero-sum conditions, in contrast to positive-sum conditions, are those where another's success predicts your loss – win-lose rather than win-win – for example, when jobs, contracts, or university places are limited. When conditions are zero-sum, you are incentivized to harm others, because their failure is your success. If someone else doesn't get the job, you might, so why would you want them to get it?

Positive-sum conditions are those where another's success predicts your success, such as during economic booms. You can get what they're getting by doing the same thing. If the coffee business is booming, you can also start a Starbucks.

Zero-sum and positive-sum conditions incentivize very different psychologies. Zero-sum conditions incentivize destructive competition as people or groups undermine one another. Positive-sum conditions incentivize productive competition as people compete and innovate to capture the abundant resources.

Excess energy, which can be measured as cheap energy, leads to economic booms. Less energy, which can be measured as a spike in energy prices, leads to recessions. These affect our psychology, behavior, and tendency to cooperate.

It's easier to be nice when there's more to go around.

Consider this analogy. Imagine yourself waiting for a bus. Let's treat the rate of buses and availability of seats as the total energy available. Let's assume buses arrive every five minutes and there's plenty of space available for everyone. As people find the service convenient, more people begin to turn up. If there are plenty of seats available, you might graciously let someone ahead of you in the queue. Even if you miss out, there will be another bus in five minutes. There may be mumblings and grumblings about the 1% with special passes that always get to the front of the line or about people favoring their friends or ingroups and letting them into the queue ahead of you. But it's only mumbling and grumbling as long as there are seats available. Eventually, the number of people begins to meet the number of seats available and it becomes harder to get a seat than it once was. Imagine now that the frequency of buses slows down – one an hour, one a day. Bus seats per person, which in our analogy represents energy per person, is decreasing, but the number of people matches the old carrying capacity. What was once mumbling and grumbling erupts into something more hostile.

In a society, such zero-sum conditions manifest as good publicly funded schools become harder to get into or less well funded, as waiting times at hospitals increase or quality of health care decreases, as well-paid jobs become more difficult to find or require more work. Local populations can become understandably resentful of these new pressures, and more resentful still when their existing resources are deployed to help newcomers.

It's not so much about inequality as equality of opportunity. Research reveals that people don't necessarily expect equality, but they do want fairness. Many people don't mind the special bus passes as long as they can get one at some point, perhaps as they get older or if they work harder. People don't mind others ahead of them in the line as long as they can get a seat if they choose to wake up earlier to get to the bus stop. But if a better life seems improbable no matter what you do; if people wake up to the idea that not being a millionaire is not a temporary embarrassment but a ceiling on their progress due to the happenstance of their birth; when who you are and not what you can do limits what you achieve – then cooperation is threatened. When this happens, the ties that bind us start to unravel.

We cooperate to compete, compete by cooperating; cooperation and conflict are two sides of the same coin. Cooperation and conflict lead to innovations and occasionally to breakthroughs in our total energy budget. These possibilities and configurations are not deliberate human designs or decisions; they are explorations in the constrained space of the possible. Explorations through the law of evolution.

The law of evolution

Charles Darwin, the father of evolution, was a pigeon breeder. And as a pigeon breeder he was well aware of how artificial selection could change the traits of an animal or plant. Through artificial selection humans turned a wolf into a poodle, grasses into wheat and other cereals, and one mustard plant into broccoli, Brussels sprouts, cabbage, kale, cauliflower, kohlrabi, and gai lan. In all these cases, human intelligence is doing the selecting on preferred traits – artificially.

In life's quest for energy and efficiency through innovation and cooperation, three simple ingredients decide the scale of cooperation

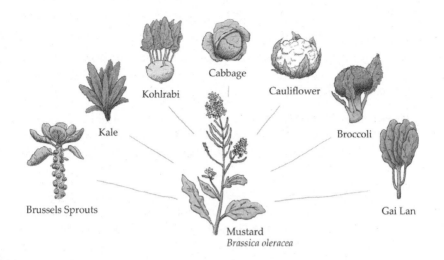

Kohlrabi

Cabbage

Cauliflower

Kale

Broccoli

Brussels Sprouts

Gai Lan

Mustard
Brassica oleracea

and the winners of conflicts: diversity, transmission, and selection. With these three ingredients, systems will inevitably converge on adaptive solutions – the law of evolution.

The genius in Darwin's theory of natural selection wasn't in the *selection*. It was in the *natural*. The realization that nature could do the selection if, in the competition for resources and energy, some survived better than others. That an organism's survival depended on its particular transmissible physiology, cognition, and behavior. But what are these traits trying to do?

Part of Darwin's great insight came from reading Thomas Malthus. Malthus argued that because animals reproduced faster than plants could grow, at some point the animal population would outstrip their food supply. We can now see that the plant population has an energy availability and EROI that imposes a carrying capacity on the herbivore population, which in turn has an EROI that imposes a carrying capacity on the size of the carnivore population. Each energy source leads to abundance and then scarcity.

Darwin's insight is that scarcity increases the strength of selection; that those who survive and thrive are those with the physiology, cognition, and behavior that leads to the highest energy returns. This might be by accessing a new food source more efficiently or by reducing energy requirements. Indeed, it is likely to be how some human populations evolved the genetic ability to drink milk and digest lactose beyond childhood – something that most humans and no other mammal can do. Famines may have provided the necessary selection pressure. Those with a mutation that led to lactose tolerance were able to access more calories from milk during these hard times.

By this same logic we can also now see how fossil fuels allowed us to escape the Malthusian trap – at least until scarcity catches up with us as our populations grow and nature's batteries deplete. We either fall back into the Malthusian trap of continual conflict once more or we move on to the next greater energy source toward abundance.

The EROI of fossil fuels is falling and availability is decreasing. The ceiling is coming down and the walls are closing in. We're all starting to feel the squeeze. Although we can't always articulate it, we can feel that there's less space and opportunity than there once was. Innovation is needed. But not just in efficiency. Our current energy is running out. The space of the possible has shrunk. To return to a win-win positive-sum world of excess energy, abundance, and high energy rewards per person, we need innovations in the energy technologies themselves. Without these innovations, our future looks bleak.

A hint of what's to come

When we first discovered oil, we only used the kerosene and wasted the rest. We literally burned off the natural gas, gasoline, and other non-kerosene products as useless waste. It's horrifying in hindsight. But over time competition led to greater efficiency. Just as our light bulbs became more energy efficient so too did everything else.

When a new energy source is discovered, initially we're not good at using it. Then organisms and organizations such as cells and societies learn to access and use the energy more efficiently, increasing the EROI of a given energy source – as is currently happening with solar panels. After an efficiency limit is reached the EROI either remains roughly the same – as in the case of hydropower (as long as the river keeps flowing), geothermal (as long as there is geothermal activity), and nuclear fusion (because the fuel is effectively unlimited). In other cases, if the energy source becomes more difficult to access then efficiency once again decreases.

Early innovations have larger efficiency gains, but then efficiency gains slow down and continued growth and progress require new, larger sources of energy or breakthroughs in efficiency. As we shall see, this arrival at carrying capacity and slowdown in innovation is where we are in the human story: the Great Stagnation as described by economist Tyler Cowen.

Entrepreneurship, engineering, and economics mostly exist to service the law of innovation. Entrepreneurs disrupt inefficiencies and, in doing so, create more efficient systems. Engineers develop technologies that allow you to do more, better, and/or more efficiently. And economists look at how to efficiently allocate scarce resources.

Innovations in efficiency are why Jeff Bezos is so rich. You can't beat two-day, one-day, or even same-day shipping, a massive marketplace, and easy customer returns. Amazon had several efficient innovations that allowed it to, first, outcompete bookshops, then strip shopping malls and high streets, and then web servers. Amazon Web Services (AWS) is now the largest web service and cloud computing platform and has become the most profitable part of Amazon's business. These innovations emerged from an evolutionary marketplace that once contained many now extinct competitors and competing products.

But all these innovations in efficiency are fundamentally different from innovations in energy. While they also expand the space of the possible by doing more with less, they are the equivalent of finding a bargain or being more frugal, which is fundamentally different from increasing your salary. Increasing your income always beats reducing your expenses.

As the sun sets on the era of fossil fuels, we need to prepare for tomorrow.

Many people push for a future based on renewable energy sources. Innovations in battery technologies and a sufficient energy surplus to pay the large start-up costs may lead to renewable sources meeting the current demands of our energy-hungry civilization. Perhaps even our immediate future demands. In the more distant future, the astonishing amount of energy released by the sun means that solar technologies in particular have the ability to far exceed our needs.

Physicist Freeman Dyson once proposed the idea of a Dyson

sphere – a megastructure encircling the sun and capturing a large percentage of its energy to meet the energy needs of our spacefaring descendants. A Dyson sphere or the more modest but still distant Dyson swarm or Dyson ring would allow us to reach fantastic levels of energy abundance. We're a long way from a Dyson anything, but the low EROI of solar means a lot of material is needed to build all the solar panels we need. That requires a massive start-up cost in energy and resources we simply don't have in Earth's discretionary budget. A fully solar future is not within reach of our current energy budget. Even if we were willing to pay the start-up cost in our life-styles, the shrinking space of the possible means that in the not-so-distant future our quality of life would continue to get worse. Proposals that require coercive population control not only require violence but will continue stagnation in innovation and further shrink the space of the possible. In effect, from where we are today, solar, wind, or any other renewable sources are insufficient to reach a new level of energy abundance.

The power of the sun

If you survey every energy source within our current or close technological capabilities, one stands out as having the necessary numbers to radically lift the human energy ceiling and enter the next level of abundance: nuclear.

Current *nuclear fission* technologies have abundant fuel and an EROI of 75. Nuclear fission exploits the enormous amounts of compressed energy created by massive supernovas and neutron-star mergers as they fused smaller elements into heavier elements such as uranium and plutonium. When these elements split, they release vast amounts of energy. As Einstein's equation $E = mc^2$ tells us, the loss of mass gets multiplied by the enormous number that is the speed of light (299,792,458 m/s), which is even larger when multiplied by itself in the squared term (9.0×10^{16}). That's the amount of energy you get back. Nuclear fission is more plentiful, better for the environment, and has a higher EROI than any fossil fuel.

Nuclear fission is a good bet for the immediate to medium future;

it ticks all the boxes in terms of availability, density, EROI, power, and start-up cost relative to return. It does, however, suffer from at least two challenges. But they're not what many people assume, the problems of nuclear waste and safety having been largely solved in modern reactors.

Nuclear waste is small because nuclear fuel is incredibly dense. The incredible devastation caused by the bomb dropped on Hiroshima was the result of just 64 kilograms of uranium. The bomb dropped on Nagasaki contained just 6 kilograms of plutonium. One ton of coal has about the same energy as 120 gallons of oil, both of which have the same amount of energy as a 1-inch enriched uranium pellet. The waste from nuclear reactors can be reused in other reactors that run on spent fuel, or can be placed in dry storage containers. These containers are lined with just twenty inches of concrete encased in half an inch of steel, which is more than enough of a barrier to stand next to a container without the need for special protective equipment. Indeed, in the Netherlands, you can sign up for a tour of the COVRA nuclear waste facility to see and stand next to the spent fuel containers yourself – such openness helps increase public trust and dispel fears. Depleted uranium, which is about one and a half times as dense as lead, is about as harmful as lead – don't eat it! Its radioactivity is lower than even unenriched uranium ore and doesn't penetrate human tissue. It's used in armor-piercing projectiles and in hospital radiation shielding. If Superman can't see through lead, he definitely can't see through depleted uranium.

Safety has also vastly improved, in a variety of ways. The three major nuclear power plant incidents – Chernobyl, Three Mile Island, and Fukushima – all used early technology in the form of pressurized, boiling-water reactors, which, as a friend in the nuclear industry aptly described, are giant, pressurized kettles where the water used to generate the power and the water circulating the heat from the react-ors mingle in a sealed unit inherently less safe than even the next generation of nuclear plants that followed. In contrast, now even old reactors developed in the late twentieth century, such as Canada's CANDU reactors, are much safer, with features such as separation of the water heated by the nuclear reaction from the water used to turn the turbines that generate the electricity. By using heavy water, it is also possible to use unenriched uranium, recycle spent fuel, or use other fuels such as weakly radioactive thorium.

Even more modern reactor designs are smaller, safer, and more flexible. There are many small modular reactor designs (SMRs), which can be about the size of a football field or two, and even smaller micro-reactors of the kind currently used in nuclear submarines and aircraft carriers. Micro-reactors range in size, from a car to a shipping container. In addition to generating power on Earth, they will play an essential role in space travel and mining other planets and asteroids. To judge the safety of present and future nuclear power plant designs based on our earliest designs from the 1950s would be similar to judging cars or airplanes as unsafe based on early models from that period.

In the United States in the 1950s there were around 60 to 70 deaths per billion miles travelled in cars. Today that number is 11 deaths per billion miles. In the 1950s, for every 10 million flights, over 400 resulted in fatalities. Today, that number is just 1 fatality for every 10 million flights.

Newer cars have become increasingly safe with the adoption of seat belts, airbags, crumple zones, ABS brakes, stability and traction control, and lane-keeping, and various accident-avoidance systems. So too in airplanes. And so too in nuclear technology.

There are, however, other real challenges to be overcome.

The first major challenge is nuclear proliferation. Given low levels of good governance and high probabilities of conflict in many countries, are Yemen or Zimbabwe ready for nuclear reactors? We will return to this thorny problem later.

The second major challenge to nuclear fission is the high cost of new nuclear reactors. Unlike many technologies, nuclear power plants have become more expensive and take a long time to build. This is in part due to regulations and in part due to insufficient innovation driven by fears of an earlier generation. We would be much better off today in terms of wealth and quality of life if we had pursued a nuclear future in the twentieth century.

Regulation is part of what ensures modern nuclear safety, but, as with the automobile industry, regulation is an area that is eminently open to innovation in the quest for reduced costs while still ensuring safety. The flexible SMR and micro-reactor designs also reduce this cost, increasing flexibility and scalability compared to current large monolithic nuclear reactor designs.

Despite these challenges, a mix of solar and nuclear fission in the future is both necessary and achievable. At a fundamental level, nuclear has the necessary physical properties to meet our energy needs this century. Greater investment and innovation in nuclear technologies alongside innovation in solar are the next step for humanity. But nuclear fission is not the next level of energy abundance. It will not substantially change our energy budgets in a way that will allow us to scalably explore the stars or mine asteroids for all those rare metals we need in the coming century.

To reach the next level of abundance we need *nuclear fusion*, the process that stars like our sun use to unlock energy. Fusion is the source of energy that has allowed for all life on this planet, that solar panels are harnessing, and that was turned into chemical energy through photosynthesis and stored in fossil fuels. It is the power of the sun.

In nuclear fusion, rather than splitting atoms, we combine them – two hydrogen atoms are fused into helium, releasing far more energy than when large elements are split in nuclear fission. Nuclear fusion has the potential to move us into effectively unlimited energy and permanent abundance. Fusion is clean and safe with no radioactive waste. It also uses the most abundant fuel in the universe – hydrogen. Although much innovation is required to increase the EROI of fusion, a nuclear-fusion-fueled future has effectively no ceiling. Were we to achieve fusion, we would have energy to desalinate our oceans to provide clean water, create new rivers and seas, mine asteroids for rare resources, and perhaps build that solar-paneled Dyson structure for our descendants. So when do we get fusion?

The arrival of fusion is perpetually somewhere between next Monday and the next thirty years. But we have little reason to believe it's anything other than a matter of when, not if. Nonetheless, there are many problems, some of which may prove more challenging than we expected. For example, nuclear fusion reactions require the rarer hydrogen isotope tritium, which has two neutrons unstably bound to the single proton. Tritium is rare. But for all these challenges, theoretical solutions exist. For example, our current conventional fission nuclear reactors, in addition to being major producers of the medical isotopes needed to diagnose and treat cancer, also produce tritium as a by-product. On a visit to a nuclear reactor, tritium was one of the particles I was constantly monitored for. Nuclear fusion reactors can also continually breed tritium during the fusion reaction through contact with a lithium 'breeding blanket'.

New advances and record outputs in the last decade have led, for the first time, to a burgeoning start-up industry in nuclear fusion firms, both private and state sponsored. Each company is pursuing different promising technologies with billions in funding. Nonetheless, energy scientist Vaclav Smil – much admired by Bill Gates – estimates that nuclear fusion is unlikely to replace current energy production any time before mid-century – 2050.

Nuclear fusion is going to require a lot of investment, innovation, and technological advancement, but, as we will see, many of the barriers to breaching then forever surpassing the next energy level are not physical or technological but rather social, economic, and psychological. Overcoming these challenges will be the largest return on investment in the history of our species.

Once we reach the next fusion-fueled energy level, we will enter a new era of peace and prosperity. It will make our current era, with all its conflicts, seem to our descendants as primitive and barbaric as we see the Middle Ages with its superstitions, witch burning, and horrifyingly brutal wars of conquest.

To solve the problem, we must first see it. To see it, we must understand our periodic table, our theory of everyone. We must understand how the laws of life created the human animal, our intelligence, innovation, ability to work together, and every aspect of our lives. Only then will it be obvious what needs to be done to reach the next levels of innovation, cooperation, and abundance.

2

The Human Animal

Humans are an astonishing animal. We are the descendants of African apes who stood up and walked across the world. We're not particularly fast, we're not particularly strong, and we're not particularly well equipped to thrive in the many places where we live. Our bodies don't have enough fat or fur for Canada's winters, enough water for Australia's deserts, nor the right proteins for processing many wild foods. And yet, despite our physical limitations, we've built towers that touch the sky, rocketed robots to Mars, and connected our planet through a worldwide communication network. How on earth did we do it?

When most animals encounter a new environment, they're forced to genetically adapt: developing powerful muscles to outrun local predators; fur and fat to prevent freezing; proteins to make plants less poisonous. But genetically, humans have changed very little. The secret to our success is not by genes alone.

Instead of evading local predators, we hunted them. Instead of evolving fat and fur, we wore the pelts of our prey. We didn't process poisons in our bodies, we processed the plants before we ate them. And we did all this by developing tools, techniques, and traditions. We did all this thanks to culture.

Civilizations and 'barbarians'

Six thousand years ago the first cities were established. Cities then, just as they are today, were a hubbub of human activity. All those people living side by side bickering and bartering with a continual supply of energy, resources, permanent migrants and temporary travelers created a hub in which ideas met and innovations emerged. City dwellers benefited from those innovations and began to consider themselves *civilized*. Rather than attribute their ways of thinking and technologies to the density of people, they assumed they were something special, in contrast to those around them – the *barbarians*.

Around 1000 BCE the Chinese of the Yellow River Valley distinguished themselves as Huá – the civilized – from the Yí – wild, uncivilized barbarians – that surrounded them. Around the same time the ancient Indians of the Indus Valley considered themselves surrounded by Mlecchá – inferior barbarians. Later, at around 500 BCE, Greeks whom both the Indians and Chinese would have considered barbarians, considered themselves civilized in relation to the surrounding foreign-speaking *bárbaros*, from which the English term 'barbarian' is derived. Those barbarians included the Romans, who later considered all non-Romans barbarians, from the trouser-wearing French Gauls to the forest-dwelling German tribes.

The Romans held an even lower regard for those further afield. Roman statesman Cicero, writing to his friend Atticus in the first century BCE, worried that there was little to be gained from invading Britain since 'there isn't a pennyweight of silver in that island, nor any hope of booty except from slaves, among whom I don't suppose you can expect any instructed in literature or music'. No doubt the Aztecs and Mayans of Central America and the Ghanaian, Mali, Yoruba, Great Zimbabwe, and other empires of Africa felt the same way about their neighbors. As people began to communicate with those beyond their immediate borders, we see the same attitude. Ironically, it was often those once labeled barbarians who, upon becoming the *nouveau riche* in innovation but lacking any memory of the past, began labeling the formerly civilized as barbarians. So the tides turn.

Writing in 1068, Said al-Andalusi, the mathematician, scientist, and qadi (sharia judge) of Toledo in Spain (then under Muslim rule), divided the world into the civilized who concern themselves with knowledge and higher learning, and everyone else. The civilized included the Arabs, Chaldeans, Egyptians, Greeks, Indians, Jews, Persians, and Romans. The rest he considered barbarians, with special mention for the Chinese and Turks as the 'noblest of the unlearned'.

Today, the terms 'civilized' and 'barbarian' have fallen out of fashion, but the same attitudes remain. Americans of the east and west coastal cities consider themselves an educated urban elite distinguishable from the rural, uneducated, fly-over states in the Midwest. Western, educated, industrialized, rich, and democratic (WEIRD) societies consider themselves developed compared to the still-developing nations of Africa, South America, and Asia – the majority world countries.

Humans in culturally and technologically more complex societies have also differed in their attitude as to whether 'primitive people' can be civilized. The Romans considered the Gauls more civilizable than the Germans. The British considered Indians more civilizable than Africans. Rudyard Kipling's *White Man's Burden* encouraged Americans to annex the Philippines and civilize the 'half devil and half child'. In contrast, Senator Benjamin Tillman of South Carolina, who was against taking up this civilizing mission, argued that the people of the Philippines were 'not suited to our institutions'. But while there was debate as to how the civilized should intervene in the lives of the uncivilized, what was agreed on was that some people were 'primitive'.

Let's start by unravelling this human universal thinking and then continue by explaining how and why people differ over time and geography.

Primitive people?

People around the world have different tools, techniques, traditions, and other aspects of culture to deal with the different problems they face. The systematic documentation of these solutions began with Franz Boas, a German physicist who would come to be known as the father of modern cultural anthropology. Boas was a child of nineteenth-century Europe. He was raised on the assumptions of the time, such as that of the hierarchy of societies from 'savage' to 'civilized' with Europeans at the pinnacle and everyone else somewhere below. These views were formalized in the cultural anthropology of the time. Another key figure was Edward Burnett Tylor, who was the first to offer a definition of culture that is often still used today: 'Culture or Civilization, taken in its wide ethnographic sense, is that complex whole which includes knowledge, belief, art, morals, law, custom, and any other capabilities and habits acquired by man as a member of society.'

At the bottom of Tylor's stages were 'savages', which literally meant people who lived in the wilds or in the woods, living off the land without agriculture. The Inuit, for example, were savage according to Tylor. Slightly better were 'barbarians', such as the Arabs – the same Arabs who would have previously considered Tylor and his fellow Englishmen to be barbarians. For Tylor, the pinnacle of civilization was of course to be found in Europe. To Tylor's credit, he rejected the Victorian belief in a hierarchy of races, believing instead in the universality of humankind. For Tylor, other societies weren't stupid in that they were lacking intelligence, but rather were lacking the knowledge needed to climb the hierarchy of civilization, which in the fullness of time they might achieve. In the late nineteenth century these assumptions about the hierarchy of humans began to be questioned. And this questioning would eventually lead to a more thorough understanding of the human animal and the critical role of society.

In 1883, hoping for adventure, with a head full of such ideas and a freshly minted doctorate in physics, the twenty-five-year-old Franz Boas boarded the *Germania* on a trip to Baffin Island in the far north of Canada, just west of Greenland. The *Germania*'s mission was to evacuate German scientists from the polar meteorological station. What Boas encountered on Baffin Island wasn't successful scientists living next to ignorant savages, but scientists who struggled to do anything without the Inuit. It shook his sensibilities. Despite coming from a society with a larger energy budget and complex technologies – it was, after all, the Germans visiting the Inuit and not vice versa – German technology and culture were ill-suited to the Arctic. In contrast, the Inuit had deep knowledge about how to live and thrive in this frozen, unforgiving environment. They knew the movements of the caribou and other animals, how to trap and hunt them, which had the best skins, and how to turn the skins into the warmest of clothing. Most impressive to Boas, the landscape of the world's fifth largest island, which seemed plain and formless to him and his fellow Germans, was to the Inuit rich in historical and navigable detail.

Boas was so fascinated that he gave up physics and spent the rest of his career trying to understand the ways in which the Inuit lived, survived, and thrived, eventually training his students, including Margaret Mead and Ruth Benedict, to do the same in different societies around the world. Rather than theorizing and speculating about faraway lands or studying long-lost ancient civilizations, these were new 'cultural anthropologists', living among people in different societies. They systematically documented how the people in each society were surviving and thriving in vastly different environments using vastly different tools, technologies, societal organizations, and traditions. Each society had different solutions for food, for shelter, for defense, for raising children, for maintaining social harmony, and for all the other problems all humans everywhere need to solve. But how did these societies figure out how to survive and thrive in such a range of environments in the first place?

One obvious answer was emerging in evolutionary biology. Charles Darwin had published *On the Origin of Species* in 1859 and *The Descent of Man* in 1871. Darwin himself knew nothing of genetics, but in 1900 – nearly two decades after he died – the careful pea plant breeding experiments of the obscure Augustinian monk Gregor Mendel were rediscovered. Mendel's work had revealed the existence and effect of genes, opening the space for a new field: population genetics.

Scientists, mathematicians, and statisticians spent the first half of the twentieth century reconciling Darwin's theory of 'natural selection' with genetics and the rest of biology in what became known as the Modern Synthesis. For example, how could it be that traits were transmitted in discrete genes, but when you looked at something like height, it was a bell curve? Mendel's experiments had revealed a mix of tall and short pea plants, with nothing in between. Shouldn't there be a mix of tall humans and short humans with not much in between, like Mendel's mix? Questions like these had to be mathematically and experimentally reconciled: the answer in this case is that most traits are polygenic, meaning that many separate genes all contribute to height, leading to what looks like a continuous bell-shaped distribution.

The Modern Synthesis forever changed the biological sciences, laying down mathematical theories, theoretical frameworks, and empirical evidence for how all animals developed solutions for food, defense, reproduction, and group living – they did it primarily through genetic evolution. Genes that helped an animal survive better persisted at the expense of genes that did not. And yet, this didn't seem quite enough to explain the success of humans in so many different ecosystems. For sure, humans in different places had different hair, skin, eyes, and noses that may have been adaptations to these ecosystems, but was this what explained their success and different solutions for survival?

Take a cheetah to the Arctic and she will shiver. Take a polar bear to the tropics and he will overheat. But Boas's anthropologists could learn from their indigenous hosts and, with their guidance and knowledge, live and survive despite having very different genes. Indeed, stories of lost European explorers, such as Robert Burke and William Wills in Australia and John Franklin in the Arctic, were appearing around the world. The difference in their survival seemed to be down not to their individual abilities or European technologies, but to their ability to access local people and local knowledge – how to find and process local foods, what plants to look for and which to avoid. There were no primitive people as envisioned by Tylor and others. People around the world were not succeeding because of what was in their genes, but because of what was in their societies. The peculiar cognitive skills and achievements of city dwellers were not due to some general difference in 'cleverness'. Instead, these specific skills and achievements arose from their access to ideas, metaphors, knowledge, and ways of thinking in larger, interconnected societies.

But how did their societies figure it all out? How did the descendants of an African ape end up learning how to survive in the Arctic? The answer to this question would finally emerge at the intersection of evolutionary biology and cultural anthropology.

The rational animal?

The human animal is like any other animal in many ways. Humans eat, poop, mate, reproduce, and then die. But in terms of other dimensions, such as brain size or size of societies, we are an outlier. How this happened or why it didn't happen for other animals isn't clear, but it seems obvious that somehow evolution selected for intelligence in our species. Philosophers of the nineteenth century described humans as the rational animal. Humans were capable of causal reasoning and understood their world in a way that other animals did not, they said. But new experiments in biology and psychology in the late twentieth and early twenty-first centuries, pitting humans against their closest cousins, chimpanzees, revealed that 'man the rational animal' might be an illusion. Economists were coming to the same conclusion at the same time, while going through a behavioral science revolution.

Let's start with the chimps.

Battle of the apes

Primatologists study non-human primates. Developmental psychologists study human children. Scientists in these two disciplines often collaborate on experiments that pit human children against young chimps and other apes. For example, one set of experiments gave children, chimps, and orangutans a battery of cognitive tests – think of it as an ape IQ test. Half the test involved physical cognition, ranging from having to mentally rotate an object to adding quantities, to causal reasoning, to understanding the functional and non-functional properties of tools. The performances of the children and chimps were indistinguishable; the orangutans did a little worse, but not by much. The other half of the test involved social cognition and ranged from learning a solution from someone else to understanding and producing communicative cues, to figuring out what someone else is trying to do. Here there was no competition, the children beat their chimp and orangutan cousins, particularly when it came to learning from others. The kids killed it when it came to *social learning*.

The results, published in the journal *Science* in 2007, confirmed what the *sapiens* in *Homo sapiens* really means: 'wise'. But we're wise not because we're particularly rational, logical, or good at causal reasoning. We're wise because wise people learn from others. Our species learned, as my collaborator Joe Henrich once put it, that it's better to be social than smart. But it's more than that – by being social, we could become smart.

One experiment that clearly illustrates this social learning difference took place in two very different locations. The first is St Andrews,

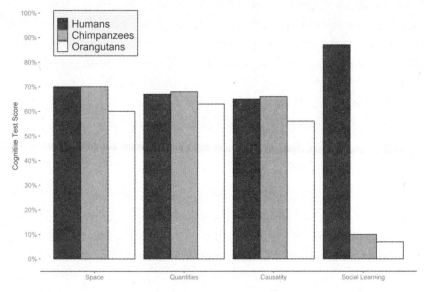

Based on graph and data from: E. Herrmann, J. Call, M. V. Hernández-Lloreda, B. Hare, and M. Tomasello (2007), 'Humans Have Evolved Specialized Skills of Social Cognition: The Cultural Intelligence Hypothesis', *Science*, 317(5843), 1360–6.

Scotland, the birthplace of golf and home of the University of St Andrews, where Prince William met his future wife Catherine (Kate) Middleton. The second is Ngamba Island Chimpanzee Sanctuary, Uganda, home to around fifty or so chimpanzees that have been rescued from across East Africa.

In 2005 two researchers from St Andrews – Victoria Horner and Andrew Whiten – gathered a group of young Scottish children and a group of young Ngamban chimps and gave them the choice to copy or think for themselves. The experimenters presented both groups with a black box. The box had a hole on the top and a hole on the side. Inside the box was a reward: a piece of fruit for the chimps, a sticker for the children. The experimenter showed the chimps and the children how to get the reward by poking a stick through the top hole and then the side hole. They then handed the stick to the chimps.

If you've ever seen videos of chimps speaking with a language board, scrolling Instagram, or whizzing through a working memory task, you'll know that they are smart. They imitated the experimenters perfectly, first poking the stick through the top and then through the side to get the fruit. Happy chimps.

The experimenters then did the same thing with the children – showing them how to poke the stick through the top hole and then the side hole. The children, just like the chimps, poked the stick through the top and then through the side to get the sticker. Happy children.

Then came the key treatment condition. The experimenters replaced the black box with a clear, transparent box that was otherwise identical. Because the box was clear, both the chimps and the children could now see that the first action – poking the stick through the top hole – did nothing. In fact there was a separation inside the box, so only the side hole accessed the reward. The top hole was irrelevant. But again, the experimenters poked the stick through the top hole and then through the side hole. They handed the stick to the chimps. What did the chimps do? Chimps are smart! They ignored the top hole and just retrieved the fruit from the side hole. They engaged in what scientists call 'emulation' rather than 'imitation'.

Then, once again the experimenters poked the stick through the

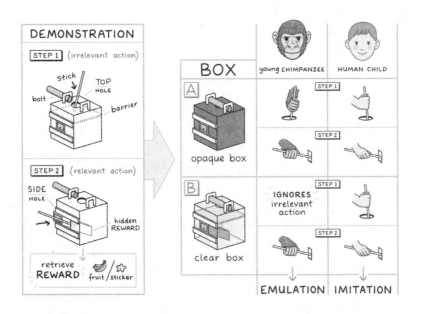

top hole and then through the side hole. They then handed the stick to the children. What did the children do? Children are smart! Or are they? The kids continued to poke the stick through the irrelevant top hole and then through the side hole.

It's not that the children didn't understand the causality – later experiments confirmed that they understood it just as well as the chimps. But the children assumed that the adults knew something that they didn't. So instead of emulating and trying to reverse engineer the reasons for the adults' actions, they simply imitated. They copied all actions because they assumed that the seemingly pointless two-poke method is *just how people do it*. Human children ape better than apes do. And in so doing, the human children weren't making a mistake, they were becoming brilliant.

By copying successful behaviors, beliefs, tools, and ways of thinking from the previous generation, even without understanding why these work, human children engage in a process unique to our species that allows them to surpass the limits of their own cognitive abilities. It is a cultural evolutionary process that has led to antibiotics, democratic governments, and nuclear reactors. It has allowed humans to build on each other's work and to take for granted the work already done in the past in order to narrow the set of things we presently need to develop and innovate on. Chimps, on the other hand, still try to figure out everything on their own, and so, as comparative psychologist Bennett Galef famously put it, still sit naked in the rain.

In this case, the children were mistaken about one thing: they assumed that the adults were demonstrating the best strategy. In fact the adults were deliberately inefficient for the purposes of the experiment (which also hints about movements and moments of mass human folly, which I will discuss later).

But in the world beyond this experiment, and by and large as far as human progress has been concerned, children learn from relatively wise adults who, when they themselves were children, imitated their adults, who when they were children imitated their adults, and so on back in time. Each generation of children, the next generation of adults, thereby acquired a head full of successful recipes – tools, techniques, and traditions. Not through understanding but by selective trust. Our lives are filled with acquired recipes, the origins of which we have long forgotten. This reliance on socially transmitted information is, in essence, a shortcut to brilliance.

When we cook, we cut the meat a certain way and add salt at the beginning or end, because that's how our parents taught us to cook, or because that's what Gordon Ramsay told us to do (no one wants Gordon swearing at them). You don't need to understand the biology or chemistry of cooking to produce a tasty, cooked meal full of pre-digested and bioavailable calories and nutrients. We check our emails on computers, a technology that, for many, may as well be magic. A sequence of clicks takes you to your inbox, crammed with more emails than you can get through. If the Wi-Fi goes down, turning it off and on again might fix it. There's no need to understand how a processor, software, and random access memory produce an image on the screen or how your router can use radio waves to show you Facebook.

The children were right to copy the adults but were wrong to assume that the adults knew what they were doing. The funny thing is, although adults often don't know what they're doing or why, they often *feel* like they do. They have what psychologists call an *illusion of explanatory depth*.

Ignorance is blasé

The illusion of explanatory depth is perhaps one of the most profound but lesser-known discoveries in psychology. It's so profound and impossible to irradicate that even after reading this section you might not believe it. But I feel compelled to mention it before we continue. The essence of the bias is that we all think we know more than we actually do. We assume we understand and have reasonable causal models for our beliefs, behaviors, and the technologies we use – it threatens our sense of self to think otherwise. The illusion is only shattered when we're tested. When we're asked to explain specifics.

In the now classic experiments, researchers first asked people if they understood how a flushing toilet works. Flushing toilets are a centuries-old technology that, with a little luck and enough fiber, you interact with every day. You probably have some sense that you understand how they work. But think about it – if someone asked you, how does the water in the bowl flush everything away and return to the same level? – could you explain it? Could you make a toilet without copying another one? In expounding the details, you might realize your understanding is shallower than you initially believed. Or take a wine bottle and turn it upside down. You'll see a punt on the bottom – any idea why? In other cases, you might even have explanations you think you understand. For example, public health messaging warns us to complete our courses of antibiotics. Why? To prevent superbugs. But surely, since not all bacteria will die, the fittest that survive will be the *most* antibiotic resistant, no? Surely finishing your antibiotics is what would *lead* to superbugs!

If you look up the answers and learn about toilet pressure differentials, the geometry and history of wine bottles, and selection pressures on antibiotic kill curves, you will once again feel like you understand the mechanisms. Indeed, the answer may feel obvious in hindsight. This feeling too is part of the same illusion. You can't shake it and some people refuse to accept it. The illusion of explanatory depth is a quirk of our psychology that is almost impossible to overcome.

Right now, you may be sitting on a chair or lying on a bed or driving a car that you could not recreate from scratch (processing wood, mining ore and making metal, creating and then weaving fibers, and so on). You are reading a book whose production process is more complicated than you think. This illusion has many implications.

One implication is that other people's problems have a seductive reductiveness. The solutions to your friend's relationship or career problems seem so obvious to you. Perhaps even the solutions to societal problems, especially if they're another society's problems. In all these cases, your psychology shields you from the true complexity of any issue.

All of this is to briefly illustrate that the world is not only complicated but even more complicated than our psychology allows us to believe. But if we don't have these causal models then who does? Where did the Inuit, Wall Street bankers, Silicon Valley superstars, or any of us for that matter, get all the information that allows us to succeed? Where did it ultimately come from? How do we know what to do?

Founts of knowledge

Three sources of information guide our behavior. First, there's what you learn from your own life experience. In the 1930s Harvard psychologist B. F. Skinner put rats into boxes and rewarded them when they pushed one lever and shocked them when they pushed another. The rats quickly learned which lever to push. Skinner laid down the laws of reinforcement learning, laws which would eventually lead to modern machine learning.

The process is simple: if you want your dog to poop in the right place, reward them when they do so and scold them when they do not. All animals learn through this process of pleasure and pain. When we touch a hot stove, we get burned and we don't do it again. When a pick-up line lands, we might try it again. But we're not blank slates. We're born with preferences. Even without reinforcement, rats and humans will naturally prefer sweetness over bitterness: bananas are tastier than Brussels sprouts. And that's thanks to genetic evolution.

Genetic evolution is the second source of information. It is also a kind of reinforcement learning, but one in which the delay on feedback from actions is too long for individual reinforcement learning to learn. Sweet compounds predict nutrients and bitter compounds predict poisons, but by the time you learn that rule, through trial and error, you'd be dead. And that's what happened – mammals that coded sweet sugars as tasty and bitter alkaloids as not tasty were rewarded with better health, longer lives, and more offspring than those that did not. And so those preferences persisted over generations all the way to us.

All animals, including us, have a genetic inheritance from our parents. Things that historically helped us survive, thrive, and reproduce feel pleasurable, and things that might harm, kill, or end our lineage feel painful. In humans, these genes affect, for example, the shape of our eyes, color of our skin, blood type, ability to drink milk in adulthood (unique to a minority of human populations and unique among all mammals), and our propensity for various diseases. These genes are also correlated with cognitive and behavioral differences.

But the law of evolution discovered a new way to use reinforcement learning in humans: not just genes, not just individual learning, but an extensive cultural inheritance.

When a baby is born, it spends the next two to three decades catching up on the last several thousand years of human history. Sometimes, we need to override our genetic predispositions – we need to learn to like the bitterness in broccoli and Brussels sprouts. We spend our earliest years acquiring beliefs, values, norms, traditions, technologies, habits, and behaviors – in short, culture.

Culture is socially transmitted information; when we communicate, we're creating and evolving culture. When we salaciously gossip about people's love lives and life choices, secretly judge or copy their parenting style, or give and withhold hearts on social media; when we learn something new or decide to try something different; when we inspire others or are inspired by others – we are changing culture. This realization that humans inherit both genetic and cultural information became known as dual inheritance theory. The rules by which culture evolves became known as cultural evolution. The mathematical theory of our dual inheritance and cultural evolution began with the question, when will natural selection lead to the evolution of learning from others?

The evolution of cultural evolution

Genetic evolution leads to genes adapting to environments, but the math says that socially learning from others will be favored not for a particular environment but when the environment is changing. Specifically, when it is changing a little, but not too much. Like in the fairy tale Goldilocks and the Three Bears, the Goldilocks zone of environmental variability is not too stable and not too unstable. This Goldilocks zone is dependent on an animal's lifespan and generation length, but when the environment fluctuates for long enough within this zone, any animal would focus on what others were doing rather than learning by themselves or blindly following their genetic instincts. We can understand one of the key insights from this model without diving into the math.

When the environment is highly stable – that is, when things are not changing over long periods of time, then genes and genetic evolution are the best way to adapt. Consider something like the amount of sunlight, which varies seasonally but is otherwise stable as a function of latitude (i.e. there's more sunshine in Australia than Austria). Here a genetic solution would be favored, as is the case with human skin color, which optimizes for UV radiation. If you're a dark-skinned person living in Europe, you risk vitamin D deficiency without sufficient supplements, and if you're a light-skinned person living in Australia, you risk skin cancer without sufficient sunscreen. Human skin color is a genetic adaptation to stable levels of sun exposure. The effects of sun exposure are too delayed to learn individually or culturally.

At the other extreme, if the environment is highly unstable then this selects for large brains and trial-and-error learning. Today the water is here, tomorrow it's there. Today we eat blue berries, tomorrow they're gone and we need to figure out if those red berries are edible. The environment is changing too quickly for both genetic evolution and the knowledge of our ancestors and so we must rely on individual reinforcement learning – trial and error. But what if the environment is changing too quickly for genes but too slowly for individual learning? Between these two extremes, where the environment fluctuates in a way that matches generation length, is where your parents and grand-parents have some knowledge worth learning.

Imagine you're a child living in an ancient society. We haven't yet invented physics, chemistry, biology, economics, or psychology. Your society goes through a cyclical drought. You may have never experienced a drought and maybe your parents haven't either, but Grandma remembers that when she was a child there was a drought and they went left of the mountain, beyond the forest, and found water, and so with this knowledge, she leads her tribe to safety. Under these conditions, past generations have knowledge worth paying attention to, and so it pays to socially learn. Indeed, this is also why respect for the elderly decreases during times of rapid technological and cultural change. When the world is quickly changing, old knowledge becomes out of date, losing the elderly their place in society as founts of knowledge.

Without this ancestral optimally variable environment, people wouldn't care as much about what others were doing. Instagram would never have been invented. This reliance on social learning moved us from relying on causal reasoning to relying on *cultural reasoning*.

Cultural reasoning

Sometimes what it takes for science to move forward is to let go of assumptions. Earth, for example, looks pretty flat and pretty stable, and our eyes clearly see a sun tracing the sky from east to west. But when we let go of the assumption of a flat earth and moving sun, we get a better model of the solar system. And with this knowledge we can go to Mars. Time feels like it ought to flow the same for all people and all places, but when we let go of that assumption, we get a better model of the universe. As the 2014 movie *Interstellar* correctly taught us, time literally moves faster at higher altitudes. And with this knowledge we can build a global positioning system (GPS). The biological world looks pretty chaotic and humans are clearly set apart, but when we let go of that assumption, we get a better model of nature. And with that knowledge we can edit genes and engineer messenger ribonucleic acid (mRNA).

The seemingly obvious assumption that has held back the human and social sciences is the assumption that human uniqueness is grounded in human intelligence. Don't get me wrong, we are bright. But not because of our ability to reason. We can reason to some degree, but without culture we can't do it much better than other animals. I imagine you're at least a little skeptical at this point – this assumption is hard to see past, but persist if you can. Because when we let go of the assumption that human intelligence is a product of our hardware and not our software then we get a better model of ourselves. And with that knowledge we not only get to a theory of everyone, but we can use that knowledge to build a better society.

Within economics, the much maligned 'economic man' or *Homo economicus* was based on nineteenth-century philosophical views about humans as the rational animal, mathematically formalized in the twentieth century using physics-inspired approaches complete with metaphors for friction and elasticity. But that assumption is wrong.

At least three Nobel Prizes in Economics have been awarded to work revealing different ways in which these rational assumptions fail to describe actual human behavior. Herbert Simon won the Nobel Prize in 1978 for his theory that rationality must be bounded by limited information, limited time, and limited cognitive capacity. We just don't ever have enough information, the time to figure out all the possibilities, or even the brain space to be truly rational. Daniel Kahneman won the Nobel Prize in 2002 for his work with Amos Tversky on prospect theory, a particularly prominent example of seeming irrationality – loss aversion – that people are more averse to a possible loss than they value the same possible gain. If I offered a flip of a coin to win $100,100 or lose $100,000, most people won't take the coin flip despite odds in their favor. Most recently, Richard Thaler won the Nobel Prize in 2017 for his work on nudging – applying our irrational biases to understand economic decision-making and applying these insights to successfully change behavior. Organ donation rates on death, for example, were close to 100% in Austria but around 12% in Germany. That's not because Austrians are nicer than their German cousins (I'll leave it to my German-speaking friends to debate this). The difference is simply that in Germany people have to opt in to donate their organs whereas Austrians had to opt out to not donate their organs. The difference was the default option. People usually stick with the default for decisions that don't have an impact on their immediate lives.

If you don't believe that you're irrational, perhaps you can at least believe that other people are. We often feel like we're the only rational person in the room. And so does everyone else. This body of work in behavioral economics is a slow realization that even if it feels like we're rational (and simultaneously that others aren't!), our behavior is not really rational, at least not in the sense of individually maximizing utility or being consistent in our preferences. Instead, we are a product of *evolutionary rationality* – fitness maximized and preferences tuned by millions of years of genetic evolution, thousands of years of cultural evolution, and a short lifetime of experience.

Herbert Simon won his Nobel Prize for realizing that we have limited information, limited time, and limited cognitive capacity. But long before Simon, evolution realized it too. Since we didn't have information, time, or intelligence to compute the answers by ourselves, we distributed the computation across the population and solved the answer slowly, generation by generation. Then all we had to do was socially learn the right answers. Using this approach, we could limit our reason and causal understanding to minor tweaks with partial causal models of the world. You don't need to understand how your computer or toilet works, you just need to be able to use the interface and flush. All that needs to be transmitted are which buttons to push – essentially how to interact with technologies rather than how they work. And so instead of holding more information than we have mental capacity for and indeed need to know, we could dedicate our large brains to a small sliver of a giant calculation. We understand things well enough to benefit from them or attempt to make improvements, but all the while we are making small calculations that contribute to a larger whole – like a wisdom of the crowds. We are just doing our part in a larger computation for our societies' collective brains. Yet sometimes our societies' wisdom of the crowds is instead madness of the masses.

This evolutionary rationality doesn't always arrive at the right answer. The answer can be mismatched, leading us to overweigh losses relative to gains, overeat cake over cauliflower, and over-rely on information we hear a lot. And these sources of information can give different signals – genetic evolution doesn't necessarily want you to be happy, your culture might lead you to fewer children, and your lifetime of experience might find shortcuts to pleasure from pornography to Pringles that simply don't optimize genetic, cultural, or even physical fitness. But on average, evolutionary rationality is or was adaptive. (We'll return to this shortly.)

Is it really true that we can't causally reason? On the one hand, experiments show that even simple feedback systems are difficult for people to disentangle. On the other hand, some researchers argue that the difference between what artificial intelligence can do and what human intelligence can do is causal reasoning. Experiments with children seem to show that they can discover simple causal patterns sometimes better than adults. And certainly it feels like we're capable of what Kahneman called System 2, deliberate causal reasoning. But at the same time, we're flush with systematic mistakes. There's a reason why students are warned that correlation isn't causation – we tend to assume temporal precedence or non-causal association is causal, which leads us to superstitions and repeating what we did last time. And there's a reason we try to make people aware of common logical fallacies like *ad hominem* – we naturally evaluate the person behind the argument as much or more than the argument itself. And there's a reason psychology gave up on Freudian introspection – we are just not a good guide to our own thinking. Most humans are not logical analytic philosophers, but we do have a limited form of causal reasoning. It's a messy kind of logic called *enthymemes*.

Enthymemes are a form of cultural reasoning: pseudo-logical reasoning with unstated premises filled by assumptions shared by others in our society or culture; assumptions that we have inherited from generations gone by. Take the classic deduction 'All humans are mortal. Socrates is human. Therefore Socrates is mortal,' as an example. This is something *Star Trek*'s favorite Vulcan, Mr. Spock might say, but most humans won't. Instead, humans reason along these lines: 'Socrates is mortal because he's human.' The unstated premise is that humans are mortal, which we learn vicariously even before we first experience a death. We do this all the time. 'She is sick since she has a fever'; 'Benjamin is a typical Canadian therefore he is polite.'

We find cultural causal reasoning intuitive in a way that formal logic is not. For example, this is difficult: 'If it is a Monday, the local band wear their uniform. The local band is wearing their uniform. Therefore it is a Monday.' Is that logical? This version with the same structure is much easier: 'If I get wet, I need to change my clothes. I changed my clothes. Therefore I got wet.' When restated in cultural causal terms, the illogical nature of both statements becomes more obvious.

An experiment that reveals the difference between causal reasoning and cultural reasoning is the Wason selection task. Given the following choice, which cards must you turn over in order to test the truth of the rule 'If a card shows an even number on one face, then its opposite face is a star'?

If you haven't seen it before, try it. No, seriously, give it a go and remember your answer.

When the same logic is stated in cultural causal terms, the answer is immediately obvious. Which cards must you turn over to test the rule 'If you are drinking alcohol, then you must be over twenty-one'?

Incidentally, the typical answer to the first case is the number 8 card and the star card. The correct answer is the number 8 card and the triangle card since you're looking for a violation of the rule to falsify it. In the second case, it's more obvious that you need to check the beer drinker and the sixteen-year-old. The difference between the two cases is typically discussed as a tendency toward detecting cheaters – that humans have an evolved tendency toward seeing who is cheating, in this case drinking alcohol when they shouldn't be, but I argue that it's part of a more general psychology of cultural reasoning, in this case about norms.

We frame our arguments in enthymemes and cultural causal logic all the time. Scientists do too and will often fail to realize the degree to which they are doing this until they try to model their theories using mathematics or computation. Suddenly all those assumptions and all that unstated and culturally shared logic in our verbal theories has to be precisely stated. The unstated invalid assumptions and leaps of logic become transparently obvious. Which is why many sciences favor formal mathematical and computational models in the development of theory, which force the modeler to write down all assumptions and logic.

Nonetheless cultural evolution has improved our ability to use logic over centuries and transmitted that improved formal reasoning ability through our education system. Everyone around us is educated, so it leads to the impression that the kind of formal rationality we value in our society is what makes us different from other animals. In other words, thanks to schools, we're a lot better at actual formal causal reasoning than we once were. To see this, we need to look at populations without schooling.

In the 1930s Russian psychologist Alexander Luria wanted to understand the effects of education on cognition. He headed to Uzbekistan, where an educational revolution was occurring: schools were being introduced to farming communities with no previous formal education. Luria ran some simple tests. What he found shocked many people.

We often consider reason and logic to be a part of what separates us from other animals. (In fact, this too is not quite true – even fish and fruit flies have some logical reasoning abilities.) Luria tested simple logical propositions such as 'If p then q', famously explaining once that 'In the far north, where there is snow, all bears are white.' As part of his test he asked participants the following: 'Novaya Zemlya is in the far north, and there is always snow there. What color are the bears there?' This was a simple question readily answered by even young children in our society and individuals with access to education in Uzbekistan. But this simple question was met with puzzlement from those not exposed to formal education. Here is a typical response: 'I don't know. I've seen a black bear; I've never seen any others.'

Luria's results are hard to believe and, in any case, the early twentieth century didn't have the fancy experimental controls and statistical methods we now have. So anthropologist Helen Elizabeth Davis and I decided to run the same experiment among the herding Himba people of Namibia and Angola. We found what Luria found.

The Himba, who never went to school, behaved the exact same way as the unschooled Uzbeks. For example, Helen asked them: 'There is a country in which boats are made of sand. I have a boat from this country. What is it made from?' The answers from those without formal education were similarly 'I don't know' or 'Probably wood'. The willingness to deal with hypotheticals and ease of reasoning in this way is a *taught* skill, culturally evolved and transmitted through our education system. The Himba who never went to school can reason, but the way they reason is different from the Himba who went to school. Today, most people we know have some amount of elementary and high-school education, and these concepts and ways of thinking are embedded, reinforced, and implicitly taught in TV shows, books, and how people around us speak. Children emerge in this world and readily learn to reason more formally. This ubiquity of more formal reasoning in our society tricks us into believing that humans are naturally rational animals.

Innovations occur as an evolutionary process, a giant calculation of the collective brains of our societies. A key feature of this system is not just social learning but knowing who to socially learn from.

Says who?

Let's return to that child living in an ancient society before modern science and technology. In that child's village lives a hunter. Let's call him Bruce.

Bruce isn't just any hunter, he's the best hunter. He's strong and well respected. Bruce brings in the most game and has the biggest house. Men are envious, women have crushes, and all the little kids want to be like Bruce. But the trouble is, they don't know exactly what it is that makes Bruce successful. There's no evidence-based, peer-reviewed study of successful hunters; no business or self-help books to tell them how to unlock their inner hunter. The kids just know that Bruce is cool.

They may use their informal cultural reasoning to hypothesize that some things may be more related to Bruce's success – the weapons he uses, the time of day he goes hunting, or where he hunts. But the world is complicated and contains many complex causal relationships. Bruce's success may be because of his weapons and training, but could also be because of the shoes he wears, the beard he sports, or the gods he worships. The best strategy a little hunter can adopt is to copy everything about Bruce and hope that maybe they will grow up to be just like him.

That psychology, a pay-off strategy, is still with us today. It's why celebrities have undue influence in areas that have nothing to do with their expertise. It's why Beyoncé's preference for Pepsi or perfume might increase sales despite neither what she drinks nor how she smells having anything to do with her rise to fame. It explains the Kate Middleton effect, where dresses worn by the Princess of Wales quickly sell out in Britain. And it's why you once believed fat made you fat (which makes sense: eat fat get fat) and then believed that sugar made you fat (which also somehow makes sense: unused calories are stored as fat) and now maybe believe it's calories overall. Or why you floss – or at least tell your dentist that you floss – despite a weak evidence base for its effectiveness. Or, indeed, it's why you hold many beliefs, about eating locally (often a higher carbon footprint because transport costs are quite low), recycling (often not economically worth it and so ultimately ending up in the landfill), supporting or not supporting big-game hunting (often an old or problematic animal is killed to provide funds that would otherwise not be there to support conservation), or even that the Earth is spherical (this one is true, or more accurately the Earth is an approximate oblate spheroid). It's also why many approaches to fake news and misinformation fail. It's not about the information; it's about whom we trust. So how do we acquire these beliefs?

Remember, any adaptive evolutionary system – artificial, natural, cultural, or even a genetic algorithm in machine learning – requires just three ingredients:

1. Diversity: things must vary.
2. Transmission: they must be transmitted without too much loss of information.
3. Selection: more adaptive traits must be transmitted or persist better than less adaptive traits.

All three are present in the cultural evolutionary system of human social learning.

Diversity is a fact of life – people do all kinds of things for all kinds of reasons. Some parents are tiger moms and dads, others prioritize creativity and free expression. Some people like the sciences, others like the humanities. Some people are extraverts, risk takers, with short-term mating on their mind; others are the opposite. The key to cultural evolution lies in transmission and selection.

Transmission needs to be faithful – not too much information should be lost. In genetic evolution, this is achieved by low mutation rates when genes are transmitted from parent to child. In cultural evolution, this is achieved by imitation rather than emulation when we learn from others. Remember Horner and Whiten's experiment with young Ngamban chimps and young Scottish children? Both animals were socially learning, but the chimps loosely emulated and the children faithfully imitated. The children copied without understanding. But no one copies at random.

We copy more when we're uncertain or when the cost of figuring it out by ourselves is high. We copy people who are successful, who are experts, who others are copying. We copy people who are similar to us. We copy less as we get older. And we copy people who meet these requirements, but especially if they seem honest and sincere. It's not a perfect process, but over generations it filters the best behaviors and beliefs into the population until the population itself becomes a source of learning. At this point, you can start following trends and conforming to the majority.

We even use these strategies to learn how to learn. You weren't born with an understanding that a suit signals success, unless you're in Silicon Valley, and then its more a signal that you're in sales. You learned that. Nor that a nice typeface meant *New York Times*-level credibility. That's a learned strategy and the reason that your grandparents often confuse weird websites with newspaper-like web templates as legitimate sources of information. When you decide who to invest with on eToro or who to follow on Instagram, you're probably looking at popularity and who is already following the person – applying conformity to infer success – a so-called prestige strategy.

The law of evolution found a new way to increase efficiency and cooperation and harness energy – by giving us rules for who to learn from that reinforced beliefs, behaviors, and technologies that worked and discarded those that did not.

Learning strategies

We can think about these strategies as homing in on who's got knowledge worth paying attention to, limited by how much information we have about them. If we've seen someone succeed by becoming a football player, we seek out those with the best football skills to learn from – a skill or expert strategy.

But sometimes, we don't know which specific skills lead to success and so, instead, we use cues of success overall such as the appearance of wealth – a success strategy. We identify those who are successful and copy a bit more generally, not knowing what exact traits led to their success.

Further, we may not even know who is successful or what constitutes success, and so we look to see who others are paying attention to and, not knowing which traits matter, over-copy these individuals – a prestige strategy.

The prestige strategy helps to explain the famous-for-being-famous phenomenon – a self-created false signal of success that becomes real. It's a bootstrapping of the prestige strategy. We copy someone because other people are copying them (or appear to be copying them).

Even our prestige strategy leads us to be more nuanced than just the number of connections a person has. We care about who those connections are. If a person on Twitter is followed by Barack Obama and Elon Musk, that counts for more than a greater number of less-known followers. We can quantify prestige using what's called eigenvector centrality, which is also how Google got so good at ranking Web pages. Before Larry Page and Sergey Brin's algorithm, search engines used the contents of pages to help you find them. In the world before Google, when you searched for 'fish' you'd get pages that mentioned fish. In contrast, PageRank effectively calculated the prestige of a website by looking at how many and which other pages that mention fish linked to it. Using the same technique that evolution also discovered led to search results that were so much better. Google completely replaced Ask Jeeves, AltaVista, Dogpile, and all the other '90s search engines.

Alongside the prestige strategy there may also be a pariah strategy – an over-avoidance of traits possessed by those who others avoid. For example, the name Adolf and the toothbrush moustache were both popular in Germany when Adolf Hitler became Chancellor. But both precipitously fell in popularity following the end of the Second World War because of their association with him. And yet, it is unlikely that being christened Hans Hitler and sporting a goatee would have prevented Hitler's evil atrocities. We have a natural inclination to avoid even associations with failure, unpopularity, or, in this case, evil.

When people in a society deploy these social learning strategies, they filter out things that don't work and make things that do more popular. And so, an even easier way to absorb a package of adaptive traits is to just copy what most people do – conformity. In WEIRD societies we think of conformity as something to avoid. But the reason we warn people not to conform is because it's such a strong human tendency. The negativity around being conformist is a mostly Western idea. Many societies assume those who can learn from and copy others are the smart ones.

We unconsciously conform and apply these strategies to all aspects of our lives. One of my favorite examples is a neuroscience study that shows that we look to others to learn who is attractive.

When our friends find someone attractive and say so, the target of their hot-or-not judgement goes up in our own estimation. We might predict that the effect is stronger if we find our friends attractive. Thus, even though attractiveness in the context of mating is a product of well-conserved cues for health and fertility, such as symmetry and clear skin, it is also subject to cultural evolutionary forces. And this explains changing standards of beauty.

Once upon a time, having more fat revealed that you had enough calories and were therefore high status and wealthy. Today it indicates the opposite; that you don't have access to high-quality foods or time to exercise. Once upon a time, being pale meant you were probably wealthy and were able to spend time indoors and being tanned meant you worked in the fields. Today being pale indicates you're stuck in an office and have less time for vacations and basking in the sunshine.

These social learning strategies are incredibly clever, as we shall see. While other animals also socially learn, humans have the most sophisticated social learning repertoire in the animal kingdom. We are the most strategic social learners.

Copying strategically

We are not blind copiers. For example, when we deploy the above strategies, we consider their relevance to ourselves. Michael Phelps, the Baltimore Bullet, credits his outstanding swimming success to copying his GOAT predecessor Ian Thorpe, the Thorpedo. Boys will tend to focus more on older males, girls on older females. And people will often pay more attention to those of the same ethnic, political, or social group, or who have things in common with them. This tendency to copy those like ourselves is why representation and role models are so important.

We also consider the sincerity or accuracy of beliefs. We hate hypocrites and pay attention to signs of insincerity. For example, if you met someone who said they took a particular supplement every day that made them super-productive and offered you some, you should be suspicious. First, you would evaluate whether they really are successful and second whether they actually took the supplement. Consuming supplements can be costly or even dangerous and you want to be sure that at a minimum you're copying an actually successful person who actually believes that the supplements are helping them. Seeing them take the supplement every day is a much stronger signal for copying. The founder of Nike, Phil Knight, intuitively understood this. Rather than listing all the features of his shoes and telling you how amazing they were, he made the brilliant business decision to sponsor sports stars. When people see sportspeople actually *wearing* Nike while competing and ideally winning, it's far more persuasive than any verbal endorsement.

None of this is to say that every weird belief of a successful celebrity is adaptive, but it is more likely to get copied. Cultural evolution is the process by which these social learning strategies over time weed out things that don't work on average and keep things that do.

Costly and sincere cues are more important when the relationship between a behavior and successful outcomes is less obvious. The transmission of religion offers a compelling example, because religious beliefs are often not directly observed and the causality is often not immediately obvious. Therefore, the transmission of religion relies heavily on costly displays, sincerity cues, and prestige.

You are more likely to become a believing Muslim if your parents prayed five times a day even when they weren't observed. If they only went to Friday prayer where they could be observed by all their friends, you might infer that their piety is a product of community pressure rather than sincere belief. This might be why many religions call for private prayer as a cultural adaptation to help transmission of the faith from parents to children, who can observe private prayer. Islam's prayer style, which involves touching one's forehead to the ground, can also create the *zebibah* prayer bump, a callus or dark spot on the forehead that serves as a hard-to-fake signal that the person engages in frequent prayer. If your parents were successful or you see successful Muslims, this adds to the likelihood of internalizing the faith.

Religion often has extreme forms of costly behaviors that signal true belief and increase the belief of others. Tertullian, one of the early fathers of the Catholic Church in the second century, was right when he said that 'the blood of martyrs is the seed of the Church'. What could demonstrate true belief in an afterlife more than giving up your current life with no obvious personal or group gain?

People also care more about some kinds of information than they do about other kinds – content strategies. For example, children will pay more attention to information about scary (and therefore dangerous) animals and avoid potentially poisonous plants. And people care a lot about gossip – information about people's love lives and information about their reputations. And because people tend to try to present their best selves, we pay more attention to negative information.

Let's say I have two pieces of information about a popular celebrity. The first is something amazing she did. The second is something terrible she did. You can only choose one. Which do you choose?

My people and what they say

Some aspects of culture are obviously and objectively true, with clear causality. For example, steel axes are better at cutting than stone axes and guns are more effective weapons than swords. And so, steel axes replaced stone axes and guns replaced swords, diffusing beyond group boundaries. In contrast, other aspects of culture, such as marriage practices, are less likely to diffuse beyond group boundaries, because each culture's wedding or marriage traditions are usually not object-ively better than those of other cultures.

A key cultural mutation in the spread of the subjective and causally obscure is a claim that it is objective. One key move that may have led to the spread of Christianity, for example, was the claim that its beliefs were objective.

Many ancient religions claimed that their god was better than your god and left it to the battlefield or prosperity to decide which was true. But Judaism was among the first religions to claim that there was only one god. Not that the Jewish God was better than your god, but that your god didn't even exist. The claim was now objective, making it easier for an evangelical offshoot of Judaism keen on converting the world – Christianity – to spread beyond group boundaries.

These are just a handful of the many social learning strategies that have been identified. Once you understand these strategies, you'll begin to see them everywhere. People trying to sell you things, special interest propaganda, psyops, or simply the nature of peer pressure and bullying. But while you can begin to recognize these strategies or even see how they're embodied in, say, the latest popular business books, we use them naturally and automatically, internalizing what we learn.

Do you know what you know?

We deploy these social learning strategies effortlessly and often without conscious awareness. Take grammar for example. You might recall the grammar lessons you suffered through in school – present and past participles, gerunds, and so on – but much of grammar, the rules of language, are implicitly socially learned.

For example, there's an order to describing adjectives in the English language that some non-native speakers are taught explicitly, but that if you're a native English speaker you might not even know that you know unless you read Mark Forsyth, who popularized this fact. It goes like this: opinion-size-age-shape-color-origin-material-purpose noun. So you can have a lovely little old rectangular green French silver whittling knife. But if you're a native English speaker, try moving adjectives around and it just sounds off: a green French old little lovely silver rectangular whittling knife – ugh!

Similarly, we acquire how to say words – our accent – from those around us, with a focus on the prestigious or majority. Immigrant children, for example, will often acquire the accent of the majority rather than their parents. Accents are thus a hard-to-fake and generally reliable signal that reveal the sources of much cultural influence. Indeed research reveals that accent trumps race when it comes to judgements. That is, people will individualize and 'forget' race if they can distinguish someone by better cues of culture, such as accent, clothing, and mannerisms. We did not evolve to pay attention to skin color. For most of our history, the groups around us had the same skin color and generally looked physically similar. But they were distinguishable through speech and behavior.

Accent reveals a lot and can be a source of discrimination. British university students face discrimination for having an accent from regions of England – or even regions of London – associated with working-class or lower education levels. A similar effect is found in the United States for certain southern accents, particularly the Appalachian accent, often derogatively labelled 'hillbilly'.

Accent is a shibboleth, demarcating communities and indicating the degree of integration and cultural transmission between communities. When groups in the same society speak with different accents, it is an indication of a group divided, a fractured society.

Implicit, internalized social learning writes the software of your mind, but its implicit nature also means that you are cursed by your own knowledge about what you think is and isn't true, what is right and wrong. You have no time or ability to recheck everything you've learned and no need if your life is going well. In fact, you're doing it right now, implicitly (or perhaps explicitly!) using these very same social learning strategies to decide if what I'm saying is sensible and useful. Testing the degree to which my claims fit your intuitions based on what you think you know and whether what you're learning will serve you well, what my politics might be, whether I belong to your ingroup, whether I'm sincere, and so on. Don't worry, I've provided a peppering of examples for you to test these claims for plausibility against your own life experience and what you think you already know.

Different people in different parts of the world have different culturally evolved and socially learned beliefs and behaviors. They have different software shaping their brains' hardware – literally changing the size and function of different parts of their brains. That software constitutes different culturally adaptive packages, filtered by generations of sophisticated selective social learning that has enabled our species to live in every ecosystem on earth. We feel like we understand why cassava should be soaked and boiled, why certain foods should be avoided during pregnancy, why bending a bow in a certain way makes a better weapon, why we should brush our teeth twice a day, eat more vegetables, stretch before exercising, or wash our hands before eating. But typically, we do not. Washing hands, for example, seems obvious to those of us who grew up in a world where everyone takes germ theory for granted, but it was less than 150 years ago that doctors refused to believe that fewer mothers might die if they washed their hands between examining a dead body and delivering a baby.

We acquire all these practices and more through social learning strategies that write the software of our brains and our societies. Our ingenuity is a product of this software, which has forced our brains' hardware to keep up, co-evolving with our genes to predispose us toward certain kinds of information, such as language, norms, and gossip. In essence, what we had to learn got so large, we maxed out on brain size and began focusing on only the most useful and relevant information, taking the rest for granted.

Our cultural package, much like energy, is part of the water we swim in – essential to our lives, but mostly imperceptible. The inability to see our cultural package, our incredible, information-dense inheritance, different for different people, leads us to judge one another as intrinsically more and less sophisticated. It leads us to assume a human nature that is instead based on culturally transmitted skills and fail to appreciate the many ways in which mental software is written by those around us and those long gone. But seeing that cultural package and where it comes from is essential to understanding who we are and how we got here.

When I think of myself as a recipient of a cultural package that generations before me have added to and subtracted from, grown and made more efficient, and of the origins of which I have little to no knowledge, I feel a mix of humility and awe. Maybe even a little unease and fear. But understanding how we acquire this package and then contribute back to it is critical to understanding how the laws of evolution, innovation, cooperation, and energy manifest in our species.

This understanding is necessary to figuring out how to maximize our chances of reaching the next level of energy abundance. Part of this requires understanding how these strategies have made us clever, because people often assume that the secret to our success is our intelligence.

3

Human Intelligence

Our society is obsessed with intelligence. What it is, what it predicts, how to measure it, how to improve it, who's smarter, whether it's caused by nature or nurture, and so on. This obsession is in part due to the implicit or explicit assumption that behind the laws of energy and innovation, and perhaps even the law of cooperation, is intelligence. That it is thanks to human intelligence that we controlled fire and fossil fuels, innovated amazing new technologies, and learned that it was better to work together than to fight and cheat. The truth is more complicated.

Intelligence is essential to who we are and how we got here. But our intelligence didn't lead us to control energy, innovate fantastical technologies, and learn to cooperate. Rather, controlling energy, innovating fantastical technologies, and learning to cooperate led us to become more intelligent. Intelligence, in other words, is not the *cause* of the laws of life; like our cultural package, it is another result of these laws. It is more a product of our quickly evolving software than our slowly evolving hardware.

But intelligence is a tricky thing to define, let alone measure or study. Attempts to study intelligence have a long history.

Francis Galton was Charles Darwin's cousin – both were grandchildren of the eminent Erasmus Darwin. Erasmus was an important historical figure in the English Midlands Enlightenment of the second half of the eighteenth century – a physician, philosopher, biologist, and poet who had died seven years before the birth of Charles (in 1809) and twenty years before the birth of Francis (in 1822). Nevertheless, his fame loomed large over their childhoods.

Francis's almost religious obsession with the inheritance of intelligence and hereditary genius began after he read his cousin's book, *On the Origin of Species* (1859). He was inspired and began to think through the implications for humans. He was perhaps also inspired by Charles and Erasmus – two apparent geniuses in the same family. Francis began to rigorously study what he called eugenics: the scientific study of *eugenes* – Greek for 'well-born, of good stock, of noble race'.

Charles Darwin wasn't much of a mathematician, writing in his autobiography: 'I have deeply regretted that I did not proceed far enough at least to understand something of the great leading principles of mathematics, for men thus endowed seem to have an extra sense.' He might have been thinking of the young Francis, who took a mathematical approach to studying almost everything, including evolutionary questions. Francis worked on the principle of 'whenever you can, count'. One of his goals was disentangling the contributions of what he called 'nature versus nurture'.

In 1865 Francis laid out his eugenic agenda in an article entitled 'Hereditary Talent and Character', writing that 'The power of man over animal life, in producing whatever varieties of form he pleases, is enormously great . . . It is my desire to show . . . that mental qualities are equally under control'. He looked at biographies of 'eminent people' and started counting, and found that although 'the children of men of genius are frequently of mediocre intellect', 'talent is transmitted by inheritance in a very remarkable degree'. But documenting this apparent inheritance of intelligence wasn't sufficient; he wanted to apply it.

Surely, Francis thought, if we could breed animals for certain traits, could we not also increase the intelligence of a society? Or as he put it, 'If a twentieth part of the cost and pains were spent in measures for the improvement of the human race that is spent on the improvement of the breed of horses and cattle, what a galaxy of genius might we not create!' And so began Galton's quest, which culminated in 1869 with *Hereditary Genius*, where he proposed policies such as arranging marriages between the 'wealthy' and the 'distinguished'.

Galton promoted his vision for eugenics with a religious fervor, writing in his autobiography, 'I take Eugenics very seriously, feeling that its principles ought to become one of the dominant motives in a civilized nation, much as if they were one of its religious tenets.' Galton's vision would become reality.

In the early twentieth century eugenics was a prominent scientific subject, supported in various forms by the most prominent scientists, politicians, and thought leaders of the time. This wasn't some fringe idea; it was endorsed and implemented in different forms and with different levels of coercion. Eugenics didn't cause the immediate revulsion many feel today. In fact, it was perceived as part of a progressive agenda – a way to make the world a better place.

'Positive eugenics' involved promoting 'good stock' (such as Francis's aforementioned proposed arranged marriages). 'Negative eugenics' involved preventing the spread of 'defective stock' (such as restricting immigration and forced sterilization of those with 'lower intelligence').

The Nazis' obsession with eugenics and their horrifying subsequent policies designed to achieve an Aryan ideal led to a decline of overt support for the idea. But eugenics lives on in various forms: population control, sex-selective abortions, prenatal gene testing, and most recently gene-based embryo selection.

Although Galton was able to convince his peers that 'talent and character' were transmitted in families and that eugenics was a progressive pursuit, he was ultimately unable to directly measure the thing he was trying to improve: intelligence. This must have been frustrating for a man who loved to count, but he lived long enough to see the first widely adopted intelligence test, published in 1905.

Measuring how clever you are

In the early twentieth century the French Ministry of Education wanted to measure how students compared to their same-age peers so that those who were *arriéré* – behind or backward – could be given special education. They tasked two psychologists – Alfred Binet and Théodore Simon – with developing a test. The Binet–Simon test, published in 1904 as 'Méthodes nouvelles pour le diagnostic du niveau intellectuel des anormaux' and later translated as 'New Methods for the Diagnosis of the Intellectual Level of Subnormals', was presented at a conference in Rome in 1905 under the title, 'Méthodes nouvelles pour diagnostiquer l'idiotie, l'imbécillité et la débilité mentale' – 'New

Methods for Diagnosing the Idiot, the Imbecile, and the Moron'. It was the first IQ test.

The three terms that emerged in the literature were not meant to be insulting but rather scientific classifications grounded in Greek or Latin roots:

> 'Moron' came from the Greek *moros*, meaning 'foolish', and designated a child who might fail to pay attention or fail to answer some harder questions.
>
> 'Imbecile' came from the Latin *imbeccilus*, meaning 'weak' or 'feeble', and designated a child who gave absurd responses, perhaps not correctly identifying an object.
>
> 'Idiot', the bottom rung, came from the Latin *idiota*, meaning 'ignorant person', and designated a child who didn't know common objects, for example confusing a piece of chocolate with a piece of wood and trying to eat both or neither.

These words are offensive today, and as such are an example of what psychologist Steven Pinker has called the 'euphemism treadmill' whereby euphemisms (indirect, milder words for unpleasant referents) become dysphemisms (derogatory words) due to the underlying reality of what they refer to. The once benign term 'retard', which means 'delayed', has become taboo for similar reasons: 'retard' is related to the inoffensive word 'tardy', both from the Latin *tardus*, meaning 'slow'.

The Binet–Simon test was later revised and translated into English by Stanford psychologist Lewis Terman in what was called the Stanford–Binet Intelligence Scales. Terman, who was a prominent eugenicist (pre-Nazi, when it was still considered acceptable), also adopted William Stern's idea of standardizing the scores as an Intelligence Quotient, introducing the concept of IQ.

You may remember from high school that a quotient is a fraction or division. For an intelligence quotient, a score of 100 means a performance at the average compared to same-age peers. Every fifteen points higher or lower represents one standard deviation from this average. So, for example, an IQ below 70 (two standard deviations below the mean) would put someone in the bottom 2.5% of the population. An IQ of 145 (three standard deviations above the mean) would put someone in the top 0.1% of the population.

What did these tests look like?

The IQ tests measured a grab bag of concepts that researchers felt all children should know, from labeling objects that were commonplace in the early twentieth century to which of two crudely drawn faces was prettier. Remember, they were initially trying to identify 'subnormal' performance, as Binet and Simon put it. As the test spread beyond France, the culture-bound nature of these questions became apparent. Different societies have different common objects and speak different languages. Some didn't learn math, reading, or writing at all. So attempts were made to remove the cultural element.

Question used in the Binet-Simon test.

Raven's Progressive Matrices IQ test was developed by John Raven in 1936 and is still often considered the most culture-free IQ test. The test has no words or numbers, only patterns to be solved. But Raven's test, too, relies on artificial, two-dimensional shapes and patterns, for example squares and triangles, that are rare in nature

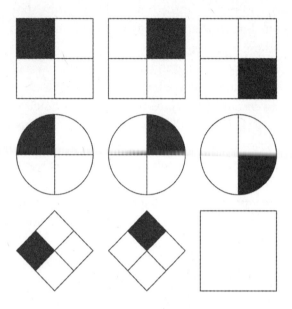

An IQ test item in the style of a Raven's Progressive Matrices test. Given eight patterns, the subject must identify the missing ninth pattern. By User:Life of Riley, CC BY-SA 3.0, https://commons .wikimedia.org/w/index.php?curid=17342989

and that we spend a lot of time training children in our society to identify. It wasn't long before IQ tests were used on adults.

The American Psychological Association began testing soldiers during the First World War. Southern and eastern European migrants had lower scores than northern Europeans and these scores were taken at face value. This led to immigration restrictions in the spirit of negative eugenics. Were these IQ tests a fair measure of intelligence?

IQ is not the same as intelligence. IQ tests are to intelligence what inches and centimeters are to length. An object has a true length, but it can be quantified in different ways and on different scales. But unlike length, intelligence doesn't have a clear, unambiguous, accepted definition. Indeed, the existence of a more general intelligence – referred to as g – rather than specific talents is statistical. What do I mean by this?

Say you're trying to measure overall fitness, let's call this f. You may care about overall fitness because of how it translates to athletic and sports performance, just as we often care about general intelligence because of how it translates to academic and work performance. But each sport is different, just as each school subject and each job is different. Nonetheless, there may be an overall fitness underlying these different specific athletic abilities. How would you know if this were true?

You might start by giving people a variety of fitness tests to measure different fundamental aspects of fitness: tests of endurance, strength, speed, flexibility, body composition, and so on. You might try the beep shuttle run test, one-rep max, 100-metre sprint time, heart rate after some set exercise, how many push-ups they can do, how far past their toes they can reach, and so on. If there were an overall fitness, you might expect some correlation between these scores – people who score high on flexibility might also be faster. Since you're correlating multiple scores, you perform what's called a factor analysis, which looks for an underlying or *latent* factor that tries to capture a kind of overall correlation between all scores as best it can – in this case, your hypothesized overall fitness f. This factor analysis also tells you how well each of your tests measure that overall correlation f factor, if it exists. In turn, you can see how f predicts performance in different sports.

This is exactly what was done to discover the hypothesized general intelligence g. Various tests for components of skills perceived to be related to fundamental intelligence, such as problem-solving, general knowledge, verbal and language ability, quantitative skills, visual–spatial processing, and working memory correlate with one another and have an underlying factor. Different tests and subtests can then be evaluated by how well they correlate with this underlying factor, g. And it is possible to look at how reliably one can measure g using these IQ tests, and how heritable it is. As it turns out, g is both reliably measured and reasonably heritable.

General intelligence, g, is almost taken for granted today among intelligence researchers. Some research focuses on explaining why these tests have an underlying factor. Could it be some undiscovered feature of our brains that differs between people (something like neural speed or efficiency)? Or perhaps there's a 'positive manifold' such that being good at different things has synergies that help improve performance in different domains? For example, being good at reading, verbal and other language skills may help you learn better and therefore become better at problem-solving and quantitative skills. Over time, these different skills will then correlate and reinforce one another.

The idea of general intelligence rests not on theory or causal experiments but on the correlation between different tests deemed to measure cognitive ability. And debate continues over the degree to which this highly valued trait is nature versus nurture, as Galton put it. But, armed with our 'theory of everyone' we can cut through Galton's debate, answer the eugenic question, and offer a more compelling and comprehensive, theoretical and empirical understanding of intelligence.

Let's start with ten facts about IQ.

Ten facts about IQ

Here are ten facts about IQ. These facts are debated and often controversial among the general public but far less so among scientists who study intelligence. The best review of the academic literature supporting these facts is a 2012 paper by Richard Nisbett and colleagues – an interdisciplinary team of leading scholars, household names within intelligence research, comprised of psychologists, an economist, a behavioral geneticist, and a former President of the American Psychological Association. Their areas of expertise include cultural and sex differences in intelligence, the effect of social and genetic factors that affect intelligence, the development of intelligence over the lifespan, the relationship between economic development and intelligence, and changes in intelligence over history.

1. IQ is a good predictor of school and work performance, at least in WEIRD societies.
2. IQ differs in predictive power and is the least predictive of performance on tasks that demand low cognitive skill.
3. IQ may be separable into what can be called 'crystallized intelligence' and 'fluid intelligence'. Crystalized intelligence

refers to knowledge that is drawn on to solve problems. Fluid intelligence refers to an ability to solve novel problems and to learn.

4. Educational interventions can improve aspects of IQ, including fluid intelligence, which is affected by interventions such as memory training. Many of these results don't seem to last long, although there is strong evidence that education as a whole causally raises IQ over a lifetime.

5. IQ test scores have been dramatically increasing over time. This is called the Flynn effect after James Flynn (also an author of the review mentioned above), who first noticed this pattern. The Flynn effect is largest for nations that have recently modernized. Large gains have been measured on the Raven's test, a test that has been argued to be the most 'culture-free' and a good measure of fluid intelligence. That is, it's not just driven by people learning more words or getting better at adding and subtracting.

6. IQ differences have neural correlates – i.e. you can measure these differences in the brain.

7. IQ is heritable, though the exact heritability differs by population, typically ranging from around 30% to 80%.

8. Heritability is lower for poorer people in the US, but not in Australia and Europe where it is roughly the same across levels of wealth.

9. Males and females differ in IQ performance in terms of variance and in the means of different subscales.

10. Populations and ethnicities differ on IQ performance.

You can imagine why some people might question these statements. But setting aside political considerations, how do we scientifically make sense of this?

Popular books from Richard Herrnstein and Charles Murray's *The Bell Curve* (1994) to Robert Plomin's *Blueprint* (2018) have attributed much of this to genes. People and perhaps groups differ in genes, making some brighter than others. But humans are a species with two lines of inheritance. They have not just genetic hardware but also cultural software. And it is primarily by culture rather than genes that we became the most dominant species on earth. For a species so dependent on accumulated knowledge, not only is the idea of a culture-free intelligence *test* meaningless, so too is the idea of culture-free intelligence.

Cultures of intelligence

Each of us possesses a diversity of skills, cognitive and otherwise. But the cluster of skills we consider to be constituent of intelligence or genius differ in different societies and at different times. Among the Inuit and many mobile small-scale societies, the most intelligent have great spatial ability for remembering how to get to different geographic locations. In early Christendom and the Islamic empires, the most intelligent were those who could memorize holy books. In the Renaissance it was scholars and artists. In the twentieth century it was the single-skilled artisans and mathematical geniuses. Even in our lifetime, the instant accessibility of knowledge through the Internet has reduced the value of simply memorizing large quantities of information and increased the value of sorting the signal from the noise, finding the right information, interpreting large quantities of data, and being able to focus in a highly distracting world. Many lament the loss of skills in mental math and memory, but each generation's focus has led to deficits in other areas. Reaction time seems to have decreased in an anti-Flynn effect. In pre-literate communities who start going to school, spatial abilities, including navigation, get worse. The point is this: an IQ test designed in the past or in a different society measures different skills and, implicitly, values. The rise of AI will no doubt reshape what we consider intelligence in WEIRD societies.

IQ tests measure what we value in WEIRD society – right now – and so they predict success in WEIRD society largely as it is. It should be no surprise, then, that IQ test scores predict school and work performance. But just as our ability to get work done on a computer depends not just on high-powered hardware – the latest Thinkpad or Macbook – but also on the right software – Excel or Photoshop – so it is impossible to think about intelligence without considering both hardware (genes) and software (culture). What is the evidence for this?

Culture is like the water Wallace's fish swim in. We can't see it when it is everywhere. We live in a thoroughly educated world and so can't see how much of who we are is a product of culture transmitted through education. Remember that all intelligence tests and the entirety of experimental psychology were developed *after* formal schooling had already become widespread. Indeed, Binet and Simon developed the first IQ tests for schoolchildren. So much of what we now consider universal core human capacities, including reasoning abilities and what IQ tests measure, are based on the empirical discoveries of psychology. But if these abilities were in fact culturally transmitted, say by an education system that was already ubiquitous by the time IQ tests and psychology itself were invented, then we wouldn't be able to tell. These ways of thinking would seem universal simply because schools are everywhere, education is seen as a right, children around the world are forced to attend schools, and most of us have never even met a person with no schooling at all. And in a world in which everyone is educated, the products of education become a new baseline of knowledge and ways of thinking that permeate every TV show, book, and conversation. If it were the case that these abilities were in our cultural software and not our genetic hardware, how would we know?

In almost all societies today a lack of education is a good indicator of extreme poverty and the other disadvantages that come with it, so it's incredibly difficult to disentangle if low IQ test performance is caused by disease, by pollution, by home, neighborhood, or by societal environment, culture, or genes. Education in modern societies has evolved as a means of efficiently downloading a specific cultural package – sit down kids, we're going to start with letters, numbers, phonemes, adding, subtracting, reading, writing, trigonometry, algebra, calculus, and so on. But this approach is new.

Many hunter-gatherers don't engage in much explicit teaching of children. Instead, children hang around adults and learn through observation. More direct instruction is more common among pastoralist and agriculturalist societies with a larger cultural corpus. Even in our society, universal compulsory formal education is relatively recent, emerging because the Industrial Revolution needed factory workers with some minimum skill set. Since then, school has become an increasingly important source of cultural transmission. How can we measure how important?

It would be unethical to randomly assign children to receive more, less, or even no education, and where this happens it's caused by other factors, such as relative wealth. But there have been some 'natural experiments' where educational opportunities are randomly assigned. The results from these natural experiments are clear and consistently the same: education increases intelligence and performance on IQ tests. I'll highlight two strong pieces of causal evidence. One experiment is from Norway, the other from southern Africa but, by complete coincidence, also thanks to the Norwegians.

Norway is an interesting country for many reasons. Once the home of horrifically violent Vikings, who shaped the early history of much of western Europe, it is now one of the most peaceful, prosperous, and least corrupt nations on earth. In the mid twentieth century Norway increased the number of years of compulsory schooling from seven to nine years. Thus children born days apart received a difference of two years in education due to a random change in policy. As a result, this reform has led to many high-profile papers measuring the causal effects of education on a variety of factors, such as number of offspring, teen pregnancies, and lifetime income. But Norway has another policy that has allowed researchers to study the effects of these additional years of education on IQ: mandatory military service. At age nineteen, as part of Norway's mandatory military service for males, all conscripts are given an IQ test. And so we can go beyond correlation and determine how much this extra education affected IQ test performance in a causal manner.

Those two extra years of education gave those who received it an average of over seven IQ points – around half a standard deviation – a massive bump in brilliance. And because this change only affected adolescent education, it was probably underestimating the overall effect of education on IQ. Earlier interventions tend to have larger effects – older dogs are harder to teach new tricks.

Not all natural experiments are as free from other potential influences as this one, but a meta-analysis of 142 tests from forty-two similar quasi-experiments like this with over 600,000 participants, conducted by Stuart Ritchie and Elliot Tucker-Drob, came to the same conclusion – the overall effect of education on IQ test performance is between one and five IQ points per year of education. The authors concluded, 'Education appears to be the most consistent, robust, and durable method yet to be identified for raising intelligence.'

How can we test the effect of no education versus some education in a causal manner? It's difficult because the peculiarly Western-style formal educational institution we call 'school' has spread to most corners of the globe, at least to some degree. This has been a boon for health, life expectancy, per capita income, and other metrics often associated with human development, but it is a challenge to scientifically studying exactly how education has rewired our brains, and, as a consequence, rewired our societies.

Two hundred years ago only 12% of the world could read and write. Today, only 14% *cannot* read and write. And that's thanks to school. But with schools everywhere, how can we know what school does to our brains and cognition compared to no school at all? If only there were somewhere in the world that had yet to receive education but also was receiving it in a sufficiently randomized natural experimental way.

My colleagues and I spent two years searching the globe to see if such a site existed. In 2016 we finally found one in southern Africa, on the border between Namibia and Angola. And by coincidence, the schooling intervention was once again thanks to the Norwegians.

The Kunene River flows from the Angolan highlands south and then west into the Atlantic Ocean. In 1886, in faraway Lisbon, Germany and Portugal decided to declare a border between its two colonies using the west-flowing portion of the Kunene as the cut-off. Namibia and Angola were born.

The Himba are a cow-herding semi-nomadic pastoralist people who live on both sides of the Kunene. You may have seen photos of them, perhaps in *National Geographic* – they are known for their use of *otjize* – a mix of red ochre and animal fat – on their skin and hair, which serves as a protective barrier against the harsh sun of southern Africa. The Himba were not consulted about the new border, but nor did it affect them to any great degree. They live in an isolated region, cut off from Angola's major cities by rough terrain and the highlands to the north and a hundred empty miles to the nearest Namibian major town, Opuwo, of approximately 20,000 people, to the south. Even today, the Himba continue to practice their traditional way of life, crossing the river to meet and marry, but largely living on one side or the other due to the difficulty in taking their cattle across. As a result of this meeting and marrying, both sides of the river are part of the same genetic and cultural population.

Schooling is compulsory in Namibia. The Namibian Himba used to send their children to school in a major town like Opuwo or to the capital, Windhoek, but many children decided to stay on after their education. Perhaps it was hard for some to return to the traditional, semi-nomadic, pastoralist life after the opportunities, excitement, sex, drugs, and rock and roll of towns and cities. Himba moms and dads were not happy about their kids leaving permanently, so they stopped sending their children to town and city schools. As a way around this, the Namibian Association of Norway (NAMAS), a Norwegian NGO funded by the Norwegian foreign ministry, worked with the Namibian government and the Himba to co-create a so-called tent school, with a localized version of the national curriculum. The school could follow children as they moved with their families to the next watering hole or grazing patch. The curriculum was co-created with the Himba and taught the Namibian national curriculum in a culturally relevant way, such as counting cows and dealing with other aspects of semi-nomadic pastoralist life. The Namibian Himba parents started sending some of their kids back to school.

What about the Himba living in Angola?

The Norwegians were happy to introduce a tent school on the Angolan side of the river too, but they ran into political challenges. There was an unverified rumor that a 'payment' was necessary. Norwegians being Norwegians refused. It's an interesting cultural clash. According to Transparency International, Norway was at the time, and even now, among the top five least corrupt countries in the world. Namibia's corruption level was, and is, in the middle of the pack, at an average level of corruption. Angola, by contrast, was at the time one of the bottom fifteen *most* corrupt countries on earth (though it has since improved and hopefully will continue to climb the anti-corruption ranks). And so unlike the Namibian Himba on the southern side of the river, the Angolan Himba just north of the river, who had little to do with the Angolan government or the rest of Angola, did not receive formal education.

So we started collecting data. What have we found so far?

First off, IQ test performance was higher in the communities with access to schools than it was in those that did not have access. So schooling does increase intelligence, at least the kind of intelligence our society values and defines through IQ tests like the Raven's matrices. But we wanted to dig deeper and also look at changes as children get older and receive more education.

IQ test performance correlates with age in WEIRD societies. We assume children get smarter as they get older. That's why Binet and Simon and everyone since them have compared test results within age groups. But here's the problem – thanks to truancy laws, age correlates almost perfectly with number of years of school in our society. Age is an almost perfect proxy for education. Are kids getting higher raw IQ test scores because they're getting older or because they're receiving more formal education?

To find out, we compared the Himba who went to school with those who did not on the supposedly culture-free Raven's Colored Progressive Matrices. As you might imagine, children with formal education had higher scores than children without formal education, but here's what was surprising: among the children without access to school, older children had roughly the same scores as younger children on the Raven's test. Their scores weren't improving with age the way they do among the schooled Himba and as they do in our schooled societies.

Older children with or without formal education are still developing, maturing, and becoming more skilled just as they are anywhere. And the Angolan Himba survive and thrive and succeed as people do anywhere. But the Angolan Himba skills would not be reflected in our society's traditional IQ test performance. Think about it – these supposedly culture-free tests still rely on colors and shapes that we take time to teach our children. And indeed, with other tasks more relevant to their way of life, such as remembering a sequence to make a bead necklace, we were able to detect developmental changes, even among those without schooling.

What about their genetically and culturally indistinguishable cousins in Namibia who have received education over the last decade? Here the pattern is the same as other schooled societies – IQ test performance increases with age (or as we can now see, with amount of formal education received). The more time they spend in school, regardless of age, the higher the IQ. IQ tests are measuring what schools in our society are delivering.

But school does so much more than just teaching you reading, writing, and arithmetic. It also trains you in things like discipline and delayed gratification – you sit down and study now for tests and rewards that are coming much later. Schools teach us how to learn through books and videos, and through other people in ways specific to our society. And they give us tools for thinking in ways that we value as a society at that moment; tools we now take for granted and perhaps even assume are genetically evolved human universals.

In the previous chapter I discussed Alexander Luria's and Helen Elizabeth Davis's and my findings on formal logical reasoning, but Luria also asked questions that resemble the *Sesame Street* song 'One of These Things'. For example, he asked which of the following are alike: a hammer, a saw, a log, and a hatchet. Here is a transcript of a conversation with Rakmat, a thirty-year-old with no education, that reveals the relational way of thinking:

'They're all alike,' he said. 'I think all of them have to be here. See, if you're going to saw, you need a saw, and if you have to split something, you need a hatchet. So they're all needed here.'

We tried to explain the task by saying, 'Look, here you have three adults and one child. Now clearly the child doesn't belong in this group.'

Rakmat replied, 'Oh, but the boy must stay with the others! All three of them are working, you see, and if they have to keep running out to fetch things, they'll never get the job done, but the boy can do the running for them . . . The boy will learn; that'll be better, then they'll all be able to work well together.'

'Look,' we said, 'here you have three wheels and a pair of pliers. Surely, the pliers and the wheels aren't alike in any way, are they?'

'No, they all fit together. I know the pliers don't look like the wheels, but you'll need them if you have to tighten something in the wheels.'

'But you can use one word for the wheels that you can't for the pliers – isn't that so?'

'Yes, I know that, but you've got to have the pliers. You can lift iron with them and it's heavy, you know.'

So with all this in mind, what is an IQ test? An IQ test is really a measure of cultural imprinting. We are taught how things in our world are related to each other and it shapes how we think about the world. But these cognitive tools change our brains in even more profound ways.

Culture makes us smarter

Crows are astonishing animals. One of the many abilities they possess is being able to fashion hooked tools. In experiments where they're given a metal wire and food that's retrievable from a long tube, they figure out that the wire can be bent into a hook, and then they use this hook to retrieve the food. It's a very human-like thing to do. 'Man the tool maker,' right?

Developmental psychologist Sarah Beck ran studies with British children using a very similar task as the one given to crows. She presented children of different ages with a long tube at the bottom of which was a little bucket with a sticker (remember kids love stickers and scientists love that stickers are so cheap) – to retrieve the sticker they needed a hook. The children had access to one of those wire pipe cleaners you probably played with in school. Crows can figure it out – could the kids?

Three- to five-year-old kids consistently failed. They poked and prodded the tube with the pipe cleaner, but just couldn't figure out that the pipe cleaner could be turned into a hook. But it wasn't that they were physically incapable of doing it (crows manage it with their beaks and claws). And when the kids were shown a demonstration, they got it immediately. That's the power of education. That's the 'Aha! Oh I see' that happens every day in classrooms across the world. Remember the Ngamban chimp experiment? We're excellent imitators.

Developmental psychologist Mark Nielsen read Beck's findings and, like many, was surprised by these results. Perhaps, he thought, children who were more used to making their own toys might be better able to complete the task? He ran the same experiment with the 'bushmen' children of southern Africa. (I grew up in southern Africa, and remain impressed by the cars and other toys the kids there create from wire, aluminum cans, bottle caps, and other scrap material.) But despite the ubiquity of homemade toys – and much to Mark's surprise – the bushmen kids behaved the same way as the British kids. The younger kids were no better at solving the hook task by themselves, but could immediately do it once it was demonstrated.

Remember, cultural knowledge of all kinds, delivered to us by our schools, parents, communities, television, YouTube, TikTok, and other media, is like software. It lets us do things with the same hardware. Imagine trying to run fancy accounting calculations without Excel. It doesn't matter how fast your CPU is or how many cores it has, without the right software you just can't make those pivot tables. So too with cultural software.

So much of what we take for granted is the accumulated wisdom, know-how, and ways of thinking of generations of humans before us – the software package we download and learn to use in the first few decades of our lives.

None of this is to say that the hardware doesn't matter or that there aren't differences between people. The hardware *does* matter – pollution, disease, lead exposure, insufficient nutrition, and exposure to smoking, alcohol, and other drugs, for example, all harm our ability to build good brains. And there are obviously genetic differences, by chance or otherwise, between people and populations. But these are unlikely to explain the ten facts about IQ we discussed earlier. Our cultural software has more capacity for change than our genetic hardware and it is the software that explains most of the difference between societies, between people, and between generations. And because it's software, it also has the most potential for change in the future. This is important for making us cleverer and helping to spark the next creative revolution we so desperately need. Let me show you how our experiences make us smarter or get in the way. Then we'll be ready to revisit those ten IQ facts.

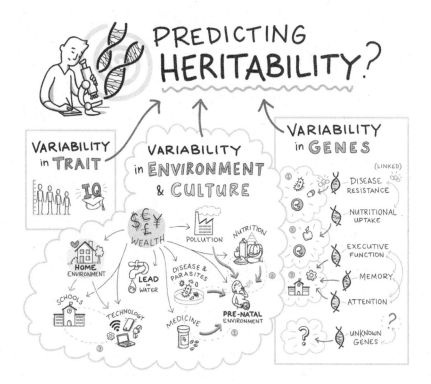

Mentalizing the physical

Your head is filled not only with words for things but with entire analogies, metaphors, epistemologies, and tools that you once learned and now effortlessly use for thinking.

When children learn how to count on their fingers, they can eventually represent those fingers in their heads without explicitly using them. Eventually this kind of concrete number representation becomes natural and you forget about the fingers completely. But there's nothing unique about fingers or indeed the number 10. Different cultures have counted on other body parts from finger joints to parts of the body using a variety of bases from 6 to 27 instead of 10. They've also found other ways to externalize the counting of objects that have helped with mathematical abilities. One example is the Soroban abacus, a fourteenth-century Japanese calculator. It looks a lot like the abacus

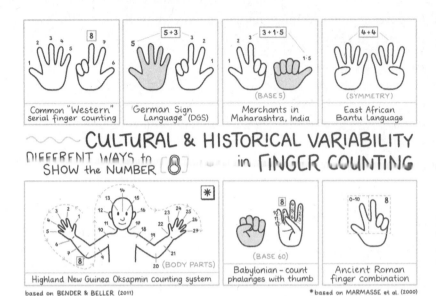

CULTURAL & HISTORICAL VARIABILITY in FINGER COUNTING

DIFFERENT WAYS to SHOW the NUMBER 8

Common "Western" serial finger counting	German Sign Language (DGS)	Merchants in Maharashtra, India (BASE 5)	East African Bantu Language (SYMMETRY)

Highland New Guinea Oksapmin counting system (BODY PARTS)	Babylonian – count phalanges with thumb (BASE 60)	Ancient Roman finger combination

based on BENDER & BELLER (2011)   *based on MARMASSE et al. (2000)

you might have played with in school, but is smaller and has five beads per row instead of ten.

The Soroban abacus saw a resurgence in the late twentieth century. Many Japanese children started taking Soroban classes to get better at mental arithmetic. But as with counting on your fingers, they would eventually stop using the Soroban and instead use the software representation of the Soroban in their heads. You can watch videos of Soroban-trained kids calculating large sums – 3267 + 9853 + 6531 + 7991 + 2641 – in seconds. It's an impressive feat and an example of how we mentally represent things we have seen and then can use these analogies, metaphors, and even mental representations of physical objects to think.

There's a second lesson here too. Our mental tools can be out of date. Those Soroban classes became less popular because mental arithmetic became less useful. My middle-school teacher warned us about the dangers of not being able to add, subtract, multiply, and divide quickly without a calculator, because we wouldn't be carrying calculators in our pocket; he didn't foresee the arrival of the iPhone. Because we view the world through the lens of all we have previously seen, these abstract mental representations can even get in the way or mislead us.

Minds are simulators

When we see something – say a rectangular, brown desk – we naturally and automatically break it up into shapes, colors, and concepts that we have learned through education. We recode what we see through the lens of everything we know. We store in our minds not the raw image of the desk, but its features in an abstract manner. You can test yourself for how you're mentally representing something by trying to draw it. We draw with our abstractions, not with a full picture of what we saw. We reconstruct memories rather than retrieving the video from memory. This is a useful trait. It's a more efficient way to store information – a bit like an optimal image compression algorithm such as JPG, rather than storing a raw bitmap image file. People who lack this ability and remember everything in perfect detail, struggle to generalize, learn, and make connections between what they have learned. But representing the world as abstract ideas and features comes at a cost of seeing the world as it is. Instead, we see the world through our assumptions, motivations, and past experiences. The discovery that our memories are reconstructed through abstract representations rather than played back like a movie completely undermined the legal primacy of eyewitness testimony. Seeing is not believing.

In her now classic experiments, Elizabeth Loftus showed people a video of a car crash. She then asked people to estimate the speed of the car and whether they saw broken glass when the cars 'contacted', 'hit', 'bumped', 'collided', or 'smashed' into each other. Different groups were given a different description of the same video scene they all saw. The participants who were asked how fast the cars were going when they 'smashed' into each other drew on everything they knew about 'smashing' and had the highest estimates for speed and the greatest likelihood of claiming to have seen broken glass. There was no broken glass in the video. It's because a 'smashed' car mentally simulates a lot more damage than one that was only 'contacted'.

These representations also allow our minds to simulate and predict the world. We see the world through anticipation of how we expect it to look. Our minds work less like computers and more like simulators. When you see a door, your brain is not calculating like a

Terminator: 'Rectangular object. Handle. Calculating . . . calculating. Door identified.' Instead it's expecting a door and would be surprised to discover that it's actually a window or a painting of a door. When your brain's predictive simulation of the world makes a bad prediction, like a painting of a door rather than a real door, parts of your brain such as the dorsal anterior cingulate cortex (dACC) fire and make you uncomfortable – a feeling that something is amiss. You implicitly and automatically check what you know and try to accommodate or assimilate the new information if you can, for example, drawing on your knowledge of paintings. But if you can't make sense of what you see through what you already know, then learning happens. This mild discomfort is essential to learning new things.

Our minds simulate the world based on what society teaches us. Societies can endow us with new skills and new natural tendencies – new culturally created instincts that can feel innate.

Cognitive scientist Peter König once created a haptic feedback compass belt, dubbed 'feelSpace', basically a belt that always vibrated north. I'm sure he received some odd looks. 'Peter, why are your pants vibrating?' 'I'm, uh . . . I'm sensing north . . .'

After Peter wore the belt for a few weeks, something strange started to happen. He discovered that his mind naturally and automatically created a map of the cities he walked in. He discovered that even when he took the weird vibrating belt off, his navigation abilities had improved. Peter had accidentally given himself and his experimental participants something close to an ability that the Australian Aboriginal Guugu Ymithirr tribe had developed through training and language. The Guugu Ymithirr language refers to relative directions in absolute terms – north, south, east, west, rather than relative terms – left, right, behind, in front. So rather than saying the lamp is to the left of the table when facing it from one direction and right when on the other side, they would say that the lamp is east of the table regardless of their own orientation. As a result, their culture teaches them to always keep track of north, much as yours taught you how to count, write, and read. All are examples of taught skills that then become a human instinct.

In literate societies, for example, reading is now an instinct. The Stroop effect involves showing people colored words where the text is either matched to the color – 'red' written in a red font – or mismatched – 'red' written in a blue font. People are asked to state the color rather than read the word, but they struggle when the words are mismatched. They just can't help but read what they see. You might think color detection is more natural, but instead, reading has become the dominant instinct.

A psychologist from Venus who knew nothing about the history of reading would assume humans have a genetic instinct for reading, but not color perception, just as we make assumptions about reasoning or counting.

These are some of the many ways in which our society writes the software of our brains and makes us better at some tasks than others. With this in mind, let's return to the facts about IQ.

Revisiting facts about IQ

IQ test performance predicts school and work performance, but there is a common variable that connects IQ, school, and work. IQ tests, school tests, and work performance are all highly valued in our society and what our current culture is training us to do. But what's valuable in our society isn't necessarily valuable in another. Despite my years of education, if I were suddenly forced to live off the land, I would struggle to survive. If I were forced to migrate to live in a hunter-gatherer society without access to modern markets and modern technologies, my hosts might consider me full of interesting knowledge about the world but thoroughly incompetent when it came to the practicalities of building a house out of mud and thatch, making a useful bow and using it to hunt, or turning animal skin into clothing. I would struggle and would hope to get by as an interesting incompetent with wild stories about theories of everyone.

Even within my society, because of intense specialization, I would struggle to perform many jobs other than the ones I'm trained in. If my job suddenly disappeared, I would not be a competitive candidate for an opening in the local builder's crew, car garage, chartered accountancy firm, massage parlor, theatre, social care provider, or on the football team coaching staff. The incredible degree of specialization in a large city like London means that I have become super smart at one thing and incredibly stupid at almost everything else. And that's just when I consider current jobs and societies. Skills change over time.

Remember the Soroban abacus and my middle-school teacher's lack of foresight when it came to the iPhone? What's valuable today may not be valuable in the future. Memory, for example, is slowly losing value as the world's knowledge becomes instantly accessible thanks to the little rectangular smart box in my pocket connecting me to everyone and almost everything people have ever known. AI will only hasten this trend.

And so it's perfectly sensible and indeed tautological that IQ test performance is the most predictive for tasks that require high cognitive skill. Those are the skills we most value in our society and those we prioritize and teach. Similarly, although IQ may be separable into crystalized intelligence (what we know) and fluid intelligence (how we think), given the way in which what we know shapes how we think, these are not cleanly separable. And our tools for thinking are getting better with every generation. Not only is education improving but so too is the complexity of our societies more generally.

The Flynn effect, the rise in IQ test performance, is too fast for genetic evolution – indeed, if anything, evidence suggests there has been genetic selection *against* higher IQs as more educated individuals have fewer children. The Flynn effect, if taken at face value, would mean that large proportions of previous generations had low enough IQs that they should have been barely functional. That obviously can't be right. Instead, the Flynn effect is best interpreted as a measure of increases in cultural complexity: widening education, better pedagogy, more complex media. Think of TV shows and films from the last century compared to those of today. Today's shows are longer, with more characters, more complex character development, and convoluted plots. Adam West's *Batman* from the sixties, with its simple storylines and cheesy 'Wham, bam, pow!' fights versus *The Dark Knight* of the noughties, with its less predictable plots; Lucy and Ricky's trope-filled 1950s humor in *I Love Lucy* versus today's *Rick and Morty*, with its complex, multilayered gags. You get the point.

The fact that IQ has neural correlates too tells you nothing about whether intelligence is genetic, environmental, or cultural. As I tell my students, if you see a headline that reads 'X Psychology Found in the Brain', don't be surprised. Where else would you expect our psychology to be manifested? Be surprised when you read a headline that says 'X Psychology Found in the Big Toe'.

Heritability, sex, and population differences require a bit more explanation. Let's start with heritability.

What is the heritability of number of fingers?

Heritability sounds a lot like inheritance and so people often confuse the two. When they hear that IQ is 50% heritable they assume that this means that 50% of intelligence comes from your parents' genes and 50% from somewhere else, like school or some other aspect of the environment. Or something along those lines. But this isn't what heritability means at all.

Heritability refers to the degree to which the variation in a population is predicted by variation in genes. That's still a bit abstract, so let me give you an example.

What do you think the heritability of number of fingers is? If you're thinking of heritability as genetic inheritance, you'd assume close to 100%. There's an overwhelming probability that both you and your parents have ten fingers. If you have more, you have polydactylism, which is sometimes dominant, but rare. If you have fewer, you probably lost them in an accident. In fact, the heritability of number of fingers is close to 0%. Why?

Almost all variation in number of fingers is predicted not by genes but by environmental accidents that cause a loss of fingers. And almost all people have genes that result in ten fingers. And so there is both very little variation in the trait and very little variation in genes. The same logic applies to heritability of more cognitive traits. Consider literacy.

The heritability of literacy differs by age and country. In Australia the heritability of literacy in kindergarten is 84% and in Grade 1 it's 80%. Roughly the same. But in Sweden and Norway the heritability of literacy in kindergarten is only 33%, rising in Grade 1 to 79%. What do you think the difference is between the kids in Australia and Scandinavia? It's obviously not genes – that doesn't somehow change between kindergarten and Grade 1 for Scandinavians, but not Australians. The answer is the cultural diffusion of literacy.

Australian children start learning to read and write in kindergarten. In contrast, the Scandinavian kindergarten curriculum emphasizes social, emotional, and aesthetic development. Literacy instruction only begins in Grade 1. And so in Scandinavia differences in literacy before Grade 1 are a product of learning going on at home, but in Grade 1 everyone starts to learn and differences are more due to genes. We're not discussing which educational system is better, but this does tell us that assessing the genetic basis of literacy without accounting for particulars of curricula on cultural diffusion is a selection bias of unknown magnitude. But note that literacy in the home environment is already shaped by cultural evolution – parents, caregivers, and other adults in the community are all literate. Children are exposed to reading and writing simply by living in a developed society. In other words, there is no 'baseline' heritability. Heritability is a composite measure that captures both genes and culture.

The lesson here is that heritability is not only a product of what is genetically transmitted – obviously, number of fingers is highly genetic – but also the variation in the trait and the variation in the predictors. For this same reason, height is not the major predictor of success among NBA players. Everyone is sufficiently tall and other factors matter more. It's also why it's such a mistake to remove standardized testing such as the SATs or GREs for college admissions based on their predictive value of performance in university. Top universities in the United States select the top SAT and GRE performers meaning other factors become more important predictors. But that's only because you have a subset of the full range of SAT and GRE scores – you've already selected on the basis of these scores. Removing that selection criteria in favor of high-school grades, reference letters, and work experience will undoubtedly lower standards and also entrench the advantages of those with the connections to secure reference letters and gain CV-enhancing experience. Removing tests that compare everyone by the same measure also removes an avenue whereby someone without those privileges can show their ability through a test given to all. That's not to say that performance on the

SAT or GRE is not affected by privilege or that the tests can't be improved, just that the alternatives are affected even more. This privilege is reflected in heritability scores themselves.

Richer Americans have higher heritability

IQ is around 50% heritable overall. But in the United States, the heritability of IQ is higher among the rich – around 70% – and lower among the poor – around 10%. How can this be?

The answer is that among the rich, the difference between schools, home environments, social groups, and neighborhoods is smaller. The difference between elite school A and elite school B is minor and so one child from a wealthy family is unlikely to have a significantly better environment than another child from a different wealthy family. Remember, it's not just genes that contribute to the differences we see even within a society, but also culture and environment.

Among richer Americans, culture and environment provide more similar input and so more of the differences in outcomes are predicted by genes. Indeed, in a utopian world where everyone had the same opportunity, same environment, same family circumstances, and access to the same resources, genetic heritability would approach 100%. One could even argue that higher heritability is a measure of equality of opportunity for that trait.

In contrast, among poorer Americans differences can be vast and often due to chance. One child from a poorer family can have a significantly better environment and set of opportunities than another child from a similarly poor family based not on their genes but on the luck of life's lottery, including the neighborhood they live in, the social services available to them, and literal school lotteries. As a result, less of the difference in outcomes is predicted by genes. To offend Tolstoy, rich American families are all alike in their environment; poor American families have poorer environments in their own way.

As a result, where interventions work, they work better when they're applied earlier. Earlier adoptions and earlier movements from more deprived to more privileged communities lead to larger gains across a range of life outcomes, including IQ. And interventions work better when they apply features that are already ubiquitous in richer communities to communities that do not have these features. For example, giving children and pregnant mothers micro-nutrients has enormous effects in poorer communities, but smaller or non-existent effects in communities who get those vitamins from reliable food supplies. Similarly, growth mindsets – teaching kids that ability and achievement are not fixed and can be developed through hard work, good principles, and learning from mistakes – often fails to replicate, but where it does show effects are in poorer communities who don't already have these beliefs or have more of a gap to close.

So what would you expect to see for the heritability of IQ in more equal countries or where educational input is more uniform – places like Europe and Australia? That's right, there is little to no heritability difference between the rich and poor in Europe and Australia.

The differences in IQ test performances between rich and poor help us understand the final two and perhaps most controversial facts about IQ: sex differences in intelligence and population differences in intelligence.

This book will tackle a few controversial topics like these. The attitude I take is as follows. Being forthright and truthful about even challenging topics is critical to trust in science. Popular books such as this can only point to the research and general conclusions and not the many nuances, and a lack of nuance more easily feeds into people's existing prejudices. Here I will try my best to faithfully and honestly offer as nuanced an approach as I can in this context. And as in all cases throughout the book, I will try my best to approach topics in as disinterested a way as is possible and will try to describe the science to the best of my ability. No one can really achieve this, but that doesn't mean we shouldn't try. Science moves past the biases of scientists as long as sufficiently different biases exist and people are free to critique one another with an agreement on how to evaluate theory and evidence. With all this in mind, let's start with sex differences.

Sex differences in intelligence

Sex differences in STEM jobs, pay, and political representation are hotly debated topics. As a researcher, I want to understand how we can get to a more gender-equal world, and as a father to two daughters, I am personally invested in figuring out a systems-level solution and not cane-toad-proximate solutions that create more problems. To fix a problem we must first understand it. One hypothesis for why men occupy more high-paying, high-status jobs is sex differences in intelligence.

Males and females differ in a variety of ways. Two obvious examples are that males are taller and stronger on average, though of course some females are taller and stronger than some males. Males and females also differ in their brains, as I'll discuss in a moment, which may be a function of both genetic propensities and cultural input, just as height and muscle mass are a function of genetic propensities, nutrition, and exercise.

Unlike strength and size differences, there are no good a priori theoretical reasons to expect a sex difference in intelligence and there is no reliable evidence of a mean difference in IQ test performance. However, mean performance on IQ tests are also a function of the design of the test, particularly the weight given to different subtests. Male and female performance differs by subtest, but these differences are unreliable in that different studies with different populations show different effects. Does that mean that males or females are smarter?

The most reasonable argument for an overall sex difference in intelligence is that male brains tend to be larger than female brains, even controlling for body size. Brain size across related taxa and even within our species is weakly correlated with measures of cognitive performance. For example, the largest dataset of total brain volume, measuring over 13,000 British participants using MRI brain scans to get an accurate brain volume, found a correlation of $r = .19$ between brain volume and fluid intelligence and $r = .12$ between brain volume and educational attainment. The perfect correlation would be $r = 1$. No correlation would be $r = 0$.

These correlations are not zero but they are weak, and as the plot that follows should make obvious, there is enormous variation. There are so many other factors beyond brain volume that affect your ability to think. Moreover, brain volume itself may be a result of other factors, such as health, nutrition, and pollutants, as we discussed.

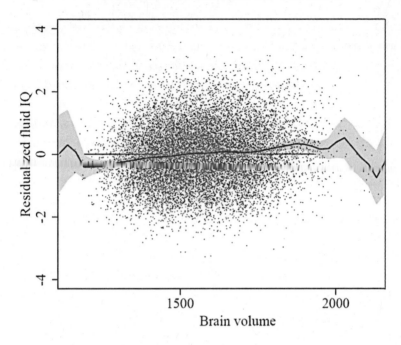

Brain size and fluid intelligence. Source: Gideon Nave, Wi Hoon Jung, Richard Karlsson Linnér, Joe Kable, and Philipp Koellinger, 'Are Bigger Brains Smarter? Evidence from a Large-Scale Pre-Registered Study' (3 December 2018). SSRN version reproduced with permission from corresponding author, https://ssrn.com/abstract=3295349 or http://dx.doi .org/10.2139/ssrn.3295349 (Final version published in *Psychological Science*).

Although male brains are significantly larger on average than female brains, the cognitive differences are small. Part of this might be because male and female brains differ in structures associated with cognitive ability in opposite ways. For example, males have more grey matter and females have greater cortical thickness, both associated with greater cognitive ability. So currently there are *no theoretical nor empirical reasons* to think that there are sex differences in average cognitive ability. There are, however, theoretical and empirical reasons to believe there is a sex difference in the *variance* of cognitive ability – the spread of the distribution.

Two out of three children with severe developmental delays are boys. For males to still have the same average cognitive ability as females, the distribution must be compensated on the other end. This so-called greater male variability hypothesis is controversial. In 2005 Harvard University President Larry Summers gave a keynote speech on diversity in engineering and science where he cited the variability hypothesis among three hypotheses that may explain the greater representation of men in top science and engineering positions. A firestorm of controversy, condemnation, and debate ensued. The controversy and condemnation comes less from the variability itself and more from the implications of greater variability in males for explaining sex differences in achievement and its use as a kind of justification for the over-representation of men in positions perceived to require greater cognitive ability. That is, greater male variability has been used to both explain and justify sex differences in STEM jobs, pay, and political representation.

If the distribution of male and female intelligence is normally distributed, has the same mean (i.e. males and females are the same on average), but males have greater variability (i.e. are more spread out in the lower and higher end of ability), we would expect males to be over-represented at both the bottom and top of the distribution. For example, there may be more men in both the highest-paid and lowest-paid jobs and this gender gap will be larger at the tails of the distribution. That is, if there really is a difference in the variance of the ability distributions between males and females, and those distributions are normally distributed, then the predicted gap between males and females at the very top and very bottom should be much larger than people assume. And that's because normal distributions are exponential functions.

Humans are terrible at understanding exponential functions. We seem to lack intuition for them. It's possible that we are actually good at it, but schooling causes us to lose the ability. For example, some populations without schooling naturally perceive the world as logarithmic, a way to make an exponential world easier to understand. Our lack of understanding of exponential functions explains why the king was tricked so easily by the court advisor, who requested a grain of rice be doubled on all sixty-four squares of a chess board. One grain on the first square, two on the second, then four, and so on. The total number of grains of rice needed by the sixty-fourth square is much more than all the rice that has ever existed – a number with about nineteen zeros. That's larger than the number of stars in the Milky Way, which only has around eleven zeros. The king received a lesson in exponential growth and the court advisor lost his head. Similarly, the lack of understanding of exponential functions contributes to people not investing more when they're younger. Benjamin Franklin allegedly described the magic of compound interest, which grows exponentially as 'Money makes money. And the money that money makes, makes money.' Exponential growth is just not intuitive.

If intelligence or ability is normally distributed, even at the tail ends, that means that small differences in the average, or in this case, variances, will have small differences near the average – like the first few doublings on the chess board – but enormous differences in the ends (tails) of the distribution. This difference will be more visible the larger the population.

Thus, a greater male variability hypothesis would predict only small differences between males and females close to the mean – the average in a population – but a large over-representation of men in the poorest-paying jobs and the highest-paying jobs (insofar as pay correlates with ability).

So insofar as the normal curve fits the distribution of abilities (and there's good reason to think that it does), even at the tails (this is not necessarily the case), then a variability difference alone (same average) between males and females would lead to small sex differences where selection is lower. Take sports for example, where both the mean and variance differ. Even so, the difference between male and female athletic performance in high school will be much smaller than in international competitions where selection is higher. And as mentioned, this effect will be even larger in larger populations (e.g. larger companies and larger countries) where there is more tendency toward a normal distribution and larger numbers at the tails of the distribution.

The greater male variability hypothesis is largely an empirical finding. Males are over-represented at the bottom end and top end in weight, height, physical aggression, brain structure, tendency to cooperate, time, risk and social preferences, as well as cognitive abilities such as verbal and spatial ability, reading, math, and IQ test performance. As with anything this controversial, sensitive, and with real implications for people's attitudes, the data is disputed. There are also domains where no variability difference is found. Moreover, there is no widely accepted evolutionary theory for why males show more variability, but most theories that make the prediction focus on the greater competition among males in the mating competition for females.

The evidence for a propensity toward greater male variability is, in my opinion, reasonably strong, but this does not mean it's unaffected by culture or it cannot be affected by policy choices. Consistent with the cultural malleability of male variability, we see differences between countries, including on the variance in cognitive abilities.

The OECD's Programme for International Student Assessment (PISA) tests children on the same exams all round the world. Focusing on just the variation, a clear pattern of significantly greater male variability in math performance was found in thirty-four out of forty countries. Some countries were more ambiguous and no discernible difference was found in Czechia, Denmark, Ireland, Netherlands, or Tunisia. Greater female variability was found in Indonesia.

Thus – and this is worth noting – insofar as the male variability hypothesis is true and insofar as ability is normally distributed even at the tails, in a large enough population we should expect to see large discrepancies between males and females. In other words, even if we were to equalize not only opportunity but encouragement, the metric of our success should not be parity – because this would need to be in all domains, top and bottom. This is not an excuse. I believe my son and two daughters largely have the same equal opportunities, but they deserve equal exposure and encouragement in domains in which men or women remain under-represented. A cultural change requires not just encouragement from my wife and me, but from society. On the other hand, policies that reduce selection based on ability are admirable in their intent to try to equalize outcomes, but in the long run and at a systems level they end up harming both males and females. Tearing down barriers that harm one sex more than the other and supporting gender parity are both important, but judging the outcomes of these efforts is not so simple.

But none of this is firmly established science – much more theory and critical testing is required. Nonetheless, mixed evidence doesn't mean that everything is equally probable. If I had to make a bet on the basis of current evidence and current theories, I would bet on there being no sex difference in intelligence at the mean, but there being greater male variability. And as I'll discuss later on, I also bet that culture, institutions, and policies can do more to equalize the opportunities and encouragement that each sex receives in pursuing the things they're good at and enjoy. This general truth is even more true when it comes to population differences. Unlike between males and females, there is a far less clear delineation between what we might consider different populations.

Population differences in intelligence

Psychologist Steven Pinker, along with many other scientists, was once asked for a dangerous idea. He selected the idea that 'groups of people may differ genetically in their average talents and temperaments', writing that 'the prospect of genetic tests of group differences in psychological traits is both more likely and more incendiary, and one that the current intellectual community is ill-equipped to deal with'.

Populations do differ in IQ test performance and many explanations for this are indeed dangerous. They lead to people essentializing these differences as unchangeable, inherent, immutable, or genetic and then defending either discriminatory policies or defeatist policies based on low expectations. But as with sex differences, to fix a problem we must first understand it with honesty and openness. Before we get into these population differences in intelligence, let's first deconstruct the populations themselves and tackle some common misunderstandings of this topic.

Ever heard someone say that race is socially constructed and not biological? I would wager most people can't explain what that means, and it seems impolite or even racist to ask, let alone dispute it. And that's true even if it seems to fly in the face of people's everyday experience with people with different ancestries, from different parts of the world, how they categorize one another and themselves, and the boxes people are asked to tick when asked about their race.

The intuitive biology of race seems to be reinforced by negative reactions to ideas of transracialism. One particular case that gained national prominence was that of Rachel Dolezal, when it was discovered that the apparently African American president of the Spokane, Washington National Association for the Advancement of Colored People (NAACP) was a blue-eyed blonde woman born to white parents. Dolezal was claiming African American ancestry despite no evidence. Similarly, when it comes to Native American or other Indigenous identification, Elizabeth Warren's claims of being Native American on the basis of Native American ancestors became an election issue. Despite race not being biological, genes and ancestry seemed to matter. Warren ultimately did take a DNA test that revealed a small percentage of Native American ancestry. So if she needed to prove her Native American status with a DNA test and Dolezal wasn't allowed to be transracial despite no one noticing that she wasn't Black for so long, how is race not biological? Both of the following are true: populations have identifiable genetic differences and race is socially constructed. Let me explain.

Let's take the case of Barack Obama, who despite having a mother of white European ancestry and a father of black African ancestry, is widely considered Black in the United States. This is a socially constructed notion of a single drop of African ancestry making a person Black, but a single drop of European ancestry not making a person White. This cultural construct now feels very natural to many Americans, but it is not a human universal. Obama would be considered neither Black nor White but a Colored person in apartheid South Africa, whereas in nineteenth-century Haiti he would be at the top of the social ladder as a light-skinned Mulatto.

Race is socially constructed in the same way color is.

Visual light is a smooth gradient of the electromagnetic spectrum. The socially constructed part in light is how we separate that smooth gradient into discrete colors. Different societies have different perceptions of where blue stops and green begins and whether cyan is in between. Like color, genetic differences are also what biologists call 'clinal' – think of a smooth incline and decline compared to an abrupt cliff.

Genetic differences between populations are a smooth gradient (with some exceptions due to geography that prevent people from meeting each other and swapping genes, thereby creating sudden but small genetic cliff-like shifts). But in the case of genetic differences, it's not a single wavelength but multiple dimensions created by a large space of possibilities of differences in a variety of different locations on the genome. In the distribution of these many dimensions there is massive overlap between all human populations and a smooth gradient – cline – as we move from Swedes to Danes to Germans to Swiss to Spaniards. But we can also reliably genetically distinguish Swedes in Europe from Swazi people in southern Africa on the basis of alleles (gene variants), which are rarer in some populations than others. While these alleles allow us to identify ancestry, they may have nothing to do with any observable physical phenotypic differences (e.g. skin color), let alone the genetic complexes related to cognition. Many forces, some entirely random and inconsequential, such as genetic mutation, drift, or gene flow, can create differences, which can be maintained in the absence of mating between people from more distant regions. In turn, these genetic differences can be used to identify ancestry without necessarily having any consequence for differences we might care about.

So the biological phenotypes that people culturally construct into races have a reality. Noticeable features such as skin color and hair type do have a genetic basis and do map on to population differences. But there is an arbitrariness to which features people use to classify one's race and the choice of these distinguishing features vary by culture.

For example, the idea that people from Africa, the most genetically diverse place on earth, are a single race is not just particularly problematic but also an obvious cultural construction. Not only because Africa has the largest genetic diversity on earth, but because people more familiar with Africans can easily identify the average differences between African ethnic groups, such as the Igbo of Nigeria, Maasai of Kenya, Dinka of Sudan, Amhara of Ethiopia, Tswana of Botswana, San of Namibia, and Zulu of South Africa, who due to lower migration and vast distances on the second-largest continent may indeed be more easily physically distinguishable than European ethnic groups. The inability to recognize these differences is a function of the culture we grew up in, not a result of biology. Not only are there many shades of black but there is also huge variation across all physical attributes.

Given this reality, categories such as Black, Indigenous, and People of Color (BIPOC) or African American in the United States or Black and Minority Ethnic (BAME) in the United Kingdom are farcical. They bind by a few arbitrarily chosen characteristics (e.g. melanin) people with massive genetic diversity, cultural diversity, and diversity in social outcomes in health, wealth, income, and education. The result is all kinds of odd inconsistencies that hide true levels of discrimination and can hinder our ability to target and help those most in need of assistance.

Group boundaries are often arbitrary, making a politics of pigmentation naive. Pigmentation politics can even hide deeper prejudices or be politically expedient without addressing actual underlying group differences. Two of my favorite examples of hiding deeper prejudices are the legal cases *Takao Ozawa v United States* (1922) and *United States v Bhagat Singh Thind* (1923).

In 1922, Japanese American Takao Ozawa applied for US citizenship, which as per the 1906 Naturalization Act was available to 'free white persons'. He argued that he was free and his skin was white and he was therefore eligible to apply for citizenship. The Supreme Court unanimously held that 'the words "white person" was only to indicate a person of what is popularly known as the Caucasian race'.

Based on this ruling, a year later in 1923, Bhagat Singh Thind, an Indian Sikh, similarly applied for citizenship under the same 1906 Naturalization Act. As per the racial classifications at the time, Indians were Caucasian and as a North Indian Thind could further argue that he was Aryan. Contradicting *Ozawa v United States*, the Supreme Court argued that although he might have been scientifically Caucasian or even Aryan, he did not meet the 'common understanding' of Caucasian; that is, he didn't have sufficiently white skin or European ancestry.

The slipperiness of racial categorization allows for inconsistencies. When material advantage is at stake – jobs, educational opportunities, and so on – it can also encourage people to identify with marginalized groups to gain these advantages. This is a common pattern around the world – when a group is offered some kind of affirmative action, the number of people identifying in that group goes up. Warren may have ticked the 'Native American' box early in her career out of genuine beliefs about her identity and ancestry, but the uptick in ethnic identification when material advantage is at stake suggests that many others would not tick the box if it conferred no advantages. The slipperiness of the categories also allows an upper-class wealthy new immigrant from Africa with no slave ancestry to claim advantages at the expense of a poor candidate with fewer privileges from both lighter- and darker-skinned native-born Black communities with slave ancestry. The next natural question is the degree to which the genetic differences between populations, clinal or not, creates the differences we see in the world.

A common retort to there being any population differences at all is that the genetic differences between populations are small and that there is greater variability within populations than between. This is true. There is large overlap between any two human populations and small differences in the relative frequency of different genes. For example, alleles for red hair are rare and more common in northern Europe than anywhere else.

But it is also true that small differences can have large effects and that when we use many different genes, we can reliably detect where people are from, especially if they don't move too much. Where populations have remained in the same place for long periods with little genetic inflow, we can map genes to geography down to 100 miles. But most populations are admixed – genetic mixtures of many migrant populations.

Migration is the norm for *Homo sapiens* and when we migrate we also mate. Of course, some populations are more and less admixed than others. Genetic isolation can also be exacerbated, not just by geography but also by cultural practices, such as when populations practice endogamy – a preference for marrying others in their ingroup or a prohibition against marrying outsiders.

In contrast, groups like African Americans and Latin Americans are new, thoroughly mixed populations, and it makes little sense to make broad generalizations about these groups on a genetic basis. So too with my ancestors from Sri Lanka, a tear-shaped island just south of India that was a central trading port halfway between east Asia and Europe. Sri Lanka's colonial past includes the Portuguese, Dutch, and English. Modern-day Sri Lankans have phenotypes ranging from fair-skinned, blue-eyed Burghers – a term meaning 'citizen' in their Dutch ancestral tongue, though it has extended to others of European ancestry – to Sri Lankans of Chinese and Malay origin – to darker-skinned Tamils and Sinhalese, better adapted to the equatorial location. But like African Americans and Latin Americans, these are not genetically discernible differences well matched to identified ethnicity. Most Sri Lankans are likely of mixed ancestry, carrying alleles for skin color and other genetic differences that are well mixed, as they are in modern America. This is not to say that those genetic differences don't matter.

Some of my ancestors, for example, are Dutch – the tallest population on earth – and it is not inconceivable that this contributed to most members of my family being several standard deviations taller than the average Sri Lankan. The average Sri Lankan male is 1.67 meters (5 foot 5 inches) and the average Sri Lankan female is 1.55 (5 foot 1 inch). At 1.85 (6 foot 1), I am one of the shorter males of the last three generations. Shorter than my 1.93 (6 foot 4) cousin, my 1.9 (6 foot 3) twin baby brothers, but also my 1.87 (6 foot 2) grandfather, great grandfather, and uncle. My sister is 1.77 (5 foot 10).

The long and the short of it is that although race is socially constructed and clines are smooth, there is also a reality to genetic differences between populations, which is why we can identify ancestry with some degree of confidence. This confidence is lower and geographic range larger when populations have mixed a lot. Which is why 23andMe, despite not knowing what I look like or my history, can identify my recent ancestors as being broadly from southern India and my more distant Dutch ancestors at the correct colonial period, broadly as French and German, which identifies a broad swathe of Europe, including the modern-day Netherlands.

Your Ancestry Timeline

How many generations ago was your most recent ancestor for each population?

1	2	3	4	5	6	7	8	8+
1960	1930	1900	1870	1840	1810	1780	1750	1720
Southern Ind								
				French & German				

The topic of genetic differences between populations is going to become even more heated in the coming decades. Not only because new DNA data from around the world will no doubt reveal genetic differences between populations – it would be shocking if there weren't genetic differences, if only by chance – but also because there are new sources of genetic differences due to local adaptations and the admixture of those older hominin populations – Neanderthals and Denisovans mentioned earlier. And we cannot rule out that just as genetic differences probably play some part in explaining group differences in height, they may also play some part in explaining group differences in aspects of cognition. It bears emphasis that there is currently no reason to believe that this is the case. But equally, we cannot base our morals or policies on the unassailability of this hypothesis. We're going to need more maturity than the current name calling and virtue signaling that these topics currently elicit. To understand some of the sources of these population genetic differences, let's start with local genetic adaptations before we dive into ancient love affairs.

Local adaptations are sometimes a result of culture–gene co-evolution – adaptations to local environments and culture. Lactose tolerance found among some European populations is probably the most well studied of these. No mammal can process the lactose in milk into adulthood, and that is true of most humans. Lactose intolerance is not some allergy, like some people have to nuts, it's the mammalian norm and the norm for most human populations. Some human populations, however, notably those who herded cows but didn't quickly discover yogurt and cheese to reduce lactose, developed a mutation that allowed them to digest the rich source of calories the lactose in their cows' milk provides. The most recent research suggests that famines may have provided the necessary strong selection pressure. In contrast, yogurt and cheese are a cultural rather than genetic solution – a way of enjoying those same nutrients in dairy with lower lactose thanks to bacteria pre-processing it for us. Culture–gene co-evolution is an active area of research, with new recent and localized adaptations still being discovered. Apart from local adaptations, another source of population genetic differences are ancient DNA.

There have been rapid advances in sequencing DNA from ancient bones and this research is quickly mapping the movement and admixture through migration of populations. This research reveals that migration is central to the human story. It also reveals that there may be some discernible differences in Eurasia through admixture with Neanderthal and Denisovan populations. These sources of distinct DNA are not absent from Africa – remember migration is the norm and some populations traveled back to Africa bringing these novel genes with them. Moreover, there may also be admixture with other yet-to-be discovered human species within the African continent imparting unique alleles – research published in 2020 hints at a yet-to-be discovered 'ghost population', ancestors to many West Africans.

All of these avenues open new possibilities for measurable genetic differences between populations. The question, though, is to what extent these genetic differences matter to intelligence, education, work performance, or anything else we care about. At the moment, the answer is that we simply don't know. But if the theory of everyone is correct then the most probable answer is that it probably won't matter all that much, if at all. For now, there is no strong case for genes explaining differences in cognitive or other outcomes between groups, but there is a strong case for the role of culture.

Culture can shape our hardware – exposure to pollution, bad diets, disease, lead (still a serious problem in many parts of the world, including the United Kingdom's drinking water), insufficient nutrition, and exposure to smoking. Culture shapes the environment of cultural acquisition – the stability of families, exposure to ideas and knowledge, value of education, and motivation for hard work. And culture ultimately shapes the software, the mental tools we acquire and then use for thinking. In all this there are vast differences between populations that do more to explain not only differences between countries but also between ethnic groups in the same country.

At various points in history and sometimes today, minority ethnic groups have outperformed the majority. Tamils had better academic performance than Sinhalese in Sri Lanka, Igbo and Yoruba did better than Hausa in Nigeria, Amhara did better than Oromo in Ethiopia, the Chinese ethnic minority did better in Malaysia and Indonesia, South Asians did better than native Fijians in Fiji, East and South Asians outperform Whites who outperform African Americans in the United States. These differences are almost certainly not due to genes. Some of these groups are probably a result of selective migration – China and India have huge populations and the barrier to migration to the United States is high and so the highest-performing members of these countries are more able to secure a competitive job or university place and migrate. Our psychology maintains cultural borders through norms and traditions better than genetic borders through preventing inter-marriage. White, Black, Latino, BIPOC, or BAME hide a lot of genetic variation and a lot of subgroup differences in outcomes.

Consider this graph of household income of ethnic groups in the United States from the US census. Many groups with large differences in income share superficial phenotypic characteristics such as skin color. In each case, history, culture, selective migration, discrimination, and institutional factors offer a better theoretical and empirical explanation than genes. Remember, different societies have dominated at different periods in history. 'Barbarians' have become civilized and the civilized have become barbarians, and these were cultural, not genetic changes. The Flynn effect and technological change are cultural changes, not genetic. With a better understanding of populations, we can now return to population differences in intelligence.

Indeed, race – in the sense of physical attributes – is not a natural evolved category that we care about. For most of human history those around us, even those from different groups, looked physically the same; remember, genes are clinal. Instead, when other, more accurate cues about culture and group affiliation are available, such as accent or clothing style, people will implicitly and intuitively prioritize these cues over racial cues. In turn, as we saw in the last chapter, these cultural-group-identifying cues become a more dominant source of differential treatment and even discrimination. In other words, race is an imperfect correlate of what we really care about – culture and group affiliation.

Returning to IQ test performance and other differences we see between populations, these are exceedingly unlikely to be explained by the genetic differences we find in our contemporary, well-mixed human populations. If we speciated, it was along cultural lines and culture is of course horizontally transmissible. The barriers to cultural transmission and, indeed, even the barriers to genetic transmission

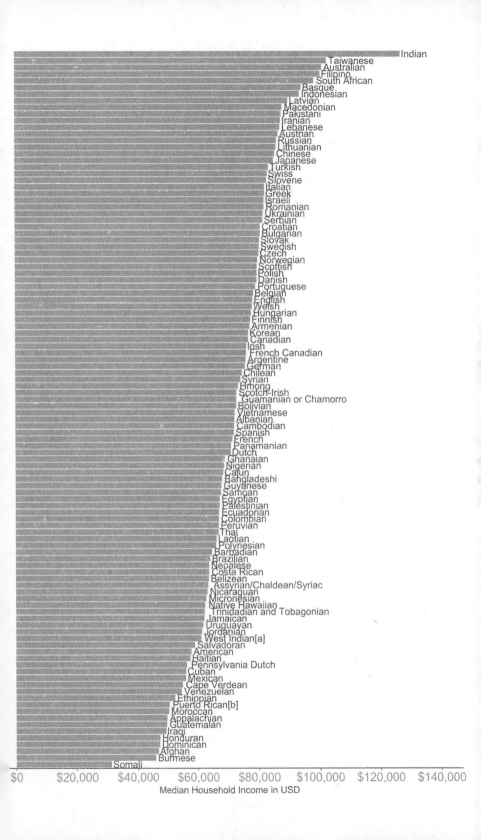

Median Household Income in USD

tend to be cultural – such as discrimination, inequalities, or preferences for marrying co-ethnics.

Pinker was right that these are dangerous ideas that we are still ill-equipped to deal with, but as we become more interconnected, deal with them we must. The science has outpaced the social discourse and that discourse needs to catch up. One way forward is denial and ignoring these possibilities, but no amount of scolding or shaming will silence everyone. Indeed, research on polarization reveals that in the face of attempted shaming and suppression, many honest and fair researchers will stay silent rather than face the consequences of career-destroying accusations or continual harassment. In such an environment the only voices left will be those committed to an agenda. Anecdotally, one piece of advice I received in writing this chapter was that I should remove all potentially controversial topics lest the broader message of the book be bogged down and lost to the controversy. Another way forward is careful, open, and honest science combined with even more careful, open, and honest science communication. Because if you can't trust scientists to be careful, open, and honest, you can't trust science. Here I've tried my best.

To summarize, intelligence is at the heart of who we are and how we got here, but there is no such thing as culture-free intelligence. Intelligence is a co-evolving product of cultural evolution. This discovery that the particular cognitive skills that have allowed us to behave intelligently in our current context are acquired and not just hardwired is important, because it means that we can structure our education systems and societies to develop the software and the hardware that will be most effective in coping with the changing world in which we live. In doing so, we can maximize our species' ability to innovate new efficiencies, new energy technologies, and new ways to work together for mutual benefit.

4

Innovation in the Collective Brain

'If you wish to make an apple pie from scratch, you must first invent the universe,' said science communicator, author, and host of *Cosmos*, Carl Sagan. Even without going back to the Big Bang, we are surrounded by foods, technologies, products, and practices that we could not possibly recreate from scratch. Almost everything we rely on – from Google to the government – were not invented by us, were not invented in our lifetime, and are typically too complicated to truly understand, let alone recreate. We simply don't know what effect a change will have until we try it.

Our success as a species has been a result of our incredible ability to innovate, but that has not been a result of individual intelligence alone; nor has it been the result of genetic geniuses who see further than the rest. Instead, innovations don't require any specific innovator any more than your thoughts require any specific neuron. Of course, specific innovators are involved, just as specific neurons are involved in your best ideas, but innovations are more accurately seen as a result of our collective brains – how intelligent humans come together to learn from one another and share ideas as a collective. Understanding these collective processes will allow us to break the next energy threshold. Indeed, it is these collective processes that have led to every innovation that surrounds us.

Even the simplest things in our lives are the product of thousands of years of accumulated knowledge, borrowed and recombined across multiple generations in multiple cultures, spanning the globe. The smartest among us could not recreate the smallest fraction of the modern world from scratch. Not even Sagan's humble apple pie.

Making an apple pie from scratch

Canadians love Kraft Mac & Cheese or Kraft Dinner as the Canuck call it. The simple three-step recipe involves boiling macaroni in water, draining it, and adding butter, milk, and the prepackaged powdered cheese. The recipe is simple because most of the manufacturing took place long before the box arrived in a Canadian kitchen. But all recipes, even those made from scratch, involve complicated histories, global supply chains, agricultural know-how, and sophisticated processing that unlike our hunter-gatherer ancestors, we no longer have to think about or even understand, let alone do.

As a simple illustration, let's *actually* make an apple pie from scratch.

First, the pastry. We're going to need some flour. Do you know how to grow the Middle Eastern grass we call wheat? Let's cheat and assume wheat has already been artificially selected over thousands of years for easier harvesting and larger grains. But even with the benefit of this selection, do you know how to plant and look after wheat? Do you know when the grains are ready for harvesting? Or how to separate the grains from the chaff? (I'll assume you know what chaff is.) Do you know how to grind the grains into flour without your coffee grinder or food processor? If you want to use those, can you find and extract the raw materials, manufacture the grinders and motors, and design and create the electrics truly from scratch?

Next, let's add some sugar. We need another grass, this time from South East Asia. Do you know how to grow sugar cane, when to cut it, and how to process it into sugar crystals? Have you got a plan for cane beetles or other bugs? What's the fertilizer? (Many of us can't even keep a house plant alive!)

Let's leave the plants aside and add some animal products. First, eggs. Those come from the South East Asian junglefowl we call a chicken. Do you know how to catch a chicken and how to keep it alive and healthy? How to help it produce more eggs?

Don't forget the butter. For that, we need the milk of a Eurasian female bovine who recently gave birth. Make sure you domesticate her and care for her, so that after she's given birth, instead of defending her calf, she'll let you milk her. Now take that milk and extract the fatty cream. Now churn that cream into butter.

Remember the illusion of explanatory depth – there's a lot of refinement and complication to each of these processes, and we haven't even got to how to make pastry or bake the pie. But once you've discovered the recipe for apple pie or any of its ingredients, that recipe is easier to spread than the process that led to it. Over time, through a division of labor and the invisible hand of efficient supply chains, the marvel of the apple pie becomes a trivial family recipe. So too with our mental software.

Learning to count

Humans can't cook without culture. And we certainly can't count without culture. We went from counting one, two, three, many, as some small-scale societies still count, to a full-blown number system. Numbers likely emerged as an innovation for more efficiently tracking agricultural inventories of cattle and crops, perhaps for the purposes of trade. (You need to know who owes you what!) Invention may be driven by necessity, but it also needs the right analogy, metaphor, or suitable mental model to make the leap. In this case, that mental model was fingers.

As briefly mentioned in the previous chapter, today we commonly use a decimal system because we count on our ten fingers, but societies have had number systems based on the twelve phalanges (three finger bones per finger) counted with the thumb or counting up to dozens with different parts of the body. There is nothing special about the number ten. But to count beyond body parts, we needed another innovation and that required a different mental model. Something like stones.

The word 'calculus' comes from 'pebble' (think calcium or limestone), and was used for addition and subtraction. It's one way to think about addition or subtraction that lets us move beyond how creative we can get with body parts. There are some stones, and you can throw down more or snatch some away. Stones are great for what we call natural numbers greater than zero: 1, 2, 3, 4, 5, and so on.

But stones are not enough to make something like zero obvious. It took millennia to get to the innovation of zero. What does zero pebbles look like? Well, it looks a lot like zero of everything else – it's nothing – and 'nothing' is hard to imagine.

But while it took the greatest mathematical minds to develop the concept of zero, the invention of the number line as a mental model helped make zero concrete and transmissible, even to children. We still use number lines in schools today. The number line also revealed the negative numbers, which 'darken the very whole doctrines of the equations and make dark of the things which are in their nature excessively obvious and simple', as British mathematician Francis Maseres melodramatically complained in the eighteenth century. Number lines work by mapping numbers not to objects but to movement and position. Without these cultural innovations, nothing about numbers is intuitive to our ape brains. But just as you can try to improve on Grandma's apple pie recipe, analogies and metaphors can help us go even further.

Sitting on my desk is a framed copy of Euler's identity, $e^{i\pi} + 1 = 0$, often referred to as 'the most beautiful equation'. Two transcendental numbers (e and π), one, and zero, all connected by an imaginary plane orthogonal to the number line. To me it signifies that the world may be complex and confusing – there is no pattern to e or π – but there are hidden rules that bring order to the complexity and confusion. Rules that once discovered can be used to do things we couldn't do before – such as solving electrical circuits or doing Fourier transformations – both of which are essential to modern technology and can be done more easily with complex numbers.

Cultural innovations literally change our minds and give us new capacities. They're like software upgrades. What is impossible becomes possible with the right innovation, and with further innovations we may even learn to reliably transmit what we have discovered. Some innovations are more general than others. For example, thanks to the invention of writing, I can convey information through straight and squiggly lines on a page. I'm doing it right now and I'm literally changing your brain. As writing became more standardized and easier to learn, it eventually extended from the purview of the elite to an essential for everyone.

How can we turn what our theory of everyone reveals about human ingenuity into actionable ways to improve our own capacity for creativity? How can we improve the innovativeness of our companies and our societies? The answer appears once we realize that innovation is a social process – a product of a collective brain. Once we realize this, we can become intentional in how we seek information and connect people to maximize the probability of good ideas emerging and spreading.

In order to get there, here is a COMPASS. COMPASS is an acronym for the seven secrets of innovation that I teach to classes and companies. After I go through COMPASS, I'll show you how the head of Uber in the United Kingdom and northern Europe, Jamie Heywood, and I used it to develop an innovation strategy to solve Uber's challenges in this tough market. (Uber encountered a clash between the American Silicon Valley approach and the European traditional approach; a battle between preferences for unregulated innovation and innovative regulation.)

COMPASS: Seven secrets of innovation

Thomas Edison famously described genius and the process of innovation as '1% inspiration and 99% perspiration'. Except that he didn't. At best, Edison popularized a description commonly used at the time, sometimes attributed to author and lecturer Kate Sanborn. Edison was good at borrowing, stealing, and recombining.

The other thing Edison didn't do was invent the light bulb. At best, Edison, along with Joseph Swan in England, developed the first commercially successful incandescent light bulb, together forming the Edison & Swan United Electric Light Company (Ediswan). It was the best way to avoid costly litigation over who got there first. But neither Edison nor Swan were singular geniuses who saw further than most. Instead, they were the first commercially successful winners in a crowded market of people trying all kinds of different light bulb designs. The successful designs of a variety of people led to at least twenty-two other patents for incandescent light bulbs at the time. So next time you see a light bulb used as a symbol for an amazing new idea, remember that it is just one of many light bulbs of different design, brightness, and longevity, only the best of which have been selected by the law of evolution. But like moths to a flame, we are drawn to the brightest bulbs – the winning innovations. We forget the evolutionary landscape full of other luminaries who just didn't quite make it into the history books.

Humans the world over tend to attribute important inventions to key ancestral, almost mythical figures. Fire is said to have been given to the Aboriginal Australians by the ancestral being Crow, to the Indians by the divine being Mātariśvan, and to the Greeks by the Titan god Prometheus. Mimi spirits taught Aboriginal Australians to hunt and cook kangaroo and Shaka Zulu, founder of the Zulu Kingdom, invented the iklwa short spear. Westerners are no exception. Edison (or Swan) is said to have invented the light bulb, Gutenberg the printing press, and Benz and Ford the automobile. Remember that our cultural learning psychology seeks out the most successful people to imitate, in order to become successful ourselves. And we do it even if those successful people are from the past. But innovations are not driven by individual innovators. The reality is more complicated. And that reality also reveals how we can become more innovative.

The key figures to whom we attribute inventions are often the first to popularize an invention, first to recombine the pieces floating through social networks where many are working on the same problem, and typically one of many inventing the same thing around the same time. At some level, we understand this.

Scientists and Silicon Valley entrepreneurs are somehow simultaneously convinced of their own originality and genius but also terrified of being scooped or beaten to market despite that originality and genius. They live in a paradox of seeing further than those around them but fearing that others may see the same thing. We can navigate past this paradox using COMPASS.

Secret 1: Collective brain thinking (C)

Inventors sometimes simply strike it lucky. You might be familiar with Alexander Fleming's prototypical story of the serendipitous discovery of penicillin antibiotics. In 1928 Fleming accidentally left a Petri dish containing *Staphylococcus aureus* bacteria next to an open window. When he came back, he discovered that the dish was contaminated with mold and that the bacteria near the mold were dying. He carefully studied the mold, identifying it as a member of the Penicillium genus. He published his results in 1929, naming the mold liquid 'penicillin'.

Fleming's careful research revealed that penicillin was able to kill a range of bacteria responsible for common, but at the time fatal, illnesses, including pneumonia, diphtheria, and meningitis. And that's usually where the story of antibiotics ends. Fewer people know that Fleming's penicillin was incredibly difficult to mass produce and was not widely used until the 1940s. What took so long?

In the early 1940s the amount of penicillin in the entire world was enough to treat around 100 patients. The miracle drug could save only a few. The urine of treated patients was saved, because typically around 80% of the precious antibiotic was passed through it and could be extracted to treat other patients. The right invention often requires the right circumstances, the right timing, and collective effort to become a world-changing innovation. For penicillin to hit prime time, it needed the Second World War.

With European powers and their allies at war, blowing off limbs and tearing bodies apart, antibiotics went from useful to urgently needed. Antibiotics for just 100 patients wasn't enough, so a team led by Howard Florey and Ernst Chain tried to build on Fleming's work by mass-producing penicillin. They moved from London to the United States to avoid the bombs and get more funding. First to New York and later to Peoria, Illinois, where the new fermentation division of the Northern Regional Research Lab was studying the metabolism of molds. The team took out ads and searched the world for different strains of mold that could be mass-produced. A suitable mold was discovered in 1943 on a moldy cantaloupe from the local Peoria fruit market. The mold contained a mutation that produced a much higher yield than Fleming's original strain. By further mutating the mold with X-rays, the team chanced upon a mutation with a thousand times the amount of penicillin. Most strains of penicillin today are descended from that 1943 serendipitous, trial-and-error, recombined, collectively discovered mold.

There are several lessons from this story, and they are echoed in many other world-changing discoveries: an initially lucky breakthrough; some recombination, followed by some trial and error and partial causal models to figure out what's going on; then the right circumstances, the right timing, and the right understanding of a problem, combined with collective effort to turn invention into diffused innovation. This is how the collective brain thinks and innovates.

Serendipity and recombination as sources of new discoveries, incrementally innovated through trial and error and partial understanding of what's going on, is still how our collective brains make breakthroughs. It's the story of the accidental discovery of Upsalite carbon nanostructures in 2013 and the intriguing discovery of ways to control a quantum computer with electrical rather than magnetic fields in 2020.

But luck isn't blind. We can make our own luck and deliberately seek out serendipity and useful recombinations. The first step is to recognize that where we are today is a result of trillions of small decisions made by the billions of people before us about where we should go and how to get there. We must recognize that not every step has been the best and that there are many inefficiencies in where we are today that can be overcome by walking off the well-beaten path.

Secret 2: Off the beaten path (O)

Path dependence refers to the role of the past and history in constraining the future; when early decisions lock us in due to the difficulty of changing to something different. For example, if you're used to thinking of temperature in Fahrenheit, it's difficult to learn Celsius and vice versa. The decisions made by James Madison and colleagues in writing the US Constitution may not be the same decisions we would make today let alone the best decisions at the time, but changing a constitution is a challenge once you have one. If you chose to study cosmetics in college, it may lead to a valuable and rewarding career but can make it more challenging to switch to civil engineering.

One of my favorite examples of path dependence concerns literal paths. Britain and many of its colonies drive on the left-hand side of the road. They've done so since Roman times. Romans rode horses and carts on the left, driving with their left hand and leaving their right hand free to use a weapon. One piece of evidence for this is groove marks in ancient Roman roads, as a load heading toward a delivery point leaves deeper grooves than with an empty cart on the way back. This reveals that Romans drove on the left side of a road. Going off the beaten path is difficult but doable. Today, Europeans drive on the right side of the road. Napoleon's empire and influence was large enough that his deliberate decision to drive on the right as a possible military strategy led to a change that eventually spread across the continent. America as a new country was able to pick a new side.

The power, vision, and influence of a new leader, the creation of new societies or new companies all offer opportunities to try something new, off the beaten path. It's why Noah Webster was able to create a brand-new form of English for America. Webster tried to create more consistency in the English language, dropping the silent 'u' from words such as 'color'. As a result pronunciation matches phonemes more reliably in America than Britain, where, I was informed, the city of Leicester is not in fact 'Lie-sester' but 'Lester' and Oxford's 'Magdalen' College is not 'Mag-de-len', as written, but 'Maud-lin'.

The tiny nation of Estonia also went off the beaten track to become the top-ranked country outside East Asia on the international PISA scores in their children's performance on mathematics, science, and reading. It achieved this not by spending more money per pupil: it spends around 8,000 US dollars per pupil, far less than the OECD country average of 11,000 US dollars and far less than countries like the United States (US$14,000), Australia (US$12,000), and Canada (US$12,000), who all had worse outcomes. Instead, these surprising results were achieved through a range of reforms that were adopted following independence from the Soviet Union, borrowing the best educational practices from around the world and combining them into an amazing new, technology-supported education system. Plagiarism is bad, but not when it comes to policy, where there should be a lot more plagiarism of what works elsewhere. We'll discuss how to innovate ways off the beaten path through start-up cities and programmable politics in Chapter 8, and the specifics of what Estonia did in Chapter 12. Estonia was using a magpie strategy.

Secret 3: Magpie strategy – steal like a magpie with a prepared mind and intellectual arbitrage (M)

In folklore, magpies are known for always looking out for shiny objects to bring back to their nests. This is a powerful strategy for empowering innovation – actively seeking recombination and serendipity, collective effort, and new contexts of application. In contrast, a 'maven' – from the Yiddish word *meyvn* – refers to someone with deep and thorough expertise on a topic. Mavens can go far, but when they are also magpies then they can go further. Take Sam Panopoulous, a magpie who created one of my favorite pizzas.

The controversial Hawaiian pizza might be the most multicultural food ever created. Sam Panopoulous, a Greek immigrant to Canada, inspired by American Chinese food (think sweet and sour) put a South American ingredient (pineapple) on an Italian dish, naming it after the Polynesian state, allegedly because the brand of tinned pineapple was called 'Hawaiian'. Hawaiian pizza is popular but divisive, and I want to take this opportunity to tell haters that they should reflect on their life choices. Hawaiian pizza is also a case study in how immigrants empower innovation.

Immigrants bring solutions from a diverse experience that can be recombined with practices, technologies, and other aspects of culture to solve unsolved problems in their new home. Their diverse experiences naturally lead them to become magpies.

The magpie strategy requires two things: a 'prepared mind' knowing the problem in sufficient depth that one's inner magpie can distinguish the shiny and useful from the dull and irrelevant; and 'intellectual arbitrage' that actively seeks potential solutions outside one's own domain of knowledge.

But you don't have to move to a different country to be a magpie.

The answers to your problems often exist in other people's heads. More often, for truly thorny problems, the answers are scattered across many people's heads. That's why they're still thorny – the pieces of the solution haven't come together yet. You can be the one to bring them together. To maximize the probability of discovering shiny new solutions, we need to talk to clever people we disagree with, people outside our disciplines and industries, and people beyond our immediate social circles. These are the conversations that make us more creative. You already know what your close friends think, but you know less about your distant friends and even less about your enemies. Having a good understanding of the problems in your nest prepares your mind. Having these diverse conversations allows you opportunities for *intellectual arbitrage*.

Arbitrage is a fancy way of describing the process of buying low and selling high. For example, finding an underpriced book at a thrift store and immediately selling it on Amazon at a higher price. Intellectual arbitrage describes this same approach applied to creativity.

Daniel Kahneman and Amos Tversky realized that economists didn't yet know about findings in cognitive psychology that undermined the predictions of economic models. Kahneman and Tversky were magpies with prepared minds who then used intellectual arbitrage to forever change economics, winning a Nobel Prize, the shiniest of all objects in science.

The McDonald brothers realized that the assembly line approach that had allowed Henry Ford to produce cheap cars could be applied to make cheap hamburgers. Rather than hiring an expensive trained chef like everyone else, the McDonald's burger chain hired people with no training at lower salaries and taught them to do just one thing – one person sliced the buns, another flipped the burgers, another added the cheese, and so on. Through intellectual arbitrage the McDonald brothers created the 'Speedee Service System', which allowed them to make hamburgers faster and cheaper than their competitors.

History is filled with famous magpies. You can google the story of Charles Goodyear – after whom the Goodyear tire company is named – with a mind prepared by the Roxbury Rubber Company to discover vulcanized rubber, or watch Howard Goodall's brilliant documentary tracing the origins of the Beatles' greatest hits to their various experiences in everything from church organs to a piccolo trumpet Paul McCartney heard on TV. But one of my favorite examples is perhaps history's most successful magpie: William Shakespeare.

Shakespeare was a master of rhetoric, a skill he likely learned at his local grammar school in Stratford-upon-Avon. A key element of the curriculum was the trivium: rhetoric, logic, and grammar. Many schools today only retain the grammar (woe betide today's potential Shakespeares). Shakespeare's grasp of rhetorical techniques like assonance (similar sounds, such as blue moon), alliteration (same starting letter, for example power to the people), chiasmus (words repeated in reverse order, i.e. ask not what your country can do for you – ask what you can do for your country), diacope (repetition with a division in between, for example Bond, James Bond), and so on is unmatched; the next most prominent rhetorical geniuses are probably British Prime Minister Winston Churchill and American rapper Eminem. But unlike these more recent figures, Shakespeare was a plagiarist, or rather a magpie, combining historical texts with rhetorical flair.

As writer Mark Forsyth recounts, Shakespeare wrote *Antony and Cleopatra* and for that he needed to look up their history. Because we know that Shakespeare had a poor grasp of Greek, we can surmise that he probably relied on Thomas North's English translation of Plutarch's *Lives of the Noble Greeks and Romans*. And when we look at North's book, we can see exactly the way Shakespeare plagiarized. But with rhetorical flair!

NORTH: . . . she disdained to set forward otherwise but to take her barge in the river Cydnus, the poop whereof was of gold;
SHAKESPEARE: The barge she sat in like a burnished throne, Burned on the water: the poop was beaten gold;

NORTH: the sails of purple, and the oars of silver, which kept stroke in rowing after the sound of the music of flutes, howboys, cithernes, viols, and such other instruments as they played in the barge.
SHAKESPEARE: Purple the sails, and so perfumed that The winds were lovesick with them; the oars were silver, Which to the tune of flutes kept stroke, and made The water which they beat to follow faster, As amorous of their strokes.

Shakespeare would have failed a college Turnitin plagiarism check. But it's hard to begrudge the Bard. He was a magpie who skillfully recombined rhetoric with history, bestowing on us not only beautiful poetry but multiple idioms that continue to be recombined by modern artists. Aldous Huxley's title *Brave New World* is borrowed from *The Tempest* ('Oh brave new world, that has such people in it'). Smash Mouth's hit 'Allstar' has the memorable line 'All that glitters is gold' – a play on the opposite from *The Merchant of Venice* ('All that glitters is not gold').

We can all be magpies by actively shaping our experiences and knowledge, who we talk to, and where we search for solutions. One way to naturally increase collective brain thinking and magpie strategies is to harness diverse ideas and ways of thinking. An old, much plagiarized saying goes: 'To steal ideas from one person is plagiarism; to steal from many is research.' A diverse team or country naturally brings together a diversity of ideas ready for recombination. But diversity is a double-edged sword – both helpful and harmful to innovation. Diversity is a paradox to be resolved. And to reap diversity's benefits without paying its costs, resolve it we must.

Secret 4: Paradox of diversity (P)

The most innovative teams are more diverse, but so too are the least innovative teams. This seeming paradox of diversity occurs because diversity offers recombinatorial fuel for innovation, but is also, by definition, divisive. Without a common understanding, common goals, and common language, the flow of ideas in social networks is stymied, thus preventing recombination and reducing innovation. But diversity is the most powerful method of becoming more innovative. Yet many companies treat it as little more than an inconvenient exercise in counting minorities using Peter Griffin's skin color chart or making sure the proportion of women on a team isn't low enough to be embarrassing. Rather than resolve the paradox, many companies opt for monoculture, 'good fit', and diversity that really means 'people who look different but still think like me'.

Resolving the paradox of diversity is in the middle of the COMPASS and at the heart of a collective brain approach to innovation. To resolve it, we have to analyze the dimensions of diversity, ignoring irrelevant diversity, ensuring we retain deep diversity, and finding common ground on divisive diversity.

Many aspects of diversity are largely irrelevant. Take food preferences for example. If you like sushi and I like schnitzels we can work it out. Other aspects ought to be irrelevant even if they currently aren't, such as the color of our skin. The key to resolving the paradox of diversity is finding common ground on things we don't share that get in the way of smooth communication. We can overcome these challenges with strategies such as optimal assimilation, translators and bridges, or division into subgroups, which retain diversity without harming communication and coordination.

Optimal assimilation means speaking the same language. Not only literally the same language, but understanding the nuances of communication – when it is appropriate to interrupt, when we should follow orders, how emails should be written, what memos should look like. Making these explicit instead of relying on unwritten norms can help quickly assimilate new, more diverse employees. But sometimes separate jargon is necessary within a team or subdivision. Things go more smoothly when sales can understand what engineering can and can't do and when engineers understand the constraints of sales and customer relations, but these groups work more efficiently within themselves by developing their own specialist jargon and unique approaches. Translators and bridges are people who can speak to more than one group, such as people with training or experience in both sales and engineering, who can help achieve more effective specialization within groups and better overall outcomes between groups.

Specialization makes it possible for society or a company to exceed the capacities of its constituent brains. To see how this works, it helps to see how specialization evolved. Imagine that there are ten things that are required to survive – food, housing, medicine, clothes, the rules of society, defense, and so on. And imagine that any individual's cognitive capacity is a maximum of ten brain units. Bigger brains can store and manage more information, but it's difficult to birth anything bigger until medical interventions like cesareans are invented. And so brain size hits a fundamental limit.

If all of us must learn all ten things to survive then each of us can achieve one skill unit on each skill; ten brain units, ten things, skill level 1. But imagine you only have to learn half those things because there are enough people that even if some die, enough others know the other half. There are enough hunters, house builders, and medical specialists that you don't have to learn it all and you know that even if one of the house builders dies, there are enough other people to build the houses. Now you can dedicate yourself to getting better at say five things and others learn a different five things. Together, our society can now reach skill level 2.

Now imagine you only need to learn one thing: society can now reach skill level 10. Your missing nine items are covered by the rest of the population. Divide it further and the sky's the limit, despite a limited ten-unit brain. Individuals become smarter at a few things but also stupider at everything else, siloing specialists into disciplines, divisions, or departments. This in turn creates a challenge for co-ordination among different specialists. This is the specialist version of the paradox of diversity and the solutions remain the same.

In a small town, there may be one general physician, but in New York a doctor may specialize on a small part of the renal system and get very good at treating that one part. Society is then able to innovate as a more intelligent collective brain. Larger, more interconnected diverse societies that learn how to share and transmit knowledge more effectively and equitably are more innovative, but there are many challenges to unlocking this potential.

Resolving the paradox of diversity is one of the great challenges of our time and so I will go into it in more detail at a societal level in Chapter 7. The reason that diversity at a societal level is a bigger challenge today than it was in the past is because people from more culturally distant places now live side by side in the same societies. This is a boon to innovation, but a greater challenge too. More often, it is easier to take advantage of culturally close diversity – the adjacent possible.

Secret 5: Adjacent possibilities (A)

Gutenberg is often credited with creating the first printing press around 1440. As with many technologies, it was a magpie recombination of screw presses like those used in wine and olive oil, replacing woodblocks with individual metal movable types for each letter, trial and error on ink recipes using various oils and black pigments, and more. It was quite the achievement, but a Gutenberg with all those same pieces in place would have failed even a century earlier. What enabled the printing press to become an adjacent possibility for Gutenberg to invent were new advances in pulp-based replacements for expensive vellum and parchment. These were simply too expensive for mass printing to make economic sense. Gutenberg's printing press needed cheap paper.

In evolution, the adjacent possible, a term coined by Stuart Kauffman, refers to the range of possibilities that can be reached with only small changes. For example, wings are not an adjacent possibility for humans – they are far too complex to reach from our current physiology. An extra couple of fingers, however, is very much in the adjacent possible. Technologies too have adjacent possibilities. New advances in one area open up new possibilities for our entire cultural corpus. For example, the falling price and rising power of computer chips led to everything from commerce to dating becoming mediated by a computer. The falling price of gene sequencing machines has made it easier to track diseases and opened up new possibilities in medicine. Tesla's creation of the modern electric car was thanks to the invention of high energy density lithium batteries built for laptops. Tesla's first cars were essentially laptop batteries strapped to motors and wheels. And today, AI is opening possibilities that we are only beginning to comprehend.

Being a magpie can help you find recombinatorial opportunities in what already exists and is well understood in a different discipline. A complementary approach is to actively follow advances in technology across different industries and consider their application to domains they were not designed for. It's very difficult to do this alone, there's just too much to learn in the modern world. And that's part of the reason why being social beats being smart.

Secret 6: Social beats smart (S)

Bigger populations are more innovative. There are more ideas floating around. Friendlier populations are also more innovative. Those ideas meet each other as people get to know one another. And big, friendly, interconnected populations? They're the best and brightest. Good ideas are no good unless the right people with the missing pieces know about them.

Archaeologists and anthropologists have long noticed a relationship between sociality – the size and interconnectedness of a population – and its cultural complexity, for example the size and sophistication of its toolkits and technologies. They've also noticed that when sociality falls, that is, say, if population size were to shrink or disconnect, then technology and culture seem to go with it. The classic example of this is Tasmania, a large island off the southern coast of Australia.

Around 10,000 to 12,000 years ago the last ice age ended and sea levels rose. Tasmania, on the southern tip of the east coast of the Australian mainland, was cut off and became an island. From this point on the inhabitants of Tasmania began to lose culture and technology to the point where they were not only less technologically sophisticated than their cousins on the mainland but also less technologically sophisticated than their own ancestors. Tasmanians lost the ability to make fishing spears, bone tools, boomerangs, and even warm clothing, resorting to rubbing fat over their bodies to stay warm. This loss of culture that follows a drop in population size or interconnectivity isn't isolated to Tasmania – there are other examples in recorded history among the Inuit and based on the archaeological record of Paleolithic Europe. Evidence that social beats smart – and indeed creates smart – can be found in the roots of the Enlightenment.

In the seventeenth and eighteenth centuries coffee houses were the source of learning, philosophical and political discussion, gossip and news. Pamphlets and other publications spread ideas between coffee houses and communities. They increased the probability of ideas meeting and recombining into something new. They forced people to engage with one another. The caffeine and sugar probably didn't hurt either.

Today's equivalents of the Enlightenment coffee house are Twitter, Facebook, Reddit, Discords, and other forums and social media. The equivalent of pamphlets are podcasts, Wikipedia, blogs, Substacks, and shared articles and videos. The effervescent, often heated debates and discussions, viral posts, and 'current thing' increase our effective interconnectedness and expose us to new ideas. The Internet and social media are where we engage with one another.

Innovation in the collective brain is empowered when we talk to each other.

Secret 7: Sharing is critical (S)

We need to talk to each other. Differences in terms of a lack of common language and hostility toward those who don't share our politics or group membership harm our ability to communicate. So too does unfriendliness or lack of opportunities to meet one another. But the final secret to human innovation is to find ways to smoothly share information.

The amount of information that is retained during communication is called transmission fidelity. Early genetically evolved and culture–gene co-evolved improvements to transmission fidelity were simple things like paying attention to others, letting others hang around without threatening them, and guessing what they might be thinking in our own minds (what scientists call theory of mind). Remember, this is what we're better at than our ape cousins.

Language would also have been a large boon to transmission fidelity, evolving as a response to having more valuable information to transmit, as we'll discuss in the next chapter.

Later culturally evolved improvements to sharing knowledge included connecting up knowledge through metaphors, analogies, and epistemologies to help people remember the world – such as the collective brain analogy for innovation – and learning better ways to teach one another.

In many hunter-gatherer societies, teaching occurs by allowing children to observe adults in action. Pastoralist societies, that typically have a larger, more complex cultural corpus, spend more time on ex-plicit and effortful instruction. In many industrial and post-industrial societies, teaching has become industrialized and specialized.

Compulsory formal education emerged as a response to the Industrial Revolution. Factories required workers to operate machinery so they needed a minimum skill level and the ability to communicate and understand instructions. To provide this education quickly and efficiently, it was necessary to formalize the delivery of a cultural package – numbers, phonemes, grammar, and so on.

Today, cultural evolution continues to increase transmission fidelity in our evermore complex cultural world, not only through improvements in education but also through technologies such as the printing press, radio, television, the Internet, video conferencing, and various iterations of social media platforms. People are freely sharing information with one another with short TikTok tip videos, longer YouTube instructional videos, and entire online courses from the world's greatest universities and greatest instructors, many freely available.

Using the COMPASS is natural and includes many things we do anyway. Cultural evolution has found many ways to improve our ability to innovate and has transmitted these throughout our populations. The difference is that once you explicitly label these secrets and codify these strategies, you can intentionally use them to become more creative and innovative. To show you one example of how COMPASS can be used at an organizational level, here's a real-world example: Uber.

Creating Uber's 5S strategy

Jamie Heywood was dissatisfied. After a successful career that included being CEO of Virgin Mobile's Chinese and Indian division and Director of Electronics at Amazon UK, he had seen a large part of the world both in terms of geography and business. He had become convinced that the very idea of a company needed a rethink. The purpose of a company, as he saw it, wasn't just to make money. Companies had to make money of course, but profits were the by-product of their true purpose, which was to solve humanity's toughest problems. The better they could do that, the more money they made.

After encountering the idea of a collective brain in my collaborator Joe Henrich's book *The Secret of Our Success*, Jamie finally found a language that helped make sense of his thirty-year career. As he describes it, companies aren't just economic instruments – rational, planned, top-down. They are also social institutions whose characteristics emerge from the bottom up and which are selected for by the company's stakeholders depending on the service they provide to society more widely. The best companies thrive over the long term because they are able to serve both their social and economic masters by aligning their interests through consistently innovating solutions to humanity's hardest problems.

In 2018 Jamie was offered the opportunity to head up Uber in the United Kingdom and northern Europe. The company was in crisis, finding that the traditional Silicon Valley belief in unfettered innovation clashed with a European preference for slow-moving and extensive regulatory protections. Jamie felt the role provided an incredible opportunity to work out how Uber could best balance the often conflicting needs of Uber's increasingly impatient shareholders with those of the cities where it operated. Critical to solving this puzzle was making Uber's collective brain cleverer.

To help achieve this, Jamie invited me to present the seven secrets of innovation and work with him and his team to develop a new innovation strategy that expanded the company's collective brain, making it cleverer. The 5S strategy we co-developed was itself a product of collective brain thinking: stealing like a magpie, having a prepared mind, solving the paradox of diversity, seeking the adjacent possible, and recognizing the importance of being social and sharing.

Every company is different and so it is important to integrate these insights into existing norms and ways of working. Over a series of meetings we boiled it down to the following five principles for Uber Europe, the 5S:

1. Simplify the problem: Be clear on what the problem is before moving to find solutions, make it engaging by using customer stories, and simplifying what we say and how we present information.

2. Socialize your problem-solving: Approach problem-solving as a collective effort and see 'socializing' the problem widely as productive time.

3. Select your meeting mode: Push the team to be deliberate about organizing specific problem-solving meetings (as opposed to decision-making or live update meetings) and plan for them accordingly.

4. Stick to the two pizza rule: For problem-solving meetings,

invite only as many attendees as can contribute meaning-
fully.

5. Seek diversity and divergence: Seek out constructive diver-
    sity and input that challenges the status quo.

Under Jamie's leadership, Uber has used these principles to help
it find solutions to some of its hardest problems, including making
peace with its regulator in London, expanding the size of its electric
fleet, and integrating bikes, buses, and trains into the app.

Jamie's story is a microcosm of what happens when we recognize
that who we are and how we got here – our brains, our bodies,
languages, societies, companies, and countries – are created by culture.
That innovation is a product not just of individual intelligence but
of our collective brains. A 10x engineer is an engineer who is 10 times
better than the average engineer. There is no doubt that 10x engineers
and other 10x workers are valuable, but perhaps not as valuable as
the 10x teams and 10x societies we can create, which, in turn, can
create more 10x engineers. Optimizing our collective brains isn't just
good for our organizations, societies, and us as individuals; it is also
essential for reaching the necessary energy breakthroughs for the next
level of abundance. Indeed, the final part of the answer to the question of who we are is that our collective and cultural brains co-evolved,
creating every aspect of us.

5

Created by Culture

Why are humans so different to other animals? When I ask people this question, they typically list all kinds of very sensible things – language, technology, art, memory, sense of humor, and so on. But underlying this list is an assumption so obvious that people don't even bother to include it – humans are just smarter.

Humans have giant brains that make them more intelligent than other animals. But as you've seen so far, human intelligence and innovation are more complicated than people often assume. Yes, we are smarter, but not for the reasons people often assume and not in the ways people often assume. Hopefully, by now, you have a better sense of how the theory of everyone applies to the human animal – how we think, the cultural nature of our intelligence, the way we learn and learn from one another, and the way we work together to innovate. Almost all aspects of our behavior and the nature of our societies are linked by this theory of everyone. With it, we can begin to understand everything from our capacity for language to the origins of the patriarchy to the existence of grandmothers and more. This will be important when, in Part II, we begin to think about where we need to go next. And it all starts with our giant cultural brains.

The human brain has tripled in size in the last few million years. Our brains are now about three times as large as a chimp's. And, you might think, that makes sense! Bigger brains are better! Who doesn't want a bigger brain? And you're right, bigger brains can store and manage more information. But if you think bigger brains are always better, you should stop and wonder why all species don't have giant brains? What's stopping them?

The answer lies in the law of energy. Brain tissue is energetically expensive, using over twenty times more energy than the same mass of muscle tissue. It's cheaper and easier to evolve brawn than brain. And so that's what most animals do; they get stronger, not smarter.

Larger brains might help you escape predators, outcompete other members of the group, and survive better in an environment, but they can only do that if they can pay their energy bills. Bigger brains have to justify their size by helping you find more food. Most animals, including us for most of our history, spent most of the time sorting out dinner. So really what an animal wants is the *smallest* brain that lets them get the job done – find food, outcompete other animals, evade predators, and so on. Too large a brain is like driving too big a car. It comes with ongoing fuel costs. And so encephalization – the evolution of big brains – needs an explanation. That explanation will help you understand how seemingly disconnected aspects of ourselves and our society fit together.

Explaining encephalization

We used to think that brains evolved for the Machiavellian manipulation of other group members or simply for keeping track of others in our groups – a *social brain hypothesis*. But we now know that that's only part of the story. Brains are not simply for tricking or tracking others; instead, they're for what you think they're for: storing, managing, and using information. Yes, for thinking! That information could be social information about other group members, but it doesn't have to be. It could be acquired from others, but it doesn't have to be.

Animals can learn all kinds of adaptive knowledge – where food is, how to evade predators, how to outcompete other individuals to secure a mate. And that knowledge can be learned by yourself through individual exploration and trial and error or by learning from others. Learning from others is by far the most efficient way to learn.

At the extreme, one could learn only the answers from others, like a lazy kid in class peeking at the exam papers of those who've studied; it's easier to copy answers than do the work yourself. Humans are like that cheating child. They arrive in the world, don't really try to figure out how it works, but instead just figure out what the past generation are doing, what most people are doing, and what the most successful members of society are doing. And then, like the Scottish children we met in Chapter 2, we just copy that. Of course, as we learned, it's not quite that simple, but as we're about to see, that general social learning approach also completely changed every aspect of ourselves and our societies. This is the *cultural brain hypothesis*.

Cultural brain hypothesis

The cultural brain hypothesis (CBH) makes predictions for the inter-connected, bidirectional relationships between brain size, group size, innovation, social learning, mating structures, and length of the juvenile period, depending on ecology and reliance on social learning across the animal kingdom. It also has secondary implications for mating strategies and the existence of grandparents that we'll get to in a moment. The theory predicts that among social learning animals, larger brains should be associated with larger groups, more social learning, more adaptive knowledge, more innovation, and longer juvenile periods. Why? Because ultimately brains evolve in lockstep with the information they can access and the calories that information unlocks.

Brains, as mentioned earlier, are for storing, managing, and using information. Bigger brains can store, manage, and use more information. You can acquire that information through some combination of asocial learning – trial-and-error reinforcement learning, building a causal model of understanding, figuring it out on your own – and social learning – copying what other members of your group are doing. But regardless of how you get that information, if you have more information that unlocks more energy, it increases the group's *carrying capacity*, which is how many individuals the environment can support.

As per the laws of energy and innovation, more information can give you more access to energy. And so with more or better adaptive knowledge, the number of individuals who could in principle survive increases.

Think of how improved knowledge about food production, from the Agricultural Revolution to the Green Revolution, or the astonishing advances in modern medicine, from antibiotics to vaccines, have allowed more of our species to survive regardless of whether

CULTURAL BRAIN HYPOTHESIS

SOCIAL LEARNING

(ASOCIAL) LEARNING

illustrated by
VERONIKA PLANT

in collaboration with
MICHAEL MUTHUKRISHNA
© 2020

ADAPTIVE KNOWLEDGE

INCREASES with / with

GROUP SIZE

LARGER GROUPS = more members/individuals to learn from

SOCIAL LEARNING

WHEN THESE INCREASE SO DOES

LEARNING

BIASES

birth → sexual maturity

JUVENILE PERIOD

OBLIQUE LEARNING?

BRAIN size

LARGER BRAINS

STORE & MANAGE more info

GROWTH LIMITS
* available calories
* physical birthing limits

extended ADOLESCENCE PERIOD

sexual maturity ↳ reproduction

DIVISION of LABOR

LARGE groups

HUMANS

HEAVY RELIANCE on social & oblique LEARNING

gaze following

joint intentionality

theory of mind

INCREASED TRANSMISSION FIDELITY

self relevance | conformist

COMPLEX LEARNING BIASES

prestige | success | expert

they understand the Haber-Bosch process for making fertilizer or how exactly a vaccine works. All that knowledge leads to more people.

But if you're a social learner, those larger groups are also useful to you in another way that we discussed. They give you more individuals from whom you can learn. In other words, a larger collective brain. As a result, brain size and group size are more strongly correlated among social learners, not directly for tricking and tracking but indirectly through the knowledge that groups offer to the social learner and through the knowledge leading to larger groups. Bigger brains, more knowledge, and larger groups are a package, mediated by the amount of adaptive knowledge.

Social learning is an efficient way to learn. Learning from someone else is far more efficient than trying to figure things out on your own. In fact, with enough information in the population, a social learner can actually get away with a *smaller* brain than an asocial learner. This might seem counterintuitive, but remember, an animal would prefer to have a smaller brain, so if you can get away with a smaller brain because the intelligence of your society helps you survive, then brains shrink as smaller brains outcompete bigger brains. Curiously, this brain shrinkage is actually what we see in humans.

The human brain, after growing for millions of years, has begun shrinking over the last ten thousand years or so. The cultural brain hypothesis predicts that such shrinkage is consistent with increased cultural innovation. Culture and our collective brain allow more people to survive even if individually they're not particularly bright – they can get away with not always coming up with the best answers simply by copying what most other people are doing. For example, we can benefit from a hospital even if we know nothing about medicine.

But this shrinkage only happens when the pressure to learn more isn't high. As the amount of adaptive knowledge that you have to learn grows, if there's more than your brain can handle then this creates a selection pressure for a larger brain. This is what happened for most of human history.

Our societies grew smarter and to keep up with all that knowledge so did we, growing ever bigger brains. But at some point we maxed out on brain size. At some point it became too dangerous to give birth to a bigger brain. Human childbirth is incredibly painful because human heads are incredibly large compared to the size of the birth canal. We can see this clearly by comparing the ratio of a baby's head size to the size of the birth canal for chimpanzees, an ancient hominin, and modern humans as illustrated in the image below. We should all be incredibly grateful to our mothers for what they went through to ensure our existence – they faced a more horrifying, difficult, and dangerous ordeal than any other great ape. More so if you have a bigger head!

A Visual Comparison of the Bony Birth Canal Vs. the Skull of the Primate infant for Primate Species

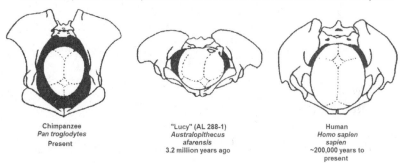

Head size vs birth canal of Chimpanzees, ancient hominin Australopithecus, and modern humans. Source: ArchaeoMouse (https://commons.wikimedia .org/wiki/File:A_Visual_Comparison_of_the_Pelvis_and_Bony_Birth_Canal_ Vs._the_Size_of_Infant_Skull_in_Primate_Species.png)

Bigger heads predict emergency cesareans and instrumental deliveries better than big bodies. Head size doesn't vary that much, but even within this limited variation, once you get to about the eighty-fifth percentile in head size, the need for emergency interventions hockey-sticks upwards on the graph. Big brains are great, but only if you can safely birth them. This variation and difficulty birthing big heads suggest that big heads are still under selection in humans. Basically, our species would like bigger heads like those Roswell aliens, but they're too difficult to birth. At least without a cesarean.

The use of cesareans is increasing. In the last couple of decades it's gone up from around a quarter to a third in the United States and has doubled worldwide. The use of cesareans and other birth interventions will likely lead to larger-brained humans in the future and an increased necessary reliance on medically supported births.

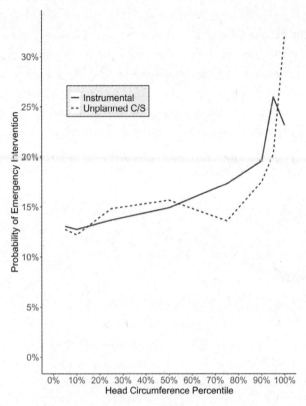

Head size as a predictor of emergency instrumental interventions (e.g. forceps) and emergency cesareans. Data from: M. Lipschuetz, et al. (2015). 'A Large Head Circumference is More Strongly Associated with Unplanned Cesarean or Instrumental Delivery and Neonatal Complications Than High Birthweight', *American Journal of Obstetrics and Gynecology*, 213(6), 833–e1.

That is, eventually, most people may have heads too big to birth vaginally. Maybe those long-limbed, big-headed stereotypical aliens are actually humans from the future!

For our species, thanks to our huge heads, vaginal births are painful and risky to both mothers and babies. Historically, up to one in every hundred mothers died during childbirth. But cesareans too come with a cost – to the mothers, who are often confined to bed for several weeks afterwards and can suffer infections and long-term complications as a result of being cut open in a major surgery – and to the babies where earlier births may be associated with cognitive and health issues. Cesareans in general can also lead to early respiratory issues. Labor leads to fluid absorption with a final vaginal squeeze removing fluid from the baby's lungs. Cesarean births also have reduced microbiome transfer, which may reduce overall health outcomes. The bacteria in a mother's vagina seeds a baby's microbiome during the birthing process. These are problems that cultural evolution may eventually solve. Indeed, new techniques are emerging that may help mitigate these risks, such as slowing the cesarean procedure – a so-called natural or gentle cesarean – and vaginal seeding, which involves a saline-soaked gauze in the vagina then being swabbed on the baby soon after birth to transfer the mother's microbiome. Birthing humans isn't the same as it is for other animals.

With all the challenges that bigger heads bring, we are lucky that big brains aren't the only game in town to deal with growing information. The law of evolution can try different strategies. For example, you can just spend a longer time learning – by extending your childhood.

If you are asked to make a face like an ape, what you often do is puff out your lips or cheeks. What you're really doing is protruding your jaw. Humans resemble juvenile chimps who don't yet have this prominent protruding jaw. Evolution extended our childhood and kept us as juvenile apes – what's called 'neoteny'.

Neoteny is a relatively easy change for evolution to make – a small change that in some sense extends features of childhood. We've done it to dogs, turning wolves into permanent puppies. We are the child-like chimp.

Neoteny might also mean that we're less aggressive, much as a younger chimp is less aggressive than an adult. It's a kind of self-domestication. Neotony might also explain why we can continue learning like a child for longer – with more to learn, we had to spend longer learning. Our childhood extended and a new period emerged: adolescence.

Adolescence is the period between the onset of puberty and full adulthood. The age at first birth and general preparedness for 'settling down' – finding a home and a job to support a family – have been increasingly delayed, creating what we could call kidults and a kind of cultural adolescence. What was initially a genetic extension of childhood has become a cultural extension where adults have no choice but to live with their parents because they're unable to afford a house of their own.

The world has also become more competitive. Even in our lifetimes, it used to be that a high-school degree was enough to compete in the workforce. Then any university degree. Then a STEM degree. Then a masters degree. Now a masters degree and one or more sometimes-unpaid internships. That's a long time to stay in school, and has created a new selection pressure: not just the ability to give birth to a big head but the ability to give birth at an older age. Those big heads and long childhoods have completely changed our societies. Starting with the relationship between men and women.

Premature babies, sexual norms, and child support

As our kids grow older, my wife Steph often marvels at how much they've grown, saying 'I can't believe they used to be small enough to fit inside me!' A fully grown human head is just too big to birth, but so too is the head of a young child. So humans solve the problem by giving birth prematurely. I don't mean some babies are born premature; I mean all of us relative to many other animals are born well before we're really ready to survive in the world. The vast majority of brain growth happens after you're born. As a father of three, I can assure you, human babies are floppy, useless messes.

We're not like a gazelle, ready to run. Or even like a baby chimp who still has a lot of brain growth left but can at least cling to their mother, easily drink milk, and will quickly mature and do even more. In stark contrast, our babies are wholly dependent on our care. We get less floppy and less messy as we get older, but we remain useless and reliant on our parents for a very long time. Some longer than others.

Our big heads, long childhoods, and protracted uselessness created new problems for our species. For one thing, our mothers need a lot more help with the kids. Today, in some societies there is still subsidized institutionalized childcare and support for single mothers providing a form of cooperative child-rearing. But another way to handle the problem is to get dad involved in protecting, provisioning, and otherwise raising children. This may seem like a natural solution, but it's highly unusual among great apes. Even among primates more generally, males typically don't have a lot to do with their offspring. It was a smart move getting dad involved, but remember genes are keen to spread themselves, so dad was happy to do it if it meant his floppy, useless kid was more likely to survive. But he also wanted to make sure that the floppy, useless kid was actually his.

One common solution to incentivizing human males to become doting dads was giving them greater control over female sexuality in return for greater control over male resources. Even today, female sexuality is more often the subject of normative control and the overwhelming majority of child support providers are male. But the control-support solution isn't the only one.

The Mosuo live at the Sichuan border with Tibet and are a rare example of a very different solution to the premature baby problem. Among this ethnic group there are no dads. Instead, your mother's brothers are expected to look after their sisters' children and there is no expectation for children within a nuclear family as we think of it. Females have complete choice over whom they sleep with and genetic fathers don't have prescribed obligations. Instead it is the job of the uncle to fulfil the role of father.

The Mosuo solution works in terms of genetic relatedness. Brothers have a guaranteed 12.5 to 25% genetic relatedness on average to their nieces and nephews (depending on whether it's their half-sister or full sister). Compare that to a value of either 0% or 50% on average depending on whether a child is yours or not. For a male, the expected value of genetic relatedness is the probability of it being yours multiplied by 50%. The Mosuo solution is also upheld by norms, but very different norms from those in our society. But this solution isn't easily applicable outside the Mosuo context. It works, in part, because of the remote location of their community – people live in the same community and it's not easy to leave or join. You also have to have large enough families to make sure you have a brother.

The Mosuo solution may be rare, but it is one of the many ways in which our big heads reshaped marriage and mating everywhere in the world.

Traditional marriages

The Mosuo are an unusual solution, but so too is the modern Western nuclear family. In much of history and still in many places, polygyny (one man, many wives) was and is the norm. President Jacob Zuma of South Africa, for example, had four concurrent wives, rejecting monogamy as a Western tradition. But even in societies where polygyny is the norm, most men had, at most, one wife. Only the wealthiest, most powerful men, like a chief or president, had more than one wife. But this created a problem.

The number of girls born tends to be the same as the number of boys born. So if one person marries more than one wife, then someone else doesn't have a potential spouse. In WEIRD societies, we still tolerate 'monogamish' behavior, particularly among the powerful. There is some degree of unfaithfulness, perhaps even long-standing unfaithfulness, but what we don't allow is for one man to legally, and often normatively, take more than one wife at a time.

Monogamy as a norm and as a marriage law is an evolutionary mystery for the following reason. If we consider it in terms of pure economic utility, what's called the *polygyny threshold model*, in an unequal society a woman maximizes resources for her children and is economically better off with half or even any reasonable percentage of a billionaire than 100% of a man who earns $20,000. Such a society offers a more efficient allocation of male resources, assuming males hold most resources and there is inequality. But it creates side effects that destabilize a society.

First, it drives down the age of marriage for females. As wealthy males monopolize the mating market, eventually there aren't any more females to marry. Therefore, in order to gain more wives, they have to marry younger females. This in turn has further effects on the relationship between males and females that are counter to our sense of twenty-first-century Western moral norms.

Another problem that it creates is a pool of young, unmarried males without the hope of finding a spouse. Males commit most of the violent crimes in every society, young males even more so. So what happens when young males are involuntarily celibate because they can't access enough wealth to woo even one spouse? They take large risks to acquire that wealth, even through unethical and violent means. Polygyny is a recipe for a violent, unstable society.

Monogamous committed romantic relationships, such as marriage, domesticate men, literally reducing their testosterone. In societies that practice polygyny, different strategies have been used to deal with the young frustrated male problem. These strategies range from elaborate rituals that force men to bide their time before finding a bride, to encouraging these young men to partake in raids on neighboring communities. These raids may result in a man finding a wife among the women, stealing sufficient resources to woo a woman from his own community, or dying during the raid. All outcomes effectively solve the problem. Although polygyny is common, polyandry – where one woman has multiple husbands – can also solve the premature baby problem.

Polyandry is rare but occurs where resources are scarce and it requires more than one man to provision a single child. To solve the paternal uncertainty problem, a woman will often marry brothers so any child has some relatedness to all fathers. Polyandry is also sometimes supported by beliefs such as partible paternity – the belief that sex with multiple men is required to make a successful child. In these societies, pregnant women will seek out the best hunters, fishermen, or other skilled artisans to endow their child with their abilities. In turn, these men may consider the child their own, providing some amount of care and resources that they wouldn't otherwise.

All of this is to say that the cultural evolution of mating practices is not constrained per se, but is affected by our biology, technology, and the environment. But ultimately, all practices are trying to solve the same problem – the big-headed premature baby problem. Indeed, it was our big heads that gave birth to the patriarchy.

Origins of the patriarchy

Matriarchy in the strictest sense refers to a society led primarily by women. By this strict anthropological definition, no matriarchy has ever existed in human history. But that doesn't mean all societies are equally patriarchal.

Even among societies that still practice traditional ways of living, some are more egalitarian than others. The Khasi people of north-east India, for example, are matrilineal and matrilocal. Matrilineal means that descent and inheritance are traced through the female line. Orthodox Jews are a well-known example of matrilineal descent. Matrilocal societies are those in which a husband is expected to live with their wife's family. Matrilineal societies represent a little less than 20% of contemporary traditional societies. These structures of our society affect our psychology.

Men are often assumed to be the more competitive sex, but experiments reveal that Khasi women are more likely to compete in an experimental game than Khasi men (54% vs 39%), showing levels of competitiveness similar to Maasai men. The Maasai are a highly patriarchal society, recognizable for their height, red robes, and impressive vertical leaps. Among the Maasai, 50% of men and 26% of women chose to compete. These numbers also reveal the enormous individual variation between people in every society around the world.

Such differences are also found in post-industrial societies. As I write (2021), the current defense ministers of Denmark, Netherlands, Germany, Austria, France, Spain, Belgium, Switzerland, Czechia, Montenegro, and Canada are all women. In contrast, the United States has never had a female secretary of defense let alone a female commander in chief. Neither have Russia nor China in their equiv-alent positions. How can we explain these patterns?

As we've seen, males and females have some reliably developing differences. One of those is strength. In fact, the average man is stronger than 99% of women. As humans transitioned from a hunter-gatherer lifestyle to agriculture, they developed different agricultural technologies and practices, such as the plow and pastoralism.

The plow was not equally useful everywhere. It's difficult to use on shallow, sloped, or rocky ground, and is particularly useful when large plots of land need to be prepared quickly. It's also more useful for certain crops, such as wheat, barley, rye, and rice. Plowing, even with the help of an animal such as an ox, requires a lot of strength. Males therefore have a natural physical advantage. In contrast, hoeing can be done by males or females. Using climate and geography as an exogenous source of variation, research reveals that not only are traditional hoe-based societies more gender equal than plow-based societies, but even long after most people have given up farming the descendants of plow agriculturalists continue to have more gendered ideas about the appropriate roles for men and women. This is even true among second-generation migrants who immigrated to countries such as the United States.

The theory of everyone offers a powerful tool for discovering the origins of sex differences. For example, it allows researchers to look for practices that exacerbate the premature baby problem. One example is pastoralism – herding animals.

Pastoralism requires men to be away from their families for long periods of time as they take their cattle to new pastures. Being away from their partners increases paternal uncertainty. Pastoralist dads are even less likely to know if a baby is theirs. This in turn heightens the compromise between controlling female sexuality and male resources. These dynamics shape the environments in which children are raised and in which their cultural package is delivered to them. Part of that package are norms such as gender attitudes.

Researchers such as economist Anke Becker have found that pastoralist practices and ecological determinants of pastoralism can cause increased female genital cutting, stronger restrictive norms around female promiscuity, and even restrictions on female mobility, such as women having to ask for permission to leave the house and requiring a chaperone when out of the house. These norms are not only imposed by men on women – indeed men are often away so cannot enforce the norm – but instead permeate the society – women are often the enforcers.

Humans became a socially learning cultural animal. This led to us having too much to learn requiring bigger brains. Those bigger brains meant that we were born premature, floppy, useless messes, which in turn meant mothers needed to spend longer caring for their babies. That in turn required more support from their communities and from fathers. Fathers needed to know that the baby was theirs to allocate resources, time, and forgo other mating opportunities. This package completely reshaped our societies in different ways, but ultimately all that cultural variation was grounded in the same reality. Having too much to learn also co-evolved with the ability to speak.

Learning to speak

As the tree of knowledge grew, we became addicted to eating its fruits. We used that knowledge to innovate better ways to unlock energy and better ways to survive, and so we could support larger numbers of people with larger calorie-consuming brains. More calories, more people, more opportunities to innovate, more to learn. And so we got better at dealing with this ever-growing body of information. One major advancement was the evolution of language.

Language is the most powerful invention for information transmission. I'm using it right now to deliver this information to you. Language is sometimes invoked as what separates us from other animals, but language is not an *explanation* for the successful human package, it's part of the puzzle. Here's the thing: there's no point having a language that only you can speak.

Language is a coordination problem – others have to understand and speak your language for it to be useful. So you also have a start-up problem. In the beginning no one knows how to speak. In fact, they lack the ability to speak – they don't have the cognitive circuitry. All attempts to teach other apes language have failed. The most we've achieved is simple sign language or language boards, which non-human apes use only to make requests, which is slightly more sophisticated than what your dog does when it's hungry. This start-up problem is sometimes called a *bootstrapping problem* – a circular dependency, in this case, language requiring others to speak it before it is useful. The term comes from the impossibility of pulling yourself up by your own bootstraps. So something had to happen to kick it all off. Before we could invent and strap on any boots, we needed to walk on two legs. For language to evolve, we needed to become bipedal.

Bipedalism may have been a critical preadaptation for the evolution of language. The nice thing about being bipedal is that it frees your hands. Freeing your hands did a couple of things. First, now that your hands are free, if you have information worth transmitting, you can supplement your crude guttural utterances with gestures. Even today, wild hand-waving isn't restricted to Italians. We can't help but gesture as we speak. The combination of a proto-language and information worth transmitting could then kick off what's called a Baldwinian process. The Baldwinian process was proposed by evolutionary biologist James Baldwin (no, not the writer and activist). It's a specific culture–gene co-evolutionary process that might currently be happening with reading. Here's how it works.

If something important and adaptive enough can be learned then genes that help you learn it better can be selected. So in this case, small mutations that improved how quickly I could learn and understand more hand gestures or guttural grunts would be selected as long as the information I was getting from those gestures and grunts was useful enough to help me survive. Today, reading might fit the same category, but back then it was just learning to speak.

With enough information worth communicating, when mutations that made us more articulate – like the gene, FOXP2 – emerged, they were selected. Some of those mutations changed our throats, giving us language. They also made us more susceptible to choking. We were literally dying to speak to one another, revealing the importance of language and communication. So freeing up our hands through bipedalism bootstrapped language by giving us another medium through which to communicate. Bipedalism and freed hands were a double win. They also led to more stuff worth communicating – like fire and tools.

By freeing our hands we could not only speak through gestures but also make better tools. Free hands also cheapened the cost of those tools. Walking on all fours makes it difficult to carry tools. A quadrupedal animal like a chimp doesn't want to invest too much time or effort in making a tool, because they'd need to make another one when they move: who wants to carry a big stone axe around on all fours! But a bipedal species can carry tools with them and so can afford to spend more time literally sharpening their stone axe – and learning how to do that.

We see the first stone tools around 2.6 million years ago. What we don't see are the bone and wooden tools that probably predate the stone versions. Wood and bones don't survive the passage of time quite as well as stones. Just as chimps fashion wooden tools today so our ancestors probably did too. Even in the absence of this direct evidence for early tools, we do have evidence for something else worth transmitting that was critical for our species – fire.

We see evidence for fire 1 to 2 million years ago, but adaptations suggest we had access to fire to cook our foods for a lot longer. The evidence is our short guts and weak jaws, suggesting the presence of predigested, softened, cooked foods. Remember, we can't survive on raw food alone. Today's raw foodists rely on large quantities of available foods and a range of supplements. Just as in the future we may need the cultural invention of cesareans to reproduce, back then, we needed the cultural invention of fire to eat.

I'm not sure if you've ever tried to light a fire without being shown a technique or without access to technologies like a lighter, matches, or Swedish FireSteel – it's hard. Really hard. And so fire-making skills would have been invaluable – indeed essential – adaptive knowledge to transmit, given that our growing brains needed the calories unlocked by cooking. Fire increased the EROI and raised the energy ceiling for early humans, and so passing on the innovation of fire-making and tool-making probably helped support the evolution of language. So we had something worth transmitting (fire and tools) and by becoming bipedal we had a new medium to speak (gestures). This would have been enough to kick off a co-evolutionary Baldwinian process that would eventually lead to our current full-blown language abilities. But who taught us? Maybe mom and dad? Or maybe many moms and dads.

It took a village

Young chimps learn from their mothers because that's who they get to spend the most amount of time with. Indeed, more females in a chimp society are associated with a larger cultural corpus. But unlike chimps, there is some evidence that humans may have been *cooperative child rearers*, helping each other raise their children. Ever heard the phrase 'It takes a village to raise a child'? It's true. And for most of human history, we probably had a literal village.

Cooperative child-rearing not only made it easier to look after big-brained children with long childhoods, but also removed a key constraint that chimps have in learning from one another – who they have access to.

Chimps learn to crack nuts with rocks, sponge water with a wad of leaves, or fish ants with sticks from their mothers, because that's who looks after them most of the time. But our ancestral cooperatively child-rearing humans had many aunts and uncles serving as many moms and dads. And so we could begin to learn who the best and smartest teachers were – Aunt May or Aunt Martha?

In contrast, parenting is harder for modern parents, not only because there is so much for our kids to learn but because we often don't have that ancestral village to provide cooperative child support. But even today, cooperation in child-rearing is essential. Lacking our traditional village, we've institutionalized that cooperation through paid childcare and schools. We've also improved our ability to educate with specialized teachers, writing, radio, television, the Internet, and online courses. But before there were specialist teachers and online masterclasses, our cultural nature evolved the first professors: grandmothers.

Information grandmother hypothesis

Humans are not alone in placing great importance on older females – elephant grandmothers lead their herd, playing a critical role in the survival of their grandchildren. Humans are also not alone in the presence of menopause. Several cetaceans, including orca and pilot whales, also have evidence of it. The common feature that seems to bind these grandmothering, menopausal societies is culture – socially transmitted information.

Grandparents and particularly grandmothers have played an important role in raising human children, teaching them, and helping them survive. Long before schools, books, and the Internet, grandparents were the major source of wisdom and knowledge accumulated over their long lifespan. Grandparents were the Wikipedias of their time. In a precarious world before we learned to write, the mere fact that a person lived to old age was evidence that they had skills and knowledge that could help children survive and reproduce.

Remember that you have been able to reproduce since your early teens, and in the past, prior to the cultural evolution of an extended adolescence, you might have done so. But you would have been a poorer parent, knowing less about how the world worked and still learning how to survive in it. This was also true in the past. It's only recently that young people can know as much or more than their parents and grandparents, a product of rapid technological change in sources of knowledge, particularly the Internet. The past was more stable with fewer places to figure out how the world works. Grandmothers of the past gathered food, cooked, and cared for their grandchildren, just as they do today. But kids weren't just getting a cook and babysitter, they were getting a chance to hang out with the professors of the past. Grandparents were the most brilliant members of their society.

Even today, grandparents naturally excel at delivering knowledge. Curiously, even with ageing-associated illnesses such as dementia, grandparents often retain the ability to be storytellers for a long time, as if making a last-ditch effort to pass on everything they know. As the large cohort of Baby Boomers retire, one way to help reduce the economic burden may be reinstating their traditional role as carers to the next generation. Intergenerational care-home-cum-childcare centers, such as Nightingale House founded and run by south-west London's Jewish community, offer a potential model with reported benefits for both children and the elderly. Every day at Nightingale House children and elderly residents come together to cook, read to each other, perform concerts, or play games. The children are supported in their learning and development and the residents seem to have better physical and mental health, including lower depression and loneliness.

Grandparents and other aspects of cooperative breeding were a way for humans to innovate and explore the space of the possible. But the law of cooperation discovered so many more solutions to get us to work together. By capturing more energy, we bound ourselves into larger bands, expanding the moral circle of whom we care about well beyond family and friends.

6

Cooperation

We are now in the final chapter of Part I. We have seen what a theory of everyone has revealed about our control over energy, our intelligence, our innovation, our brains, our bodies, and our societies. But all of this is contingent on one thing: our ability to work together. The four laws of life interlock. Our control over energy and ability to innovate are empowered by our ability to cooperate.

Over the last two and half centuries we have seen an explosion in energy control, innovation, and population. Former enemies have become friends, and the previously oppressed now work together with former oppressors. We are cooperating at heights that would have been unimaginable to even our recent ancestors.

How did we achieve cooperation at the scale of large unions of nation-states like the United States and European Union? Answering this question is essential to ensuring that cooperation and all the progress we have achieved thus far doesn't come crashing back down. In Part II we will zoom out to see the threats to cooperation and what we need to do to overcome them, but first we need to zoom in and understand the specifics of how we got here.

The puzzle of cooperation

I'm often asked to give talks at universities, for companies, and in public settings around the world. As I'll remind my audience, putting large numbers of strangers together in the same room, or indeed inviting a stranger into their midst, is unusual.

It's unusual from a cross-species perspective: a room full of strange chimps is a room full of dead chimps. It's unusual from a historical perspective: even a few hundred years ago, a stranger was a potential threat and in danger themselves from those who felt threatened. Even today there are geographical variations: I would be a lot safer giving a talk in Switzerland than in Somalia. And yet there are many places around the world today that have created sufficiently stable, large, and diverse populations, peaceful for the most part, and with well-connected collective brains that support innovations that leave all of us better off. How did this happen?

This question is puzzling enough that even after decades of work, in 2005 *Science* magazine listed 'How did cooperative behavior evolve?' as a top twenty-five big question for the coming quarter-century. To understand the essence of the puzzle, we can go back to a now classic paper written in 1968 by Garrett Hardin. He called the puzzle the 'tragedy of the commons'.

Hardin asked us to imagine a common, shared field that farmers use to graze their cows. The number of cows getting enough calories to thrive and grow is constrained by the size of that grassy field – its carrying capacity, if you recall. But how the field is shared requires mechanisms that support working together. Why? Because it is in the best interests of all the farmers to be careful with their grazing so the field will exist for many years. But it is in every individual farmer's best interest to graze their cows as much as possible so their cattle grow as large as they can. Cooperation requires finding ways to suppress that selfish urge. If your policy is to rely on goodwill alone then it's a bad policy, because between selfishness and altruism, all else being equal, selfishness wins in the end. Taking advantage of others is an easier and more efficient way to gain more resources. And so by the laws of evolution and innovation it is selfish mutations that will dominate over altruistic mutations.

You can see this dilemma in many spheres. The person in the office who free-rides by not doing their fair share or by taking credit for work done by others. It would be great if those behaviors didn't lead to career advancements, but they often do and that's why such selfishness persists. At an international level, climate-change mitigation is an example of managing the commons of our world. Yes, we would all be better off in the long run if we all agreed to cut back our carbon output. But as I mentioned at the beginning of this book, it was always unlikely that we would slow the economy to save the planet. In the absence of a global government or credible ways to enforce the carbon commitments made by other countries, every person, every company, and every country uses the energy they can afford. Even if a few cut back, they would be outcompeted by those that did not.

The tension is always between what is best for *me* and what is best for *us*. Or what is best for a smaller us over a bigger us. A society succeeds – and arguably only becomes a society – when it suppresses the tendency to be selfish and moves to a new equilibrium that incentivizes people to be more altruistic. In working together, if there is energy to be exploited, a society can unlock more energy and resources to expand. It can not only manage the field Hardin imagined, it can grow it.

To study these cooperation dynamics and the behaviors of people in different places and under different conditions, scientists often use economic games related to the tragedy of the commons. Each game changes the pay-offs of different decisions to capture different facets of the cooperation puzzle. Take for example the public goods game.

In the public goods game, people are given some money which they can either contribute to a public good or keep for themselves. The money contributed to the public good is multiplied and then shared equally between all players. You can think of it like paying your taxes for things we all enjoy, such as clean air and water, roads, firefighters, and police officers. In these cases and in the game, you are personally better off by not contributing, not paying taxes, and instead free-riding on the contributions of others, even if we would all be better off if everyone paid their fair share of taxes.

Data from public goods game experiments reveals that people play close to the cooperation norm in their society. At least at first. In WEIRD societies, people's first instinct is to cooperate, and they often only play selfishly after thinking about it and realizing there's more money to be made by not contributing. Instinctively being cooperative makes sense when we're surrounded by other cooperators. But if you grew up around people who were trying to exploit you, you would instead be intuitively skeptical and perhaps intuitively selfish.

Cooperative behavior can easily be overwhelmed by selfish behavior. Unchecked selfish behavior is the Nash equilibrium – the optimal, highest pay-off strategy that will dominate with no counteracting forces.

Even in these games, after an initially cooperative decision, players slowly realize that just by being a little bit more selfish they can make more money and eventually, over subsequent rounds, they reduce their cooperative contributions to the public good. They slip backwards into selfishness. These experiments are a sped-up version of what happens in our societies.

Even in the most cooperative, wealthy industrialized societies, we are always in danger of slipping backward toward selfishness and conflict. Many people prefer to avoid taxes if they can – money under the table, tax loopholes, offshore havens – which in turn leads to more people avoiding taxes if they can. No one wants to be the chump contributing for others to benefit when others are not paying their fair share. So the question is always, what's stopping everyone from doing this? The answer is the various mechanisms of cooperation that have been discovered in answer to *Science* magazine's challenge.

Mechanisms of cooperation

Even before *Science* magazine laid down its challenge, biologists, economists, and psychologists had identified various mechanisms that incentivize cooperation over selfishness. These mechanisms had limits to how much cooperation they could achieve and with whom. Only today do we have a more complete picture of how animals cooperate and how humans have reached the heights of cooperation we've achieved. For humans, the lowest level of cooperation is between family members.

Loving families

A common cliché is that love is a mystery. That the bonds of family are hard to explain. Perhaps this was once true, but we now have a deep scientific understanding of both love and the bonds of family.

This explanation is what's called inclusive fitness or kin selection, and it's the most basic level of cooperation. It explains why grandparents are willing to teach their grandchildren, why we love our kids, and why all animals, if they favor anyone, favor their kin. It's the reason a lion might kill another lion's cubs but rarely their own and the reason your own baby crying is tolerable while another baby crying is miserable. The basic idea is captured by a joke made by biologist J. B. S. Haldane.

A friend asked Haldane, 'Jack, would you lay your life down to save your brother?' Haldane only had a sister, but nevertheless responded 'No.'

'But,' he continued, 'I would save two brothers or eight cousins'.

What Haldane was getting at was an evolutionary logic later formalized by Bill Hamilton in 1964. It's the $E = mc^2$ of evolutionary biology: $rb>c$. The basic logic is as follows.

At a genes-eye level, genes that can make more copies of themselves will outcompete genes that make fewer copies of themselves. That's what Richard Dawkins meant when he described genes as selfish. One way for the genes to make more of themselves is to convince you to have children – that's the standard logic of natural selection. Your children have 50% of your genes. But other people around you don't have 0%. My brother, Daniel, was surprised when his gene sequencing results revealed that he had a half-brother who shared 25% of his genes. That 'half-brother' wasn't a half-brother at all but my son, his nephew, Robert.

So, genes can also spread themselves by identifying and favoring other less-related individuals who carry copies of the same genes, such as your nieces and nephews and also more distantly related relatives. But evolution also shapes the amount of support you will provide to relatives. Daniel should provide a lot of support to his identical twin brother, Chris, and more to his own children than to his nieces and nephews. More formally, the rule is that when the relatedness (r) times a benefit (b) to the family member is greater than the cost to yourself (c), then you should cooperate. And more so when either the relatedness and/or benefit terms are much larger than the cost to yourself.

Inclusive fitness is the mechanism that gets humans to the level of cooperation among relatives – those hunter-gatherer bands that persisted for thousands of years. But it doesn't explain the kind of cooperation we see among humans today. We regularly cooperate with strangers. Inclusive fitness and kin selection can be trumped by direct benefits to yourself.

An eye for an eye, a tooth for a tooth

A more powerful mechanism for larger-scale cooperation is referred to as direct reciprocity, reciprocal altruism, or peer punishment. It can be summarized by the adage 'You scratch my back and I'll scratch yours' or 'An eye for an eye; a tooth for a tooth'. Remember the laws of life: ultimately, we're not cooperating for no reason, we're cooperating to efficiently access energy and resources. And often it pays to network and trade favors, helping those who will help you in return.

Direct reciprocity gets you to cooperation at a village level or the level of a workplace. It's what explains friendships. Other animals also cooperate through direct reciprocity. As long as everybody knows everyone else and they regularly interact, they will help those who help them and harm those who harm them. You don't even have to like the other villagers or your office colleagues to cooperate with them. The promise of returned favors or the threat of retaliation is enough for people to get along. But direct reciprocity suffers from some problems.

First, it requires an ongoing relationship. You have to have some reasonable probability of having a favor returned or you're being exploited. Con artists don't try to con the same community – they'd get caught; they have to move on and find another mark.

A second problem is that the direct punishment aspect can lead to cycles of retaliation for punishment – as Mahatma Gandhi put it, 'an eye for an eye makes the whole world blind'. And a final problem is that because direct punishment often comes at a cost to the punisher, it also suffers from the second-order free-rider problem. Not that people are unwilling to contribute to the public good, but that they're unwilling to pay the cost of punishment when others do not.

Imagine you're waiting in a queue. Someone skips the line ahead of you. It makes you mad. Someone should tell them off! But hopefully someone other than you . . . Do you really want to face potential retaliation or harm? But if no one is willing to enforce the norm, eventually queues collapse into throngs as queue-jumping becomes more frequent.

But even when these problems are overcome – which, as I mentioned, they mostly are in small communities, including non-human communities – direct reciprocity still doesn't get you to the level of cooperation in a modern large-scale anonymous society where you don't know or regularly interact with everyone in your country, city, or even neighborhood. We need something more.

Reputation is everything

Beyond *direct* reciprocity, we can use *indirect* reciprocity – cooperation conditional on a good reputation. For direct reciprocity you personally need to know someone and regularly interact with them to know if you'll have your favors returned. But for indirect reciprocity, you don't need to know everyone, you just need to know *of* them. You need to know their reputation so you can conditionally cooperate with those who have a reputation for cooperating back and conditionally avoid those who do not. In doing so, you can improve your own reputation.

Think of gathering together a team for a company or project. Ideally, it's people you know and have worked with before (direct reciprocity), but to expand your network, you rely on whether someone has a good reputation. You ask around, listen to gossip, and in the worst case read their LinkedIn endorsements. This is actually why we evolved to love gossip so much – it's about tracking reputational information.

Indirect reciprocity requires that you know *of* people and have reliable information about them. If reputational information is uncertain or untrue then cooperation collapses, just as, if a review platform offered fake restaurant reviews, you would stop using it. Once again, at least prior to online reputation management, it's still not a powerful enough mechanism to scale up to a large-scale society of *anonymous* strangers.

Leviathan

The mechanism we encounter most often in the modern developed world is what we might call *institutional punishment*. Rather than relying on our genetic relationships, punishing people directly, or relying on reputation alone, we bypass all the challenges and difficulties of these mechanisms by instead paying our taxes to an institution that does the punishing for us: our governments, police forces, courts, and judiciaries.

Institutional punishment with the right rules is incredibly effective at stabilizing large-scale cooperation. But, while the right institutions can stabilize the high scales of cooperation we see today – and experiments reveal that people, at least WEIRD people, prefer them to these other mechanisms – anyone who's traveled, knows their history, or is keeping up with current geopolitics knows that institutions can be, and often are, undermined.

Institutions securitize trust. Rather than trusting one another directly, instead we place trust in our institutions to protect our interests and act fairly. This in turn increases our trust in one another, knowing Big Brother is looking out for us. But the trouble starts if our governments, regulatory bodies like the Food and Drug Administration (FDA), Food Standards Agency (FSA), European Medicines Agency (EMA), or Centers for Disease Control and Prevention (CDC), police forces and justice systems are not perceived as unbiased and impartial in their decision-making. The trouble starts if it feels like the law is selectively applied based on who you know, future favors, or direct financial rewards; when legal systems are undermined by lobbying, political patronage, and personal connections. Under these conditions, the power of institutions to sustain cooperation collapses. We call it discrimination, racism, and corruption. But what would cause institutions to become biased? What undermines this high level of institution-mediated cooperation? The answer may surprise you. The answer is *cooperation*.

Institutions are ultimately made up of people with competing priorities and cooperative commitments. All mechanisms of cooperation exist alongside one another – we have our family, our friends, our reputations. Altruism at one level is selfishness at another. Favoring your family over your friends; your friends over your local community; your local community over your state, country, or world – all are both cooperative and corrupt. And so higher scales of cooperation – such as nations – can be undermined by lower scales – such as family and friends – if these are not suppressed. Examples abound.

Family dynasties in unstable developing countries that enrich themselves at the expense of the common good are ultimately kin selection undermining institutional punishment. From a Western vantage point it may seem like such corruption is a failure to be explained, but it is not. Corruption is far more natural than impartiality. The puzzle is how we overcame it in some places, but not in others.

From cooperation to corruption

There is nothing natural about democracy. There is nothing natural about living in communities with complete strangers. There is nothing natural about large-scale anonymous cooperation. But there is something very natural about prioritizing your family over other people. There is something very natural about helping your friends and others in your social circle. And there is something very natural about returning favors given to you. These are embodied in cultural obligations such as Western old-boy networks or Eastern *guanxi*. These are the mechanisms of cooperation found across the animal kingdom.

When a president negotiates his son a government contract, we call that nepotism. But it's also inclusive fitness undermining institutions. When a manager gives a job to a friend or a friend of a friend not because of private information but because of the relationship, we call this cronyism. But it's also direct or indirect reciprocity undermining our meritocracies. Bribery is a cooperative act between two people, and so on. It's no surprise that India, China, other parts of Asia, as well as much of Latin America – all family-oriented cultures – are also high on corruption, and particularly, you guessed it, nepotism.

We often think about supporting our families as a virtue, but when it really is the case that *la famiglia è tutto* – when family is everything – that might also be at the expense of a more impartial and fair society. The norm of looking out for your kin over others prevents countries from reaching a better outcome for everyone, an outcome where even those without strong connections can thrive. These are some of the aspects of culture that are the invisible pillars that support successful institutions.

Institutions rest on invisible cultural pillars

It doesn't matter that the law says you must be impartial if the norm is to favor your friends and family. It doesn't matter what the constitution says if you don't have a norm around the rule of law that enforces the idea that not even the leader is above the law. Successful institutions require that we are ruled by principles and not by people. And these norms vary around the world.

A dilemma posed by Dutch psychologist Fons Trompenaars captures this normative difference.

> You are a passenger in a car driven by a close friend, and your close friend's car hits a pedestrian. You know that your friend was going at least 35 mph in an area where the maximum speed was 20 mph. There are no witnesses. Your friend's lawyer says that if you testify under oath that their speed was only 20 mph then you may save your friend from any serious consequences. What would you do? Would you lie to protect your friend? What right does your friend have to expect your help? On the other hand, what are your obligations to society to uphold the law?

When people around the world were asked this question, their answers differed dramatically. The majority of Koreans, Russians, and Chinese said that they would lie for their friend. But over 90% of Swiss, Canadians, Americans, Swedes, Brits, and Dutch said that the friend has no or only some right to expect support and that they would not help.

Impartiality and rule of law are two of the many norms that are essential to well-functioning democratic institutions. When a country discovers the all-too-common phenomenon of a corrupt leader absconding with billions in national funds that could instead have been used to build better schools, hospitals, and roads, they question why less-corrupt leaders can't be found. But corruption is not a function of bad leaders that can be replaced by better leaders; it's a function of entire cultures where the same behavior of favoring friends, family, and close connections occurs at every scale: from the manager giving her friend a job, to an official allowing a connection to skip the usual bureaucratic process, to the minister giving his nephew a government contract. The difference here is not in the behavior but in the scale of the implications.

Even within Europe, we can see differences in impartiality from surveys of the method people used to find their current job. In Switzerland, Germany, and Norway, people primarily found jobs through job adverts. Job adverts level the playing field and create a larger, fairer competition by being available to all. In contrast, in Portugal, Italy, Greece, and Spain, which are all higher in corruption, people primarily found their job through friends and family.

Norms such as individualism and impartiality rather than familial obligations co-evolved with democratic institutions, largely in Europe. But as these within-Europe results make stark, they are hard to fully implement and countries are always in danger of slipping back to these more natural, lower scales of cooperation. This is also why it's such a challenge to try to export democratic institutions and fairer, impartial, non-family corporations to places around the world that lack these necessary norms.

Liberia, for example, founded by formerly enslaved Americans, took more than its flag from the United States, but is now by almost all metrics on the other end of the spectrum in the strength of its democracy, human development, corruption, and violence. The institutions alone were never enough. You also need those prerequisite cultural pillars. But like Wallace's water, those pillars are invisible to those from successful, less-corrupt countries. Unless you've lived in a country without these pillars, it's hard to fathom the difference in psychology, norms, and institutions. This failure to understand the diversity of people, how they handle relationships, and the different normative obligations in other nations has had disastrous consequences for foreign policy. Take, for example, the failure to transplant democratic institutions to Afghanistan.

Lessons from Afghanistan

In the wake of the 2021 Taliban takeover of Kabul, President Biden defended his decision to withdraw US troops from Afghanistan saying, 'We gave them every chance to determine their own future. What

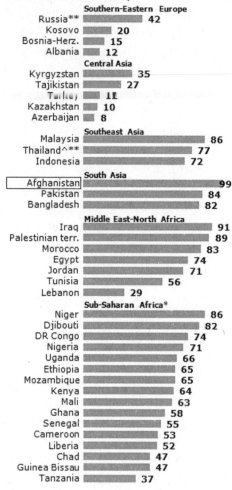

Support for Sharia

*% of Muslims who favor making sharia the
official law in their country*

Southern-Eastern Europe
- Russia** — 42
- Kosovo — 20
- Bosnia-Herz. — 15
- Albania — 12

Central Asia
- Kyrgyzstan — 35
- Tajikistan — 27
- Turkey — 11
- Kazakhstan — 10
- Azerbaijan — 8

Southeast Asia
- Malaysia — 86
- Thailand^** — 77
- Indonesia — 72

South Asia
- Afghanistan — 99
- Pakistan — 84
- Bangladesh — 82

Middle East-North Africa
- Iraq — 91
- Palestinian terr. — 89
- Morocco — 83
- Egypt — 74
- Jordan — 71
- Tunisia — 56
- Lebanon — 29

Sub-Saharan Africa*
- Niger — 86
- Djibouti — 82
- DR Congo — 74
- Nigeria — 71
- Uganda — 66
- Ethiopia — 65
- Mozambique — 65
- Kenya — 64
- Mali — 63
- Ghana — 58
- Senegal — 55
- Cameroon — 53
- Liberia — 52
- Chad — 47
- Guinea Bissau — 47
- Tanzania — 37

*Data for all countries except Niger from "Tolerance and
Tension: Islam and Christianity in Sub-Saharan Africa."
^Interviews conducted with Muslims in five southern
provinces only.
**Question was modified to ask if sharia should be the law
of the land in Muslim areas.
This question was not asked in Uzbekistan.

PEW RESEARCH CENTER Q79a.

Source: 'The World's Muslims: Religion, Politics and Society', Pew
Research Center, Washington, D.C. (2013), https://www.pewresearch.org/
religion/2013/04/30/the-worlds-muslims-religion-politics-society-overview/

we could not provide them was the will to fight for that future.' But did Biden really understand what type of future Afghan people actually wanted? The little data we have, supported by deeper historical context, suggests that it was not the one imagined by those in charge of US foreign policy.

The United States occupied Afghanistan in 2001. At the height of this occupation, in 2013, a Pew poll suggested that 99% of Afghans favored making sharia the official law of the land – a figure much higher than any other Muslim country.

But the word 'sharia' simply refers to Islamic law and can mean different things to different people. So what did the specific policy questions tell us?

81% of Afghans favor corporal punishment, such as lashings and cutting off hands, for theft.
84% favor stoning as the punishment for adultery.
79% favor a death penalty for leaving Islam.

Pew claimed the data was representative of the population, but in certain areas women were under-represented. It is possible that even if the interviewer were a woman interviewing an Afghani woman with no man present, women may nonetheless have answered with the perceived norm, as may have Afghani men. And neither men nor women may have had a good concept of alternative laws and norms. Indeed, in larger cities, we can see more (though still a minority) of women lobbying for greater rights. On the other hand, these numbers from Afghanistan are not small majorities. And moreover, many other norms and conditions in Afghanistan also pose challenges to successful democratic institutions.

Afghanistan, a nation of over a dozen tribes with different languages and histories, cooperates primarily at the level of kin. People rely on their kin for survival through support and favors. Afghanis even marry among their extended family – the rate of cousin marriage in the country is 46%. Kin-based obligations undermine the kind of impartial institutions that liberal democracies require. Instead, partisan, tribe-based politics dominates. It becomes critical that *your person* and not necessarily the *best person* is in charge. These dynamics unsurprisingly undermine good decision-making. And unfortunately, these norms are self-sustaining.

People rely on their friends and family and place great importance on these relationships, because they can't rely on governments and other institutions to support them. In such places, people prefer cash, take loans from family rather than banks, and are less likely to engage in broadly pro-social altruism such as donating blood or impartial charitable donations to strangers. For many in such places, charity starts at home, but that's also where it ends. These countries are trapped in a self-sustaining equilibrium that prevents them from reaching higher scales of cooperation.

The aforementioned Afghani data can be hard to understand let alone accept if we've never met people who hold such views. When this data is shared with people in WEIRD countries, it is often met with instinctive incredulity and sometimes moral outrage at the idea of such cultural differences. The very idea that people might want something so drastically different to WEIRD sensibilities sometimes invites patronizing and paternalistic attitudes – that people in other societies just don't know better. Regardless, this attitude contributes to failed foreign policies in these distantly different cultural contexts. If you read this section with surprise or incredulity consider that even if one accepts a diluted version of the data, the point still stands – the rights and rules commonly found in WEIRD societies may not be what everyone everywhere wants. And successful foreign policy requires full understanding of the norms and values of people in other nations. But of course, even if we narrowly focus on economic development, the WEIRD package of norms and institutions may not be the only one that works for large-scale cooperation and economic development. The law of evolution just happened to discover these ways of cooperating as an adjacent possible of Europe's specific historical trajectory, as we will see in the next chapter. In other words, other solutions are also possible. Many nations, most notably those in Asia, have borrowed and recombined elements of successful WEIRD institutions and adapted them to local cultures and circumstances.

South Korea, for example, has successfully and in their own way integrated WEIRD-style corporations and institutions. Hong Kong, as a former British colony, was an engine of development for China and is culturally halfway between China and Britain. These examples represent WEIRD-style corporations and constitutions tailored to more collectivist cultures. The overall message is that cooperation and conflict are two sides of the same coin.

. As novelist Nafisa Haji quoting an ancient Arabic proverb put it, 'I, against my brothers. I and my brothers against my cousins. I and my brothers and my cousins against the world.' This is the evolutionary dance between cooperation and competition, the duality of the human condition.

Humans cooperate in groups – overlapping and embedded within one another – and then these groups sometimes cooperate and sometimes compete, depending on contexts and conditions. It is through strong competition in the presence of large energy sources that higher scales of cooperation are reached. The mechanism by which this happens is called cultural-group selection.

Cultural-group selection

Cultural-group selection describes competition between cultural-groups, which are defined as groups of cultural traits rather than groups of people who possess those traits, because people can change and acquire different traits over their lifetime. I use a dash in cultural-group to avoid ambiguity and the mistaken view that we are referring to some cultural form of group selection on groups of people. These cultural traits might include belief in democracy, female empowerment, hard work, arriving on time, giving to charity, which sports should get more funding, and much more. It requires a psychology to represent normative behavior, a psychology of reputation to reward those who follow norms and punish those who do not, and a psychology to identify groups and subgroups who may have different norms.

In the strictest version of cultural-group selection there may be a perfect overlap between the cultural-group of traits and the group of people, such as in an ethno-linguistic group. A small New Guinean tribe's unique beliefs may be completely correlated with its unique language. But this is rare. In most cases, we belong to multiple overlapping and embedded cultural-groups. For example, cultural-groups of liberal democracies and of shared religions; of Googlers and of Department of Defense employees. There are British Catholics and Spanish Catholics; Americans who are also New Yorkers; Amazon employees who span the globe.

There are a few well-studied mechanisms by which these groups compete, though there are probably many more yet to be discovered. The best-studied mechanisms include the following:

Direct competition: Groups outcompete one another through conflict or simply surviving at the expense of another group – e.g. war or corporate bankruptcy.

Selective, assortative migration: Individuals carrying cultural traits move to some places at a greater rate than others – e.g. more people move from South Africa to North America than vice versa. America as a cultural-group will grow at the expense of South Africa to the same degree these migrants or their descendants acculturate to dominant American values. If these migrants don't acculturate but instead change the local culture – perhaps South Africans Trevor Noah and Elon Musk brought values, norms, beliefs or behaviors that other Americans now embody – then they represent cultural mutation or recombination (think Hawaiian pizzas). If they segregate as separate communities, a satellite to their group of origin, then they represent a smaller cultural-group, potentially competing within a larger one. If you choose to work for one company over another or people choose to stay or leave during a merger or acquisition, these too are examples of selective, assortative migration.

Demographic swamping: Some groups grow faster than others. For example, agriculturalists at the expense of hunters and gatherers; pro-fertility religions emphasizing large families at the expense of religions that did not emphasize fertility; companies that secure more investment or larger profits.

Prestige-biased cultural-group selection: Groups copy the cultural traits of more successful or prestigious groups en masse. For example, Americanization or Westernization of many traditional communities, or the spread of hip-hop culture beyond its African American origins. Watching a Japanese hip-hop crew in Tokyo, I was struck that they had not only absorbed the musical style but also the hairstyles and fashion. And of course, companies borrowing policies and practices from more successful companies.

One of the clearest examples of cultural-group selection and the duality of cooperation and conflict is the evolution of religion.

The evolution of religion

To some non-Muslims in recent years, the Takbir, a short prayer of praise meaning 'God is the greatest' – '*Allahu akbar*' – has instead become associated with violent terrorist attacks. Like the Takbir, religion itself, especially since 9/11, is often seen as a source of conflict. Indeed, it is a source of conflict. But it is also a uniquely human source of cooperation that bridges reputational psychology and Leviathan governing institutions. Religion was the ladder we used to climb from reputational-based systems to impartial government institutions. But like all mechanisms of cooperation, it is also a means for creating cooperative groups who can compete with one another. Here's how this mechanism of cooperation works.

Religions often use ethnic markers, for example a cross around your neck, a hijab around your face, a vibhuti on your forehead, a pirit string around your wrist, or a yarmulke on your head. These markers allow co-religionists to identify one another. To the degree that the marker is maintained by a community and linked to beliefs, such as being good to others who share your religion, your brothers and sisters in Christ or those in your *ummah* (community of all Muslims) – for fear of God's punishment or karmic retribution or wish to please Allah – then two individuals sharing those markers may be slightly more likely to trust each other when they meet than they would otherwise. In a world in which hijabs or zebibahs (the calloused, discolored prayer bump Muslims can develop on their foreheads from touching the prayer mat) are rare and indicative of Muslim beliefs, Muslims who don't know each other (direct reciprocity) or even know of each other (indirect reciprocity) may still trust and help each other by recognizing a co-religionist through their hijab, zebibah, facial hair, or clothing style. They do so because they know that those markers indicate a belief that requires them to be good to their co-religionists, norms which are often enforced or encouraged by their community. But remember the ultimate–proximate distinction. Although the world's major religions do share these prosocial beliefs, they didn't have to and indeed many religions over history did not. The question then is how did these beliefs evolve?

Religion may require basic cognitive biases, such as mentalizing, the ability to represent other minds in your mind; teleological thinking, the belief that things happen for a purpose or that there is a reason behind everything; or intuitive mind–body dualism – that the mind and the body are separate. But these biases alone are not sufficient to explain the variety of religious beliefs we see around the world. These cognitive biases are at best proximate building blocks upon which norms can be understood and enforced by the belief in a supernatural punisher. But, of course, different religions have, and have had, different beliefs, ranging from loving enemies to sacrificing children. A clue as to how prosocial beliefs became more common lies in comparing the religious beliefs of societies of different sizes. The gods of small-scale societies are very different to the big gods of the major world religions. The gods of small-scale societies don't want you to cut down the trees or desecrate the water. They are less interested in your sexual habits or whether you're nice to one another. They are limited in the scope of their interests, their power to punish, and their 'goodness'. In contrast, the gods or supernatural punishing forces of large-scale societies, like the Christian God, Allah, or karma, are all-seeing, all-knowing, powerful in their ability to punish, and all-good, with cooperation paid back in this life or the next.

People often marvel at the commonalities between major world religions – the golden rule, prioritizing family, not lying or cheating, and so on. Major world religions share these beliefs thanks to an evolutionary process that winnowed winning traits helping groups to grow and thrive. Or to put it another way, any major world religion today is a major world religion because it has been able to sustain large amounts of cooperation among co-religionists. Indeed, this has allowed religion to serve as a *super-ethnic* category, binding people of different ethnic groups that might otherwise be in conflict, for example, the many tribes of Europe under Christianity and the many tribes of Arabia under Islam. As these religions grew, they unified larger numbers of diverse groups. You can be Catholic or Muslim regardless of your ancestry or geographic location. Religion is as much a commitment to a group of people as it is to a set of cultural-group beliefs. Religion supports cooperation, but then, of course, these religious cultural-groups can compete with one another creating higher scale conflict.

Religion is not unique in creating conflict. Cooperation and conflict occurs through all the mechanisms of cooperation. Families against families, Montagues and Capulets or Hatfield and McCoy; villages against villages; regions against regions; nations against nations; and now large unions – think NATO or the EU – often bound by a common cultural and religious heritage against other large unions bound by another common cultural and religious heritage. But while religions share a lot in common, different beliefs matter, and these are under selection.

A handful of religions now dominate the globe: Christianity (2.3 billion), Islam (1.9 billion), Hinduism (1.2 billion), and Buddhism (500 million). Why do these religions share beliefs in a powerful, good, supernatural punishing power; the importance of obligations, such as to family; and the importance of helping others? From a cultural evolutionary perspective, the reason is simple: any major world religion that has spread to this degree has spread because its features have facilitated that spread.

Being nice to one another and being pro-fertility means stable and often large families that look after each other. These are not beliefs that are universal to all religions that have existed. It's just that some beliefs don't lead to growth and spread and so those beliefs and those religions are no longer with us. They're like the preference for banging your head against a rock.

As one example, you may have Quaker friends but are unlikely to have Shaker friends. Shakers are an offshoot of the Quakers that believed in celibacy for everyone. That's right, no sex, not just for a priestly class, but for everyone. So there are no Shakers left.

These kinds of beliefs don't have to be religious, but they do have to affect action. Setting religion aside for a moment, if the idea of the American Dream leads people to take risks and work harder and there are sufficient resources and energy for these behaviors to pay off, even if only at a country level, then the belief persists because the country can access more energy. The American Dream or America as a 'shining city on a hill' need not be true, but only believable enough to lead to behaviors. And if those behaviors lead to a stronger America then they will persist. The same is true of beliefs about equality, freedom, consent, honor, or patriotism.

It takes a group of individuals working together to accomplish both our greatest triumphs and our darkest tragedies. Greater scales of cooperation allow for greater scales of conflict between larger, more cooperative groups. Although we tend to attribute responsibility to individuals (because of our tendency to seek out models to learn from and avoid, as we discussed earlier), no one acts alone.

Vladimir Putin doesn't carry oil in his pockets. His control over Russian resources is contingent on those who benefit from some share of those resources (oligarchs) and in turn the supporters of those oligarchs who get some smaller share, then their supporters, and so on, in a network of political patronage, favors, and financial benefit that stretches even to the politicians and media in other countries. Putin can do this because innovations in efficiency have meant that Russian energy can be controlled and exploited by a smaller cooperative group than was required to access it in the first place. But despite this kind of corruption and even conflict, overall violence has declined.

Long peace?

Is the human heart kind or cruel? Are we fundamentally cooperative or competitive? Are we racist or can we see past our differences?

As both human history and your never-ending newsfeed make clear, the answer to these questions is both. We are capable of great kindness even to those far away. We are capable of great cruelty even to those close to us. We cooperate in groups but those groups compete.

The long arc of human history has seen a decline in cruelty and destructive competition. Blood sports were once mainstream entertainment – gladiators fought each other and large animals for everyone's entertainment in Roman amphitheaters. Today we are entertained by mostly bloodless sports. We used to torture cats for amusement. Today, we are amused by cute cat videos on YouTube.

The centuries and millennia have seen an overall decline in deaths through violence, as Hans Rosling, Steven Pinker, and others have documented in detail. That includes deaths from both homicide and war. Our probability of dying almost anywhere in the world today is lower than it was in centuries past. But that decline has been jagged, punctuated by great evil and periods of intense violence: the two world wars that scarred the twentieth century; the New York crack epidemic in the 1980s and 1990s; the Rwandan genocide of 1994; various Balkan and Middle Eastern conflicts, and more.

There is small solace in reminding people that the arc of history bends toward peace and justice during such times. In telling them that if they were given a choice and had to pick a moment in history to be born without knowing their sex, skin color, sexual orientation, or disabilities, it would probably be around the last few decades. In pointing to statistics or identifying punctuations when, even today, vast inequalities, deep unfairness, and terrible violence still permeate our world, creating much suffering. And there is small solace in reminding us of an overall decline in violence, if we are about to lose a generation to another drawn-out violent world conflict or civil war, or face the economic aftermath of these events.

To continue to make the world a more just, peaceful, and safer place for everyone, we must answer the question of why we see this pattern of an overall decline in violence with large variation over time, geography, and social groups. To know if peace will continue or even expand, we need to know how we became peaceful in the first place.

As we alluded to earlier, various popular answers exist to explain the long peace. Yuval Harari attributes it to the likes of our imagination and intelligence. Steven Pinker attributes it to a variety of factors, such as the rise of states, commerce, Enlightenment values, and the power of reason. But we can imagine many things both peaceful and warlike; we can use our intelligence and reason for both good and evil; the Enlightenment produced many ideas, and the rise of the state itself needs an ultimate explanation. To say that specific ideas produce peace is like saying genes allow for cooperation. Genes are the fodder for genetic selection and ideas are the fodder for cultural selection. The question is why did some ideas spread while others did not?

Take the Enlightenment. German philosopher Immanuel Kant gave us laudable ideas such as 'Freedom is the alone unoriginated birthright of man, and belongs to him by force of his humanity', but also deplorable ideas such as 'Humanity is at its greatest perfection in the race of the whites'. 'Enlightenment values' is less an explanation and more an exaltation of values we now possess. It is circular reasoning. The expansion of values we now consider 'laudable' and rejection of those we now consider 'deplorable' is not an explanation; it is another example of the way in which the world has become more peaceful. We need an ultimate explanation.

The laws of life can create a pattern of an overall decline in violence punctured by small and large conflicts. To demonstrate this process, Eric Schnell, Robin Schimmelpfennig, and I formally modeled cooperation in the context of multiple sources of energy with different EROI requiring different levels of cooperation to unlock and different carrying capacities created by the unlocked energy. We built on a class of cooperation models called the stag hunt.

The stag hunt game captures the dynamics of two people deciding between two scales of cooperation. They can choose to either (1) hunt a hare, which they can do on their own to get a guaranteed energy return of one food unit regardless of what the other person does, or (2) work together to catch a stag with a larger energy return, say six food units: three food units each. The trouble is, it is not certain they will catch the stag. Our ancestors and even present-day hunter-gatherers return from most hunting expeditions with nothing. And so the other person's willingness to cooperate for an uncertain reward rather than go catch their own guaranteed hare is also uncertain. And you can't catch a stag on your own – trying to do so leaves you with nothing. This is the basic stag hunt dilemma. To capture multiple scales of cooperation and a decline in violence, we needed to modify the math.

In the real world, there aren't just stags but also larger animals like buffalo and whales. Not just wood but coal, oil, natural gas, nuclear fission, and fusion. And in the real world it's not just two players but many. In fact, each reward requires a different number of cooperators with different levels of uncertainty depending on the number of cooperators. When you model these complexities, you see exactly what you find in the real world.

Each energy source unlocks a large energy surplus, which creates a larger carrying capacity. Our societies run on this excess energy. This excess energy in turn means more potential cooperators. Larger energy sources require a larger minimum number of people. For any given energy source, the probability of successfully capturing the energy source can increase with more people, but with more people the energy per person decreases. This leads to several interesting dynamics.

First, it's easier to reach a higher scale of cooperation if you are already cooperating at a high scale. An industrialized society can more easily reach nuclear fusion than a pre-industrial agricultural society (even with access to the right technology), and it is easier for an agricultural society to industrialize than a hunter-gatherer society. Consider the ease with which you might put together a team for a project that requires 5 people over one that requires 50 or 500. Doing so from scratch is difficult. For these larger projects or companies, it helps to have an existing large and cooperative group that requires just a few more people – it's easier to expand an existing team.

This model showed an overall decline in violence. With each new, more available, and high EROI source creating a larger space of the possible, violence declined. But what was interesting was that as the carrying capacity exceeded the necessary number of cooperators, smaller scales of cooperation could dominate, creating what we might call corruption. Smaller groups could work together to capture a larger amount of excess energy with a higher energy return per person than a larger group. Moreover, as the number of people grew or the energy ceiling fell, abundance turned to scarcity, leading to conflict between large groups. To put it simply, the presence of sufficient energy leads to an overall increase in cooperation and corresponding decline in violence within these large cooperative groups, such as countries. But this peace is punctuated by violence either from smaller cooperative groups with a higher energy return per person (exacerbated by efficiency innovations that let fewer people capture the same energy) or occasionally between large cooperative groups competing over a larger scarce resource. The decline in violence punctuated by internal conflict and larger-scale conflict are part of the same pattern. But because the punctuations to peace are rare, they would be difficult to detect statistically in the real world, making them look like noise. This explanation is more consistent with everything else we know about our theory of everyone, including the mechanisms of cooperation. Conflicts like the First and Second world wars were not noise. And that means as EROI falls and energy becomes scarce, future internal conflicts and large-scale conflicts are all but inevitable unless we address their underlying cause – energy scarcity and other threats to large-scale cooperation.

To summarize, people cooperate in ever larger groups to access available energy and resources. Within these groups there is more peace, cooperation, and kindness. But between these groups there is often cruelty, exploitation, and destructive violence. Larger groups discover values and norms, such as ideas of equality of all people under the law, stigmatization of discrimination, or valorization of meritocracy, which spread and are enforced through reputation and institutions. These are not self-evident, but in the presence of sufficient resources, they can support higher scales of cooperation. But our cooperation, innovation, and intelligence are a result of these evolutionary forces selecting among possible worlds with different sets of interconnected norms. Successful beliefs persist not through reason, causal understanding, or knowledge, but by their effect on the world and on people's outcomes.

The Great Divergence refers to the way in which the Industrial Revolution catapulted Europe past all pre-industrial empires across the rest of Eurasia and elsewhere. Many other countries have since caught up or are on their way to doing so, using their energy stores and the innovations unlocked during this period in what's called the Great Convergence.

The sudden post-industrial rise in wealth, energy capture, population size, size of countries or polities, child survival rates, human rights, or just about any other indication of progress and social development, as Ian Morris puts it, 'made mockery of all the drama of the world's earlier history'. That rapid rise is all thanks to fossil fuels, cultural-group selection, and the ways evolution has found to get us to work together, innovate, and become brighter. Yet this astonishing progress may be but a temporary break from harsh Malthusian logic.

All of what we have achieved requires continued access to abundant, dense, high EROI energy sources, without which we start crashing back down. Energy bills rising and discretionary budgets falling; home ownership slipping out of reach; rising prices diminishing our ability to travel, enjoy restaurants with friends, provide for our families, and do all the things that we think of as being part of a good life – are all a result of falling EROI and energy abundance. Our excesses are all dependent on excess energy. As energy abundance turns to scarcity, what we are all feeling in our bones is the beginning of a slow descent before a societal freefall.

Returning to the laws of life

When energy is available, life harnesses it by working together in larger, more complex units and discovering innovative ways to do more with that energy. Geothermal energy and the heat of the Sun with the Moon stirring warmed water full of potential were enough for Earth to evolve simple self-replicating unicellular life. But once simple unicellular life evolved to utilize this energy, a package of stored energy existed that could be exploited. And so multicellular life – single cells cooperating and working together – could evolve to eat these readily available stores of energy. This process continued where plants specialized in harnessing the energy of the Sun, herbivores specialized in eating the stored solar energy in plants, and carnivores specialized in eating the stored energy in herbivores.

For most of human history the energy return was a one-to-one return on our time. As a hunter-gatherer, the amount of food you gathered was a function of how long you spent gathering food. If game was large and easy to find then populations grew to meet this energy ceiling until abundance once again turned to scarcity. In turn, these larger populations with larger collective brains might innovate more efficient hunting, gathering, or food processing to increase excess energy. Two major innovations were fire and cooking. But then once more abundance turns to scarcity. To break out of this requires a larger energy ceiling.

Think of a household budget. You can do a lot more if you work very little for a lot of money. And increasing income always beats reducing expenses. More revenue beats greater efficiency. Only with excess can you go beyond the basics in a household or society. Only with excess can a company grow or conquer new markets. Those with larger budgets can beat those with smaller budgets.

After burning wood and learning to cook, the next major energy innovation was agriculture. It was the first major energy revolution for humans – deliberately and efficiently harnessing the sun's energy for a reliable food source, outcompeting smaller, less energy-rich hunter-gatherers. The reliable food source led to larger populations, which led to scarcity but also to further innovations that allowed for even more efficient use of energy. There was now sufficient surplus food to feed and domesticate animals. So instead of driving the plow by hand, we could drive it with an ox, effectively multiplying the work we could do through the solar energy we captured in the plants we grew and the livestock we looked after.

The next major unlocking of energy was burning the densely stored solar energy of ancient organisms – fossil fuels. This led to the Industrial Revolution, which massively multiplied the amount of work that could be done. Again, our populations grew and so too did our innovative capacity. Instead of driving the plow with an ox, we could use a fossil-fueled tractor. As a result, industrial societies outcompeted non-industrial societies. When energy is available, the processes of innovation lead to more efficient use of that energy, the ability to harness more resources and even to live in places that were previously unlivable (think of cities like Dubai or Phoenix, Arizona).

As with every energy revolution, our populations grew, but the energy of fossil fuels was so abundant it has taken two centuries for abundance to turn into scarcity. That is where we are now.

With the right conditions, our current energy budgets are enough to lead to the next breakthrough in the harnessing and use of energy. But this isn't inevitable. As energy declines, the probability of conflict between large, energy-rich cooperative groups grows. And lower scales of cooperation are always present. These smaller cooperative groups will always try to access more energy per person with fewer people. The cost of corruption and civil unrest are greater when energy is scarce. The threat of international war and civil conflict comes ever closer as the energy ceiling descends.

All complex life is forever in a battle with lower-order cooperation, namely bacteria, viruses, cancers. The COVID-19 pandemic has made this point more obvious than ever. Likewise, all developed societies still deal with lower-order cooperation, such as corruption, cheating, and favoring one's own groups. But the threat from these lower orders depends in part on energy availability.

Organisms and societies become sick when they don't look after themselves; when resources and energy are limited. If you arrive at a parking space and another car unfairly takes it, how will you react? If there's plenty of open spaces, you might graciously carry on. But if everything is full and you've been driving around for thirty minutes, things may be different. The cracks that always existed in a society – the smaller groups based on race, ethnicity, politics, or economic status – may begin to polarize and fracture. The moral circle of who we care about becomes smaller.

It's easier to be nice when there's more to go around.

Expanding Hardin's field

By the laws of life we traverse the space of the possible created by abundant high EROI energy-dense sources. We scramble to do more with less and we transition between scales of cooperation. A good way to think about these dynamics is to return to Garrett Hardin.

Hardin's tragedy of the commons describes a single field of fixed size that can support a fixed number of farmers and their families. We can sustainably manage the field with the mechanisms of co-operation instantiated through principles such as those documented by Nobel Prize winner Elinor Ostrom. That's where most readings of the familiar story stop. But a theory of everyone takes that story further.

If we manage to preserve that first field, and experience bounty and stability, something happens. First, by the law of evolution, if there is excess then we don't just sit on that same field stagnating with our families forever. Rather, some of those farmers may go off and find new fields. Some of those fields may be larger, together supporting a larger number of farmers. Some of those fields may create such an excess of people that it's worth trying to take over the fields of others. In the absence of this between-field competition, a smaller group of farmers may try to take more for themselves and their families at the expense of their village. This process of evolution is an exploration of different beliefs, behaviors, norms, and social organizations, between and within fields, increasing through innovation, cooperation selected through competition.

With enough fields, enough farmers, and a surplus of food, some of those farmers' children can now spend their time doing things other than farming. Some might figure out ways to drive a plow with a domesticated ox or eventually a fueled tractor. These innovations can more efficiently farm the fields, even expanding the fields or making them greener. Innovations that increase efficiency – by unlocking energy – mean better food returns on the number of hours spent farming, requiring fewer farmers to feed a larger village.

More people and more energy mean more people to refine other processes and develop other skills and knowledge – not only in science and engineering but also in art, literature, and entertainment – which lead to more opportunities for new ideas that lead to breakthroughs that are – inevitably – about energy and the control over it.

This pattern of growth and expansion, abundance and scarcity, driven by innovations in efficiency and greater energy control – originating in our sensibility to not burn through a single field we share – is at the heart of the human journey.

These patterns embodied in a theory of everyone and the laws of life are lenses through which to understand our genetic and cultural inheritances, to grasp the manner in which we work together to learn and to know and to innovate, to appreciate where we have been as a species, and glimpse where we are headed. Because by looking ahead, we can better steer our ship.

In the second part of this book we'll discover where we're going. Where we need to go is the next level of energy abundance. To get there, we need to solve several puzzles. For example, one puzzle that plagues economics is why, despite large innovations in computing and the Internet, production has increased but productivity – the rate of production – has been slow to increase. As economist Robert Solow put it, 'You can see the computer age everywhere but in the productivity statistics.' Part of the answer is that the Internet and computation is less like the Industrial Revolution and more like the Enlightenment that preceded it.

The Enlightenment was the launch platform, if you will; the apparatus and even the initial burn that then quite literally shot the human rocket into space. The knowledge it unlocked eventually led to physical innovations in the control of the energy-dense fossil fuels that wildly accelerated what humans could do. We are still living off the back of that revolution, which truly increased not just production but productivity.

There are barriers that block us from reaching this next level. How do we get past them? How do we reunite humanity, develop government institutions for the twenty-first century, create a fairer world, trigger a creative explosion, and maximize the potential of all people? And, by corollary, how do we overcome the forces pushing us in the opposite direction – tearing us apart, disenfranchising communities, increasing inequality, slowing innovation, and wasting human potential?

PART II

Where We're Going

Biologist E. O. Wilson once wrote that 'We have created a Star Wars civilization, with Stone Age emotions, medieval institutions, and godlike technology.' He was eloquently juxtaposing our individual limitations against the astonishing achievements of humanity.

In Part I we discovered who we are and how we got here; the four laws of life that describe all life on this planet and the theory of everyone that describes our species. Thus armed, we can begin to understand ourselves and our societies, our intelligence and our creativity, our capacity for both cooperation and cruelty.

Rousseau believed human nature was naturally good but corrupted by society.

Hobbes believed human nature was nasty and brutish but civilized by society.

Hobbes was wrong. So was Rousseau.

They weren't even asking the right question, because we now know that there is no *single* human nature. Human nature is deeply nurtured. How we nurture comes from our nature. We now know that the nature versus nurture debate for human behavior makes about as much sense as a right leg versus left leg debate for human walking. We have a dual inheritance, inextricably entwined.

Human nature has co-evolved with our norms and institutions, all of which has been molded by the laws of cooperation, energy, and innovation. From this vantage point we can marvel at the space of the possible created by the energy we've unlocked and put to work for us. Our productivity increases when we marshal vast energy budgets to do our bidding. Fossil fuels astronomically expanded our energy budgets. Excess energy fueled the evolution of technologies and social innovations in efficiency and cooperation. The future of our energy budget will determine what comes next. Where we're going is not inevitable. It is a choice. Who will make it and how?

Armed with our theory of everyone and the laws of life, we can bring new solutions to old problems. We can understand ultimate, systems-level causes, look ahead to the challenges coming our way, and apply the science to create new solutions. We will go as far as the science can take us and then go a little further.

It has been ten generations since the Industrial Revolution. Up until now, our energy ceiling has been in the rising phase of growth and abundance. That ceiling has been so high for so long that almost every generation alive today has lived through a period where it has felt limitless. Nobody alive today can remember the before times. The data is too abstract to truly appreciate.

Instead, our economic systems, invented after the Industrial Revolution, are focused almost entirely on innovations in efficiency – how to do more with less energy – the law of innovation – *ignoring the total available amount of excess energy* – the law of energy. But our energy ceiling is now falling. The era of growth is over and we are living through a Great Stagnation in productivity as we run out of ways to improve efficiency through non-energy-expanding techno-logical innovations.

The energy ceiling matters far more than the technological innov-ations in efficiency.

There is a limit to how much more efficient we can make the heating system of a house. At the end of the day, some minimum amount of joules of energy are needed to keep it warm.

There is a limit to the efficiency of our ability to hunt with better weapons or better techniques. At the end of the day, the calorie returns are a function of the animals we kill.

There is a limit to the efficiency that new technologies can squeeze out of a given amount of energy. Ford's assembly lines created more efficient factories, but they still required a minimum amount of joules of energy to operate.

Energy has been abundant for so long that we have taken it for granted while our larger, smarter collective brains innovated greater efficiencies that did more with less.

Now, as the energy ceiling falls and innovations in efficiency hit limits, the space of the possible shrinks. The squeeze is cracking and even breaking societies. This coming century will determine if we support and then raise the ceiling, clean up our planet, and set the stage to become a spacefaring galactic civilization, or if the ceiling crashes down on a failed species unwilling to look up and now forever stuck on a chaotic, climate-changed Earth being slowly depleted of the highly available, energy-dense high EROI sources needed to spring forward to the next energy level.

As EROI falls and high-density energy sources become scarcer, we are scarring our planet as we dig, frack, and scramble over the little that's left. Climate change has created new challenges over limited resources as we cycle through droughts and floods and sudden shocks to essential supplies. In the first decade of the twenty-first century, droughts in Syria turned previously fertile land into desert. Failing crops led people to move from farms to cities. Insufficient resources and the sudden influx of migrants led to dissatisfaction. Dissatisfaction led to protests. Protests turned into civil war. It wasn't long before the troubles spilled beyond Syria's borders into Europe. Like hosts who hadn't bought enough groceries for unexpected guests, Europe scrambled to accommodate the refugees. Not everyone was pleased at the influx of newcomers, leading to a rise in prominence and power of right-wing populists across the continent. At the height of the crisis, in 2015, 45% of Brits said that the refugee crisis on the Continent made them more likely to vote Leave in the Brexit referendum. Xenophobia was a strong predictor of voting Leave. The crisis was not the cause of Brexit, but it may have been enough to tip the scales. In 2016, Britain voted to break away, declaring itself an economic and political island. What happened in Syria was not unique.

In Africa, 'unprecedented' and 'record high' have become the climate catchphrases of the new century. Both insufficient rain and flooded rivers create food insecurity, increasing violent conflict over water, pasture, and land between farmers and herders and different regions. The instability and disasters have led to millions of migrants spilling over into countries like Uganda and Sudan. It was easier for the West to ignore what wasn't on its doorstep. We are not living through temporary bad times; these are all signs of the challenges ahead.

It is only thanks to the sacrifice of life long dead, fossilized as fuel, that we are able to live in a technological wonderland today. In just a few centuries we burned through these carbon batteries that had taken millions of years to charge, becoming more globalized and more diverse. That globalization led to greater efficiency through specialization, but also centralized production. Centralized production has made our supply chains less resilient and flexible. Take for example the resources and technologies at the heart of our information economy.

Taiwan alone manufactures nine out of every ten computer chips powering the world's cellphones, laptops, and Web servers. Lithium is critical to battery technologies among other electronic items essential to modern life. Almost half the world's lithium reserves are in Chile. Australia and China have a further 40%. This kind of dependency is found in just about every essential mined metal that is needed to power our solar panels and build our technologies. China alone manufactures seven out of every ten solar panels. A further 20% are manufactured in East and South East Asia.

We are not prepared for shocks to these suppliers.

The cultural diversity of our societies empowers innovation but also creates division. The world over, we have less trust in our institutions and in each other. Technological efficiencies have allowed a small few to accrue vast wealth. In turn, wealth inequalities and power imbalances are biasing our political decision-making. When that power is passed on from the original investor or innovator to their heirs, it leads to inefficient allocation over our remaining still-vast energy budget. And so the allocation of our energy budget has become less efficient with each generation.

If these social challenges weren't enough, our more diverse, unequal, and divided societies are tasked with dealing with sudden shocks, from droughts and dry summers drying up hydropower in Brazil and Europe to gas shortages leading to lower food supplies and to a global pandemic and all its consequences. And thanks to social media, we're creating new tribes based on common interests, all of whom are more aware of each other.

Our public spaces, both real world and online, are battlegrounds. Here cultural-groups promote ideas and visions for the world, all vying for dominance and in turn creating cooperation at different scales and competing over energy by the law of evolution. Some of these ideas are a reason for hope. Others are a reason for despair. As an author, my hope is that, having read this book, you will find the messy, confusing, and chaotic human world a little less messy, confusing, and chaotic. That you will feel better equipped to push for better decisions moving forward. Because the decisions we make today will determine what our future will look like and what remaining choices are available to us. Which of those futures is *our* future?

In some futures we live in perpetual zero-sum conflict, forever trapped in the Malthusian dystopia of the past as EROI continues to decline, leaving us without sufficiently large and accessible energy sources to cooperate at the current large scale of diverse societies of strangers. In these futures we are polarized into ever smaller co-operative groups that circumstances force us to pick, and become entrenched in our positions, unable to think clearly or in ways that cross ideological lines. In this future we fight with one another in an ever-escalating conflict. As you may have noticed, we are at the beginning of this shift.

Collapse doesn't happen overnight.

Collapse is a gradual decline.

Our bills rise as energy becomes more expensive. As energy becomes more expensive so too does food, transport, and everything else. For the first time, children have harsher lives than their parents, and we are seeing the beginning of more people sliding down Maslow's hierarchy of needs, from the creative pursuit of our full potential down to basic concerns over food, water, and housing. No amount of sustainability or cutting back can prevent the inevitable. The progress we have made in reducing poverty over centuries is being reversed in a matter of years, and our higher ideals are becoming lost as we struggle with forces beyond our control. Liberal democracy, freedom of speech, and pluralism become the ideals of a more abundant age and are seen as irrelevant to the realities of ever-present scarcity.

In this future inequality continues to rise and innovation continues to stagnate, immune to all government attempts at stimulation. We cooperate in smaller tribes of those we trust, against those we do not trust. And like the Tasmanians, we go backwards and lose the advances in technology and intelligence of our own ancestors. We begin violently tearing ourselves and each other apart over the limited energy and resources left on our planet, no longer possessing sufficient energy and cooperation to unlock the next energy level.

We are currently heading down this cold, dark path, but it is not inevitable.

In alternative brighter futures, we use our theory of everyone to scale democracy to deal with large and diverse populations of competing cultural-groups. In these futures, concerns around inequality dissolve as each of us has the opportunity to compete in a fair competition for wealth and ultimately for control over how we allocate our vast energy budget. A competition that is not rigged by the arbitrary circumstances of our birth. We reinvigorate innovation through a creative explosion and redirect all our current efforts at energy control toward bets that rocket us to permanent fusion-fueled abundance. In this future we head to the stars and, if we're lucky, become the first generation of a civilization that spans the galaxy.

Which of these futures will our descendants inherit?

That depends on what you and I decide today.

7

Reuniting Humanity

Our societies are being torn apart. The United States grows ever more politically and socially polarized. Europe has greater support for right-wing nationalist parties than we've seen in a century. Across the globe we see similar patterns.

Social problems on this scale require systems-level solutions. Without an understanding of how the system works, the problems will remain or even lead to new problems. Proximate solutions are like applying duct tape to a leak instead of identifying the leaky pipe and repairing it. As philosopher and writer Robert Pirsig put it,

> If a factory is torn down but the rationality which produced it is left standing, then that rationality will simply produce another factory. If a revolution destroys a government, but the systematic patterns of thought that produced that government are left intact, then those patterns will repeat themselves.

The ties that bind us are the very same ties that tear us apart. We are quick to blame greed and selfishness, but these vices have been our companions throughout human history. They are not explanations in themselves and certainly not systems-level explanations.

What destroys the high scales of cooperation evident in a peaceful, prosperous society is lower scales of cooperation. The president who steals from the people to enrich her family; the mid-level manager who gives jobs and other perks to his friends. These temptations are ever present, but when there are not enough jobs to go around, when the world feels more zero-sum, we enter a vicious feedback loop that incentivizes these lower scales of cooperation. When that zero-sum switch is flipped in people's heads then fractures widen, cooperation falls, and the few increasingly benefit at the cost of the many.

In our society, those living paycheck to paycheck are the first to feel the effects of zero-sum circumstances: stagnant wages, lack of quality education, unaffordable health care. These people are like the canaries in the proverbial coal mine. The societal-level forces from which they suffer will eventually hurt us all.

The goal of a successful society should not be just tolerance, but harmoniously working together in friendship and comity. Ever higher levels of cooperation is a secular aspiration that aligns with the preachings of the major world religions. To achieve these higher scales of cooperation there must be enough to go around such that fair competition is incentivized. Humans do not necessarily expect perfect equality of outcomes, but they do have a desire for fairness in the competition that leads to those outcomes. In practical terms, this means that when a population's size is increased, often through immigration, we must also invest in infrastructure to ensure there's enough for everyone.

Access to housing, health care, education, and other essentials must keep pace with population growth. Racism, discrimination, nationalism, tribalism, and polarization are not ultimate causes – they are proximate symptoms of the circumstances we find ourselves in when this doesn't happen. This in turn can lead to vicious feedback loops that decrease the circle of those we care about. When the frequency of buses slows down, when car parks are harder to find, we increasingly favor our own families, friends, class, and ethnicities, and are unable to work together for the common good. To see how easily this can happen and the vastly different futures it can create, let's look at Norway and Britain.

Norway versus Britain

Every Norwegian is born with an inheritance of around $250,000. The 5.5 million citizens of Norway start life with access to over $1 trillion through the Norwegian Government Pension Fund Global, also known as Oljefondet or the 'oil fund'. The oil fund was founded to manage the profits from the vast oil reserves discovered in Norway's part of the North Sea in 1969. The fund has since reinvested profits in the largest companies in the world. This small country, with half the population of Los Angeles County, now owns more than 1.5% of the world's stock market. It is the world's largest sovereign fund. In 2021 alone, returns from just the United States contributed more than $100 billion.

The Norwegian government uses this vast wealth to ensure a high quality of life for all Norwegians. For example, health-care costs are capped at around $200 per year, after which everything is free. Education – elementary, secondary, and university – is free, even for foreign students. Norwegians are a lucky people, but their enviable situation was not entirely down to luck and nor was it inevitable. It was the result of specific decisions. To see the path not taken, we need only look across the North Sea at Great Britain.

Britain also discovered a similar amount of oil on its side of the North Sea at the same time, and has since produced about the same amount of oil as Norway. Britain's population is larger of course, but Britain has also had access to large amounts of coal.

Rather than create a sovereign fund with these resources, which would benefit all, the British government of the time prioritized the profits of private companies and the pockets of a few wealthy people. Rather than plan for the future or consider how to maximize the potential of *all its citizens*, it took a smaller share of taxes on those profits and used them to bolster political support through tax cuts that primarily benefited the wealthy. Much of this money went into more property for the already rich, inflating the UK housing market and raising the cost of living for everyone else.

So, instead of helping all British citizens, the bulk of Britain's oil profits passed into the hands of oil company shareholders and wealthy citizens. Those closest to the oil benefited from higher salaries and a boost to the local economy, but relative to the value of the oil, these were small spillovers trickling into their bank accounts and the Aberdeen economy. Without strategic investment, the rest of Britain received mere droplets.

And British governments continue to make decisions that favor the few. Today, Britain's poorest are the poorest of all the major countries of Western Europe, with the lowest income share. Norway's are the wealthiest. Norwegians and their descendants will be wealthy for many generations to come. Perhaps forever. Britons today and their descendants will not.

How did this happen?

In 1965 Britain and Norway divided the North Sea shelf by a median line. In September 1969 oil was discovered in British waters. Three months later it was discovered in the Norwegian waters. The discovery of such a vast resource might seem like a boon but it doesn't always lead to increased wealth for the country concerned. Or at least not wealth for all. Sometimes the sudden discovery leads to what's called the resource curse – lower development, greater inequality, increased corruption. The resource makes everything worse than if it were never discovered at all, because people fight over it, the winners use the money to suppress the losers, and the country spirals deeper into poverty. The laws of life allow us to understand these dynamics.

A resource curse happens if a country is not cooperating at a sufficiently high scale to exploit the resources for its citizens' collective benefit. Typically, when we talk about a resource curse, we have in mind failed states like the tribally diverse and incredibly unequal Democratic Republic of Congo, a massive country in the middle of Africa blessed with diamonds, gold, oil, and rare metals but whose people remain the third poorest in the world. Countries like the UK are not directly comparable with the DRC but they do have a milder version of the resource curse – let's call it a resource hex – with similar causes.

In both cases, there is a small cooperative group based on class and/or ethnicity who do not feel they owe anything to the country as a whole because of the large cultural gap between them and everyone else. And so, they ask the question, can we exploit this resource without involving anyone else? The answer is often yes. For example, an oil company is often a more cooperative group with more shared goals than a large yet divided nation and can work effectively in collaboration with a relatively small number of people within the nation, such as the ruling political party. The returns from the fields can then be used to maintain those people in a position of power at a smaller cost to the oil company than if they were sharing the benefits of the oil with the whole country. In this way a small number of people cooperate at a lower scale to control the resource, making themselves wealthy but leaving the rest of the population poor. Indeed, the few who control the resources are not only wealthier but also now have the means to control the rest of the country and put their own interests first. Recall from the last chapter that Putin doesn't carry oil fields in his pockets. His power is contingent on cooperating with those who benefit from some share of the resources he controls (such as the oligarchs). In turn, those supporters derive their power and influence from those who get some smaller share from them (their supporters), and so on all the way down to the local policeman, judge, or small-business owner. Historian Rutger Bregman once accused Fox News host Tucker Carlson of being 'a millionaire funded by billionaires' (the interview unsurprisingly never made it on air but can still be found on YouTube). I don't know enough about Carlson's income sources to comment on the accuracy of Bregman's claim, but, in general, corrupt cooperative power structures undermining the general public's welfare are often a case of billionaires funding millionaires. And the millionaires in turn funding ordinary people. Ordinary people who are happy to accept a decent income, putting food on their table and a roof over their head, especially given the lack of alternative options in a captured economy. Everyone else not connected to this corrupt network of cooperation? Too bad.

Norway understood the resource curse and actively attempted to avoid it. So far, they have succeeded.

In 1971 Norway laid out the 'ten oil commandments' to ensure that this vast new wealth would benefit all Norwegians then and into the future. These commandments served as the beginning of the oil fund, a formal institution – codified norms – around which Norwegians could coordinate and compel one another to ensure the vast wealth was used for the greater good. Norway continued down this path, accumulating so much money that during the 2008 financial crisis the oil fund was able to buy up half a percent of shares in all the companies in the world. When the market recovered, over 60% of the oil fund was now made up of stock market returns. As current CEO of the fund, Nicolai Tangen, eloquently described it, 'Norway found oil twice, first on the continental shelf and second in capital markets'.

Britain never even got to that point. Long before the 2008

financial crisis and long after, the profits were pilfered away by the few at the expense of the many.

What was the difference between Britain and Norway? An obvious answer is that British politicians lined their own pockets and the pockets of the wealthy well connected and Norwegian politicians did not. But this is a proximate explanation and just moves the question back a step. Why did British politicians act in the interest of the few and Norwegian politicians act in the interest of the many? An ultimate explanation emerges from considering the laws of life.

Britain launched the Industrial Revolution on the back of cheap and available coal. The energy was plentiful but needed innovations and people to access it. Through cultural evolution and cultural-group selection, social and technological infrastructure and cooperation evolved to exploit that available energy. The efficiencies didn't yet exist to exploit the resource with a smaller group so vast numbers of people received education and training that in turn gave them access to the energy and improved their lives. Britain was a society divided by class, but this mass empowerment shook up the old order. The energy unlocked was enough to springboard Britain toward higher scales of cooperation. It used that energy and cooperation to exploit other wealthy but less cooperative targets and created an empire that colonized and dominated the world.

Eurasia was connected by trade routes, such as the Silk Road, swapping technologies and ideas across the more easily traversed east–west latitude. The more climatically and geographically variable and difficult north–south orientation of Africa and the Americas couldn't connect into as effective a collective brain and so was outcompeted. But even within Eurasia, once Britain and later the rest of Europe harnessed a vast new energy source and later industrialized with the power this energy provided, other parts of Eurasia, including China and India, were also outcompeted. From the eighteenth century to the beginning of the twentieth century Britain was a powerhouse. Then the world wars began.

During the Second World War, despite harsh conditions, Britain was an even more united kingdom. The aristocracy had begun to weaken in the late nineteenth century and the classes became more permeable as the children of former aristocrats married wealthy industrialists. In any case, when facing an existential threat, people tend to bind together and corruption decreases as groups are forced to put the best people in power rather than 'their people'. When you're not at war, there's no harm in your brother-in-law running the armed forces. But when the enemy is at your door, you want someone with competence and experience of warfare in charge – someone who can actually win.

At a proximate psychological level, shared suffering also leads to increased cooperation; it forges strangers into bands of brothers. The suffering of the Second World War led to unprecedented levels of national camaraderie. Old fractures in class and culture gave way to common purpose. Britain was swept by reforms that helped a larger number of people.

Class boundaries, which were already weakening, weakened further as estate taxes (so-called death taxes if you're American, which we'll get to) were raised from 65% in 1940 to a peak of 75% in 1945. In 1942 the Beveridge Report, a social manifesto, laid out a plan for universal social security, free health care, employment benefits, and ultimately the creation of a social welfare state. In 1946, after the war's end, the revolutionary recommendations were implemented, turning Britain into a welfare state with social safety nets and a free National Health Service (NHS) that at the time was the 'envy of the world', as is sometimes dubiously claimed today. But the end of the Second World War was also the end of the empire. Britain was deeply in debt and the EROI and availability of coal in its mines was falling. The middle of the twentieth century is remembered by many Brits with fondness. It saw the final throes of an empire in decline. Life was still good, just as it is for many in America today, despite the feeling of unease and impending decline. But collapse is gradual. Britain's energy ceiling was descending. The sun would finally set on the largest empire the world had ever seen.

The end of the British Empire was the beginning of a new Britain. After 1945, the country that had once sent its sons and daughters to the farthest reaches of the empire now saw the sons and daughters of the farthest reaches of the empire come to Britain, escaping difficult circumstances – often created by British interventions – or in search of a better life. Now, in addition to the class divide, groups of people with large cultural distances between them, with psychologies culturally evolved under very different conditions, were living side by side. Such diversity doesn't always spell disaster, but it can be a challenge when it exists under conditions of resource scarcity. The paradox of diversity was at play. It was an opportunity for greatness, but also division. And division won. So what happened?

Fractured countries make bad decisions. When there isn't enough energy and resources for everyone, the scale of cooperation collapses. Just as a malnourished organism becomes weaker and sicker as lower scales of cooperation – cancers and bacteria – dominate, so too does an under-resourced society as class and ethnic divisions dominate. Britain faced diminished resources, inequalities of wealth and class divisions, and quickly rising diversity. It could not meet the challenges of cooperation.

The challenges of integration are also more complex when an already established culture welcomes large numbers of newcomers. When large numbers of European migrants settled in the countries of the New World such as the Americas and Australia, they didn't have to think about integration, they just violently displaced the indigenous populations. As a result, immigrants with culturally similar origins became the majority. But the countries of Europe already had large established cultural-groups and this created a very different dynamic for the small groups of culturally distant migrants arriving from Africa, the Caribbean, and Asia into the UK.

In post-war Britain migrant populations were growing by around 20% per decade. The various migrant communities at the time numbered around 3 million people, just over 5% of the population, but numbers were rising rapidly. The country had not yet forged a new unified multi-cultural British identity and old economic class lines, which had never really gone away, began strengthening their internal bonds. Without a booming economy or a government willing or able to address the issues, and the feeling that there was not enough to go round, it was too much for the country to handle. Rapid change led to racial conflict in the form of discrimination and violence in communities, political conflict in government, and policies that once again favored the wealthy.

Exemplifying the cultural and political divide, Conservative MP Enoch Powell's infamous 1968 Rivers of Blood speech expressed the zero-sum and diversity concerns of many at the time. As he described it, local communities 'found their wives unable to obtain hospital beds in childbirth, their children unable to obtain school places, their homes and neighborhoods changed beyond recognition, their plans and prospects for the future defeated'. His speech called for an urgent and immediate halt and reversal of immigration. Even for the time, it was an argument made using incredibly racist and emotive language, and Powell was shunned. But he had tapped into real concerns that were unaddressed and unexpressed by other parties. Many argue that it was Powell's speech and the importance of the issues he expressed that explain the unexpected Conservative Party victory over the incumbent left-leaning Labour Party in the 1970 general election.

In the 1970s and 1980s Britain was racked by race-based violence. Black British rioted against police harassment and South Asian communities were routinely targeted by anti-immigrant groups. This violence was a symptom of a failed integration policy and served to further polarize communities, widen fractures, and encourage people to self-segregate within their own ethnicity – if only for safety. Lower scales of cooperation dominated.

Britain also continued to fall apart along old lines of class and wealth. The estate tax, which peaked in 1969 at 85%, was soon replaced by a weaker inheritance tax with more loopholes that would continue to fall to pre-war levels of 40% in the 1980s, where it remains today.

Democracy and large-scale cooperation are easier when goals and values are shared. When a population agrees on the fundamentals, they can put the best person in power to implement that shared vision. But when there is disagreement on fundamentals, people fight to put *their* person in power because the opposition represents such different values. For example, if you're in Denmark, until recently an extremely culturally homogeneous society, and you agree on the desirability of universal health care, you can vote for the person who can best implement it. If you are in the United States and are divided on whether universal health care is fundamentally desirable in itself, you will be voting for who will or won't implement it and each group needs to put *their* person rather than *the best* person in power.

If wealth inequality is low then more people have a shot at a position of power and a broader range of people in power will ensure policies that benefit more people. But if inequality is large, a small proportion of the wealthy will control political decisions to their benefit.

Britain's political class run quite different cultural software – they speak differently and see the world differently – to the rest of the country. It's a product not just of privilege but of a specific and incredibly effective institutionalized privilege honed over centuries. It's a system that creates a cultural bottleneck and perpetuates the interests of the powerful, who are the product of and in turn feed their own children into the private school pipeline.

(In the UK the term 'public school', somewhat confusingly, refers to an elite private school, originally called 'public' many centuries ago when such schools first emerged because they had no religious, guild- or other group-based prerequisite for entry. The two most well-known public schools are Eton and Harrow. For ease of understanding of those less familiar with the British terminology, I will refer to the problem as the private school pipeline, but will continue to refer to these specific elite schools as public schools.)

The private school pipeline traditionally starts with prep boarding school at age eight, though sometimes children are sent as young as four. No longer living with their parents or within the broader community, the private school pipeline immediately starts programming children's brains with homogeneous cultural software. From here, most continue to private secondary schools, again, typically living in the school as boarders, limiting outside cultural input. It's a good academic education, highly successful at getting pupils into Britain's two most elite universities, Oxford and Cambridge. Nonetheless, it's remarkable how the mannerisms, accent (remember, accent is an important cue of cultural identity), and ways of thinking of children who attend these schools, differ from those of the rest of the population. Former chancellor and current prime minister Rishi Sunak, a private school student, was once interviewed for a documentary as a twenty-one-year-old. Asked about the diversity of his friends, he said: 'I have friends who are aristocrats, I have friends who are upper class, I have friends who are working class . . . Well, not working class.'

For many private school pupils, like Sunak, the pipeline ends in positions of power, often in government or finance, always as a result of being part of an exclusive, self-sustaining network of power built on shared experience and world view. As a result, Britain's leaders and many members of the elite are not the best and the brightest, who have fought their way to the top on merit. Rather, the elite and particularly the political class, future prime ministers, finance ministers, heads of newspapers and major public and private corporations are where they are because they were all once friends in the same exclusive playgrounds. Of Britain's fifty-five prime ministers, forty-five went to private schools. Twenty prime ministers went to Eton alone.

Returning to the North Sea in 1969 and the decades that followed, an entrenched class system with a ruling elite far removed from the experiences of ordinary people, personally unaffected by and disinterested in addressing the rising challenges created by significant cultural diversity, meant that when Britain found the treasure, the spoils were never going to be shared by all. That oil boom still temporarily rescued Britain from a decline caused by falling EROI in its failing coal mines, but it was not invested in a way that secured Britain's future and so the decline continues.

In contrast, 1969 Norway had a population that was less than a tenth that of the UK, at under 4 million people, was ethnically homogeneous (remember the paradox of diversity), more economically equal, and already enjoyed greater social mobility. Today, Norway has enjoyed a positive-sum productive cycle. Its energy budget and oil fund continue to benefit all Norwegians, who now have one of the highest levels of income, household wealth, standards of living, and the highest level of social mobility in the world. This last statistic reveals the power of truly equal opportunity.

The differences between the way Britain and Norway handled their oil booms are a result of both path dependence and the specific political decisions made within each country. They are the same dynamics playing out today in an increasingly divided and polarized United States and European Union. Although decline is gradual, the same vicious vines are penetrating every aspect of these and other societies too, tearing people apart. We face the paradox of ever-increasing diversity. The lives of our children will be more difficult than our own, so resolving the paradox becomes evermore urgent but also evermore difficult. So much so that many are afraid to even address the topic for fear that pointing out the elephant in the room might wreck it.

Resolving the paradox of diversity

Some topics are hard for scientists to discuss because they fear their words can be twisted or misused to support hatred, cruelty, and xenophobia. None of us wants our work to make the world worse. Of course not all topics are equally controversial or have the same potential to be exploited for political gain. Knowing the mating behavior of butterflies is very different to identifying the sources of inequality between groups or how to sustainably manage immigration and maintain a harmonious multicultural society.

These topics are so difficult to discuss because they involve real implications for real people – mothers, fathers, children; their livelihoods, freedom, even their safety. Discoveries in these spheres are the equivalent of what engineers call dual-use technologies – technologies that can be used for both peaceful and violent aims.

Nuclear technology gets you power plants. And bombs. Rockets can launch satellites. And warheads. But while there are a lot of stages and resources involved in moving from the theory to developing and launching actual nuclear warheads, in the human and social sciences, mere discoveries and words alone can cause harm. So what should we do?

One approach takes the view that unwelcome hypotheses should not be explored and unwelcome results should be suppressed, denied, and condemned for fear of possible societal harms. Noble lies must be enforced and a fuller discussion never takes place because the topics themselves become taboo. This is an intuitively tempting but dangerous path.

Censoring science is self-defeating. It feeds into the narrative that scientists are not to be trusted because they are suppressing the 'truth' in an attempt to protect themselves. That in turn creates a space for less scrupulous scientists or even non-scientists to fill with questionable work, driven by specific social and political agendas rather than the pursuit of truth. These become the sole sources of information on random Google searches, reported in dark corners of the Internet and sometimes by the media. They leave people with a sense that governments, academics, and the elite are not on their side, which in turn feeds into misinformation, conspiracies, and falling institutional trust. Remember, it's not about information, it's about who we think is on our team and whom we think we can trust.

Here's a recent example. Early in the COVID-19 pandemic, the public were told that masks were ineffective if not worn correctly. This was a short-sighted attempt to preserve masks and other then scarce personal protective equipment (PPE) supplies for health-care workers. It was an attempt to influence the public, treating them with a disrespect that was costly when supplies were secured and the messaging then flipped to the importance of wearing masks. And that short-sighted decision prevented people from being told that the masks they should have worn in the first place were proper PPE supplies – N95/FFP2 or N99/FFP3 – the kind of masks physicians working with contagious patients wear and the very same masks people were initially told were ineffective for public use. It was a catastrophic failure of a panicked public health messaging system. How could anyone trust what they were being told? The mistrust created by this expedient messaging will cost lives in the decades to come as people ignore well-evidenced advice on other public health matters.

Being forthright and truthful about even challenging topics is critical to trust in science. If you can't trust scientists, you can't trust science.

And so in a world where public messaging emphasizes that 'diversity is strength' or that 'diversity is destroying us', depending on which channel you're watching, it is important to acknowledge that, in reality, diversity is a double-edged sword. As the collective brain teaches us, diversity is a fuel for innovation and economic progress as diverse ideas recombine into new innovations. Immigrants have been the super-serum that led to America's super-strength in innovation and technology.

In April 1924 the *New York Times* declared 'America of the melting pot comes to end', a reference to the newly introduced 1924 US Immigration Act that created immigrant ethnic quotas based on national origin. It was a restriction on 'less favorable' Europeans – those from southern and eastern Europe. Restrictions, such as the 1882 Chinese Exclusion Act, already reduced migration from non-white parts of the world. America was losing some of its lifeblood.

Recent analyses suggest that the 1924 Act led to a massive 68%

baseline decline in indicators of innovation, such as patents, in industries where these migrants, such as Italians and Jews, worked. There was also an overall decline in innovation across other industries – innovations spread and are built upon, and so there were fewer innovations everywhere. To what degree can we generalize results like these from a particular population at a particular time?

The first and perhaps most obvious issue with generalization is that 'immigrant' is not a useful category. To ask if immigrants contribute more or less to the economy, commit more or less crime, or are happy or unhappy in their new homes is as crude as asking if citizens contribute more or less to the economy, commit more or less crimes, or are happy or unhappy. When we talk about our ingroup, it's more obvious that we need to disaggregate broad, sweeping claims, be they positive or negative. We need to disaggregate by factors such as the nature of the economy at the time of immigration, the characteristics of both immigrant and local populations such as culture, age, economic conditions, the cause for migration, and level of education.

Humans are a migratory species. Not only have we been marching across the globe for thousands of years, but we have done so back and forth, replacing, mating, cooperating, and fighting with one another on the way. Few populations are guilt-free in replacing a population that was there before. But until the late nineteenth century's age of mass migration, migration was primarily from geographically and culturally closer places. Without large ships and airplanes, we lacked the ability to quickly and cheaply traverse the planet. That is no longer the case.

Today, more people from more culturally distant societies increasingly live side by side. And at a global level, their culturally distant countries of origin are forced to coordinate on global issues as never before. Our world is smaller, but diversity remains. This new form of culturally distant migration and cooperation has enriched our societies but also created new problems.

Remember that diversity has been central to the success of all complex life on earth. Diversity provides the new traits needed to make life evolvable. The recombinatorial power of sex increased evolvability and the speed of genetic evolution. Today, diverse societies recombine diverse cultural traits to empower cultural evolution. Beyond Hawaiian pizzas, businesses started by foreigners, for example, tend to be more profitable and more likely to expand. But there are many barriers to cultural traits meeting and recombining. These are often challenges in communication, coordination, and cooperation – barriers created by diversity itself.

A successful immigration policy requires population-level thinking that considers cultures not as homogeneous blobs (Chinese are like X, Canadians are like Y), but as distributions of different traits. As you know from your own country, state, or city, we can talk about New York culture or British culture, but not everyone is even remotely the same in either of these places. Those differences reflect differences in the relative frequencies of different beliefs, values, and behaviors – not all Americans are laissez-faire about norm-breaking, but they're more laissez-faire than Germans, not all of whom insist that norms be followed. Immigration policies are ultimately policies that enable the host country to sample these different distributions and traits. And there are many ways to do so.

Numbers alone are important. Immigrants from a particular population arriving in small numbers are more likely to integrate with local populations. It's difficult for the only two Norwegian families in a Nova Scotian town to not integrate with locals. But when immigrants arrive in larger numbers, they can represent a cohesive cultural-group, sometimes preserving a fossilized version of the cultures from which they came. An Indian friend of mine was shocked to discover attitudes toward dating in a London Indian community, which seemed to her to be like those of her parents' generation. She was right: it was members of her parents' generation who had migrated and preserved the values in the communities she visited.

When immigrants arrive in large numbers with no restrictions, such as during a humanitarian crisis that leads to mass movement of refugees, it is as if the host country were randomly sampling from the whole distribution of their countries of origin. Not everyone will follow the norms common in their own country, but with unrestricted migration, you are more likely to see a fuller representation of the entire distribution – the whole variety of people in the proportions found in a country.

New migrants bring new cultural values, norms, practices, and psychologies that can differ from those of the host country. At a noticeable level, these differences culturally enhance our society, allowing us to enjoy a good taco on Tuesday and a spicy curry on Thursday. Not every Mexican can make a good taco, but the proportion of people with good taco-making skills is higher in Mexico than in the United States. In the United Kingdom, it's nearly impossible to get a good taco – there just aren't enough Mexican migrants. But immigrants bring more than just food.

Some valued cultural traits might be present at higher proportions than in the local population – valuing hard work, education, or entrepreneurship. But other less-desirable cultural traits can also be present at higher rates than in the local population – reduced tax compliance, less support for gender equality, reduced belief in the rule of law. Even with smaller cultural differences, these migrant communities shape societies.

The present regional differences in the United States can be traced to their founding immigrant populations. The Puritans, for example, brought an emphasis on education to New England. The Scotch-Irish brought an honor culture of politeness, avoidance of offending others, maintaining reputation, and condemning 'improper conduct' to the Deep South. This trend has continued, with cultures remixing and finding more and less compatible previously arrived traits. For example, migrants from agricultural societies who used plowing rather than hoeing are, even in the second generation, more likely to agree with statements such as 'When jobs are scarce, men should have more right to a job than women' and 'Men make better political leaders'. It's not just beliefs. Fewer women from these groups are in the workforce.

It should be emphasized that these are always overall average effects that mask the distributions. There are many New Englanders even of Puritan ancestry who don't value education, Southerners even of Scotch-Irish ancestry who don't possess an honor culture, and those from plow-based agricultural societies who have a highly developed sense of gender equality.

New methods we've developed for measuring these cultural differences between groups based not on averages but entire distributions of cultural traits are revealing. One metric, the cultural fixation index (CFst), gives a score between 0 (identical cultural trait distribution) and 1 (completely non-overlapping cultural trait distribution). Using the United States as a comparison culture, for example, we find that most countries range from 0 to 0.3 in their cultural distance from the United States, suggesting large overlaps between all peoples in the world. But those cultural distances combined with that range predict many broader cultural and psychological gaps in personality, values, corruption, and even the tendency to donate blood or return a lost wallet. The following figure shows how distant different countries are from the United States. The height of the bars are just labels, but greater distances are found as you move further to the right.

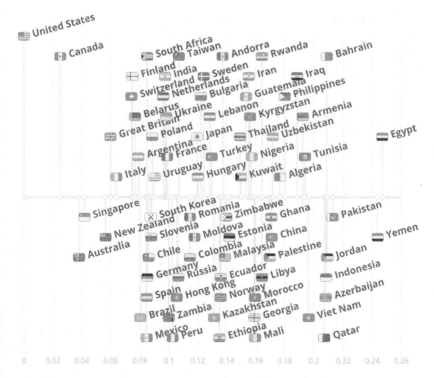

Cultural distance from the United States calculated using CFst on the World Values Survey. Source: Muthukrishna, et al. (2020).

Unsurprisingly, the most culturally close countries to the US are Canada, Australia, New Zealand, and Great Britain. Singapore is culturally close to the US as well, a result of many Singaporeans receiving an American education – what you might call a cultural download. The most culturally distant countries are Egypt, Yemen, Jordan, Azerbaijan, Pakistan, and Indonesia. We also find large cultural distances within countries – in the figure, the bigger the polygon, the larger the cultural distance. Note that similar cultural distance from the United States does not suggest cultural closeness to one another. Britain and Bolivia are similarly geographically distant from the US but not geographically close to one another. Similarly, Colombia and Bulgaria are similarly culturally distant from the US but not culturally close to one another.

Non-metric multidimensional scaling (NMDS) plot of cultural distances between regions in China, India, USA, and EU, with a larger shape revealing a larger cultural distance.

The United States is highly diverse but regionally much more similar than the other largest populations – China, India, and the European Union. The diversity in the United States is not primarily regional.

And so new migrants bring with them cultural values both desirable and less desirable. In other experimental work, we find Canadians who have lived in countries with more corruption lead experimental groups to also behave in more corrupt ways, presumably because they have internalized or been exposed to similar behaviors. The entire group becomes more corrupt because those who take the bribery option end up with higher pay-offs and the others don't want to be seen as gullible. We find no such effect based purely on ancestry – the tendency to offer or take bribes is not a trait intrinsic to Canadians as far as we could tell. But such traits can spread.

In research on the spread of the use of tax loopholes, a large predictor of use is not only whether you would benefit from the loophole but whether you know others who use such loopholes and get away with it. The actual probability of getting caught by tax authorities like His Majesty's Revenue and Customs (HMRC) in the United Kingdom or the Internal Revenue Service (IRS) in the United States is unknown, and so many societies rely on norms of compliance perhaps supported by unknown risk of punishment rather than actual punishment of tax violations.

In almost every country, most people do not commit crimes and as such, most immigrants also do not commit crimes. But with no selective migration, migrants will possess cultural traits at a rate comparable to those found in their countries of origin. Some populations have higher levels of educational attainment matching the educational attitudes found in their countries of origin. Some populations have higher rates of violence matching their countries of origin. These can also be quite specific, such as sexual crimes being associated with gender attitudes which see women as less equal in their countries of origin. Much trumpeted and troublesome statistics are that there have been hundreds of bombings in Sweden over the last few years, primarily in foreign-born or second-generation migrant communities from regions with higher rates of violence. Similarly, around 50% of rapes and attempted rapes in Sweden are committed by foreign-born residents, most from countries with corresponding negative attitudes to gender equality and higher rates of female harassment than those found in much of Europe. These trends for foreign-born crime more generally are also found in Norway, Denmark, Finland, and Switzerland. This pattern is not unique to Europe. In Australia, for example, Sudanese migrants, many of whom experienced high rates of violence in Sudan are over-represented in crime statistics, including violent crime.

This is an emotionally heated and polarizing set of statistics that is often ignored on the left and exaggerated on the right. Even mentioning these statistics can reinforce existing prejudices and biases. Moreover, these statistics can lead to self-reinforcing feedback loops – if the migrant group is perceived to be problematic, this can lead to increases in violence and anti-immigrant targeting of the group, further widening the cultural fissures. It is also important to remember that we are talking about distributions – most migrants do not commit crimes nor necessarily excel in education, but the rates of both may match rates found in cultures of origin. The point here is not about specific crimes but that immigration is a process of sampling cultural traits and assimilating both the desirable and the undesirable. And culture isn't just about identity or country of origin – it is the beliefs and values in your head and how you behave. Those can also be shaped by education or wealth. For example, foreign-born Germans commit crime at a higher rate than other Germans living in Germany, but rates are similar to native-born Germans with comparable education and wealth. Education too is a form of cultural difference.

What should we do with all this information? Diversity may be double-edged, but it is too powerful a sword to put aside.

The first thing is to recognize the importance of the sampling strategy you choose. Are you selecting migrants based on education, on wealth, on language, or at random? How do these correlate with different cultural traits?

The choices are not always easy. One example of sampling at random is when there is mass migration by refugees fleeing war or famine (as was the case of Syrian migrants to Europe and Sudanese migrants to Australia). When a million displaced people are at your border, economic framing is the wrong choice. Mass migration is a humanitarian crisis that we navigate as best we can. But we must recognize that it is ultimately sampling at random from all the cultural traits present in migrants' countries of origin with the heartbreaking additional trauma of forced displacement.

Samples taken from illegal migrants, whatever their circumstances, correlate with the willingness, motivation, and ability to overcome the barriers to entering a country. Perhaps those traits include high tolerance for risk and willingness to start new enterprises, but perhaps also a willingness to break the law. Within the United States, where economic integration is high, immigrants commit fewer crimes than local communities, but illegal immigrants commit more crimes than legal immigrants.

Legal migration offers the host country better opportunities for selection on the basis of cultural traits, including education. For example, countries can develop policies that select for cultural traits associated with greater personal success for migrants and greater economic benefits for the host country. Encouraging high-skilled immigration into industries where home-grown skills are lacking is a good general heuristic. Countries such as the US, Canada, UK, and Australia have benefited from migrants with relevant training for industries such as science, engineering, and health care. The US tech sector is a particularly notable success story, using high-skilled immigration to meet the growing need for engineers.

Once people migrate, ideally through a selective policy, the next step is to introduce policies of *optimal acculturation*. Both acculturation and integration sometimes have negative connotations based on the fear that people will be forced to change, they will lose their sense of identity, or that the process involves value judgements about which culture is better. Optimal acculturation focuses on cultural differences that impede communication and coordination, much as one might want to do when two companies merge. The aim is to close cultural gaps to the benefit of both migrants and locals, and for cultural cohesion. It involves greater understanding and communication on both sides, migrant and local. Closing the cultural gap is essential to strengthening harmony against a falling energy ceiling, tightening the bonds between us during times of economic growth and recession.

Optimal acculturation

Ineffective immigration and multicultural policies are bad for both immigrants and existing populations. Poor policies lead to immigrants not doing as well as they perhaps expect, and when there are limits on migration, they displace other migrants who may have had more success to the benefit of themselves and the host country.

If the country has a strong social welfare system, it means that locals and migrants who contribute more are subsidizing locals and migrants who contribute less or even cost the system more than they contribute. Further, support for the welfare state itself is threatened by lower contributions and more fractured societies.

Getting immigration and multicultural policies right is worth it. As Lee Kuan Yew, Singapore's founder, is said to have told political scientist Joseph Nye,

> China could draw on a talent pool of 1.3 billion people, but the United States could draw on the world's seven billion people and recombine them in a diverse culture that exudes creativity in a way that ethnic Han nationalism cannot.

Lee would know all about how difficult resolving the paradox of diversity can be but also the large rewards from achieving it. He was a controversial leader, but under his stewardship Singapore knitted together multiple cultures, languages, religions, and classes of largely Hindu Indians, Muslim Malay, and Buddhist, Taoist, and Christian Chinese into one of the most successful countries on earth. Remember, big, friendly, interconnected populations become the best and brightest. It should be no surprise that Singapore is at the top of many tables for education, low crime, business, anti-corruption, and more.

So how do we get to a successful multicultural society that allows us to resolve the paradox of diversity and benefit from its rewards? Let's begin by evaluating three common multicultural strategies in terms of the degrees to which they achieve optimal acculturation. I want to emphasize that every country has different policies that continue to change over time and these categories are merely broad generalizations about their overall philosophy.

No hyphen model

In 2018 France won the FIFA World Cup. Its twenty-three-member men's football squad was made up of fourteen players of African ancestry, prompting South African-born host of America's *The Daily Show*, Trevor Noah, to joke that Africa had won the world cup. 'Look at those guys,' he quipped, 'You don't get that tan by hanging out in the South of France.'

Diplomats don't normally pick fights with comedians, but French Ambassador to the United States, Gérard Araud, was outraged. He chided Noah, claiming that the joke denied the Frenchness of the players and fed into far-right talking points, legitimizing 'the ideology which claims whiteness as the only definition of being French'. Conveying the French ideal, he wrote that,

> France is indeed a cosmopolitan country, but every citizen is part of the French identity and together they belong to the nation of France . . . Unlike in the United States of America, France does not refer to its citizens based on their race, religion, or origin. To us, there is no hyphenated identity, roots are an individual reality.

No Algerian-French or German-French. No Muslim-French or Christian-French. Just French. France, at least on paper, is prototypical of what we might call the *no hyphen model* of multiculturalism.

France has a long history of this assimilationist model. Even as a brutal colonial power, its official policy was that those who adopted the French language and French culture were truly French and part of the Republic. The morality of not recognizing different ethnic groups as subcategories is up for debate, but pragmatically if such high levels of integration could be achieved, it would resolve the paradox of diversity. But this is a big challenge, and in practice, France often falls far short of the ideal.

Take North Africa for example, where the French no hyphen assimilationist model led to policies that suppressed local culture and

traditions. Violence, including mass killings, torture, and forced reloca-
tion, was used to enforce policies and suppress dissent while exploiting
the wealth of nations. Although North Africans absorbed the French
language and some French norms, the unequal status of those from
North Africa and those from France affect present-day group relations
and the experiences of North African migrants in France. The colonial
experience of French culture no doubt made a pathway to integration
easier, and, indeed, many North African migrants continue to contribute
to present-day French culture. But cultural and socioeconomic gaps
remain and discrimination, prejudice, and the legacy of the colonial era
have led to a North African identity and subculture within France. The
present reality, in other words, is far from the no hyphen ideal.

A no hyphen model can be achieved, but it requires either small
numbers of migrants from any particular origin such that cultural
enclaves are unlikely to form, culturally close migrants (including
socioeconomic cultural closeness) who can more easily integrate, or
selective migration for those who are motivated to integrate. Ideally,
it requires all three conditions.

It also requires local populations to welcome newcomers to social
gatherings and other forms of bonding critical to cultural transmission.
Typically, this requires sufficient resources and people to share things
in common as a basis for friendship. These many conditions are often
not met. Migrants often don't assimilate to French values and norms
as measured by everything from beliefs and behaviors to jobs and job
types, and the composition of friendship groups. These interpersonal
dynamics are overlaid against France's ongoing struggle between a
no hyphen model and inequality between groups, on the one hand,
and the human tendency to cooperate and affiliate with similar others
at a scale that maximizes personal benefit, on the other. The symptoms
of this struggle manifest in everything from rife discrimination and
lack of integration to public discourse on the wearing of a hijab.

At the other extreme of the no hyphen model is an approach where
there is little to no encouragement for migrants to integrate at all,
an approach that encourages communities to coexist as separate cooperative groups. This model is sometimes called a mosaic model (or
the less alliterative 'salad bowl').

Mosaic model

On entering Canada you will be greeted with the iconically Canadian 'Hello-bonjour', a reminder that independent English- and French-speaking colonies found a way to cooperate and forge a common country. This history led Canada to approach diversity as a mosaic. Each community is like a separate piece of glass, representing cultures across the globe, together making a coherent montage. In the mosaic model, different cultural groups coexist within a country; separate, but connected.

The mosaic model has its benefits. Communities serve as satellites to their countries of origin, pathways through which ideas, capital, and people can flow. Insofar as groups truly do cooperate and communicate, the mosaic model can encourage one of the greatest and under-utilized benefits of multiculturalism – the ability to borrow the best cultural traits from across the globe, and share and recombine them to strengthen a whole new society. If one group is doing well, others should seek to find out why and copy whatever it is that helps them succeed.

But while Canada is mostly peaceful today, mosaics are more fragile than glass melted into a single pane. Especially if those mosaics are put under pressure. If resources reduce or zero-sum perceptions are triggered, people perceive the success of other people or other groups as predictive that they are taking from a limited pie and thereby reducing people's own success or the success of their own group. Separate communities living side by side are natural fractures that can come apart under stress. That is to say, mosaic societies may be fundamentally fragile, waiting to shatter into tiny shards under the right economic conditions. Canadians are reminded of this threat when there are occasional sparks between majority French-speaking Quebec and majority English-speaking Ontario, as well as occasionally between some ethnic communities and others. Anti-immigrant sentiment increases during recessions, including in Canada. Only time will tell if the Canadian experiment is stable. A theory of everyone would suggest that it is a fragile model.

Somewhere between the no hyphen model and the mosaic model is the melting pot model, most often used to describe the United States.

Melting pot model

A successful melting pot is integrationist in a different way to the no hyphen model. Rather than encouraging assimilation to a pre-existing culture, the melting pot is supposed to promote a new, mixed, American identity drawing on all people from around the world. The idea is that no single culture dominates but all contribute to the creation of a uniquely American culture. American society is not fully melted together – it does retain a strong multicultural element – but a melting pot model is in principle an effective middle ground for multiculturalism, if it can be achieved.

America's melting pot is helped by a long-running history of migration from so many different places. This has shaped many aspects of American culture. For example, this long history of migration has made America a deeply expressive culture which puts a lot of emphasis on explicit emotional expression – thinking about how you feel and the feelings of others. Many Americans may take this focus on expressing emotions for granted, but it is not universal.

Americans are known for their broad smiles, obvious displays of anger, and other clear emotional expressions. These features are common in countries with long histories of migration. When your neighbor doesn't speak your language or share your culture, emotions serve as a common ground for communication. Clear expression is critical to being clearly understood.

In 1990, after the fall of the Iron Curtain, the iconic American burger chain McDonald's opened its first restaurants in Russia. One of the first challenges was teaching Russian workers to smile as part of that authentic McDonald's experience. Both workers and customers initially found this difficult. In Russia people who smile when something isn't funny are considered crazy. But with sufficient training, workers – and customers – accepted the new smiling norm. They came to understand that people smiling without a joke might be crazy or they might just be American. They came to accept a local norm at McDonald's. Of course, this didn't change the overall culture. Prior to hosting the 2018 FIFA World Cup, Russia once again ran smile training sessions for service workers lest tourists from countries with long histories of migration leave with the impression that Russians are unfriendly.

Within cross-cultural psychological research, the emotiveness of Americans is often contrasted with other, more monocultural countries like Japan, known for its more muted emotional expression. Japan's homogeneous culture allows for any Japanese person to know what any other Japanese person feels from the context alone. A Japanese person would immediately recognize a shameful situation or one that would provoke anger without anyone displaying emotions. Outsiders, on the other hand, may be oblivious to contextual cues, not realizing when they've offended their hosts.

This emotional control has further downstream effects – for example, eyes are often used as a focus of emotional expression more than the mouth in many similarly homogeneous East Asian cultures. By corollary, Asian immigrants to America are often surprised by explicit 'I love yous' and the 'Thank you; you're welcome' routine, even for family members. In places like India or China, being so explicit with family members would be considered odd or even insulting. Implicit communication can be more efficient when everyone shares the same norms and understanding, but it makes it difficult for newcomers who have to discover hidden norms and rules through faux pas – ideally someone else's. A broad policy of explicit communication may be uncomfortable and unfamiliar initially, but is part of what helps newcomers assimilate more easily.

The melting pot model has led to a largely successful immigrant story in America, albeit one marred by a history of slavery and discrimination alongside more noble melting pot ideals. But the success of the melting pot may in part be due to abundant resources. People form multiple overlapping groups naturally and it takes work to de-emphasize smaller group affiliations. The shrinking space of the possible is making things more difficult in today's America. And in reality, of course, different-sized populations and levels of wealth and power may lead to Ankh-Morpork. Ankh-Morpork, as fantasy novelist Terry Pratchett described the fictional city state, is 'the melting pot of the world, which occasionally runs foul of lumps that don't melt'.

All these models and metaphors are simply ideals. The extent to which there is integration, separation, or fusion in countries like France, Canada, or the United States varies across provinces, states, cities, and communities. For example, the United States has effectively hyphened, segregated communities (often distinguishable by a noticeable accent), but is also broadly integrationist. Indeed, its strict immigration policies help maintain its melting pot with a high proportion of migrants who make large contributions. This is a large part of the political division over the status and pragmatics of illegal immigration. Illegal immigration notwithstanding, all three multicultural models have a blind spot.

The no hyphen model, mosaic model, and melting pot model all focus entirely on relations between ethnic groups, between different immigrant communities, between immigrants and local populations, and between communities and the state. All ignore the broader context of the interaction – which is shaped not only by culture and policies but also by the space of the possible that they share.

The same culture and policies may succeed with plentiful resources but fail in times of scarcity. A better model and metaphor that encompasses these complexities is what I call the *umbrella model*.

Umbrella model

The umbrella model uses a successful company culture as a metaphor for a successful national culture. In any great company, people need to be able to work together. To have common purpose and common culture. And so cultural fit along dimensions that matter for cooperation, communication, and coordination is essential. People have to be willing to work with others of different religions, status, or ethnicity, but they also need to have shared goals, coordinated behavior, and some common moral values. They need to speak the same language; they have to drive on the same side of the road. Other cultural traits are up for debate.

We can think of a successful immigration policy as similar to building a successful large, multinational corporation. In such a multinational, there may be many different elements or companies, the central authority or government may be hands off or more hands on, the central brand or identity may be more obvious or a more diffuse collection of brands and identities. Regardless of the particulars, a great organization knows how to hire the best people and support their growth to benefit both them individually and the organization as a whole. We're looking for the missing skills we need with a purpose in mind – greater innovation, efficiently accessing resources and energy, growth, or profit.

Within this umbrella model the sister companies support each other, creating supply chain alignment and vertical integration for synergies that support all companies. But you can also imagine it as people standing under a real umbrella, knowing the umbrella needs to be held up so that everyone can stay dry. The umbrella must be made large enough – with sufficient resources and investment in infrastructure so that there are enough school places, jobs, resources for hospitals, and other public goods. People's training must match the jobs that are needed.

Too small an umbrella and people get wet and fight over who stays dry. The wetter people get, the more they grumble and try to take control of the umbrella. Ultimately that conflict can destroy the umbrella and everyone is left standing out in the rain.

In the umbrella model, just as in a successful large company, people must see themselves as being part of the same team, have a shared vision, and see their futures tied together. If a company wants to grow, it must have sufficient resources and investment in infrastructure – people need space to work.

If people don't have enough work, they end up fighting over the few valuable jobs. That conflict will eventually destroy the company and everyone will be out of a job.

Metaphors aside, there are challenges to achieving this harmony.

Just as in recruitment, we need *sustainably managed migration* – too many people at once are difficult to onboard. Too few and companies can't grow. Recruitment is best done through a fair competition for employees or migrants who can best develop ways to improve the company or country. Selective migration policies involve working out the right number and how newcomers can work with existing employees and citizens. In turn, current citizens must themselves understand that they have to help onboard newcomers. It requires a cultural shift in how we think about immigration.

As with a great company, it means hiring the best people with the skills that are most needed. Great companies can be built with a monoculture, but even greater companies can be built when the conditions are established for diversity to flourish. When diversity becomes inclusive and people feel like they belong, more people can contribute what they have to offer. Even when migrants have much-needed skills – engineers, nurses, teachers, care workers – often greater investment is needed to help more culturally distant migrants maximize the use of their skills. A great software developer from Brazil can immediately apply their JavaScript knowledge in Vancouver, but still needs to know how other aspects of work and life differ in the Canadian context.

More culturally distant groups represent a greater challenge to national harmony, but also potentially greater benefit. Language and other cultural translation programs are needed to help with the transition, as is an expectation that locals play a role in welcoming the newcomers.

Ethnicity isn't the only form of cultural diversity. Education is often a greater source of cultural diversity. Remember that education is a powerful mechanism for downloading a cultural package with certain assumptions and ways of thinking. Psychologist Cindel White and I measured the cultural distances between highly educated people from across the world. The cultural distance between highly educated people is smaller than the distance between those with less education. If a global culture is emerging, it is mediated by education and probably movies, TV, and other media. Indeed, even within a country, education is often what separates what some have referred to as nationalists versus globalists, somewheres versus anywheres, the labor class versus the laptop class.

For all these reasons, large numbers of low-skilled immigrants and refugees with unknown skills and cultural backgrounds should be welcomed on humanitarian or charitable grounds, but with full recognition of the greater challenge and economic cost they may present. As German Chancellor Angela Merkel discovered, we cannot succeed on willpower alone.

Early in the 2015 European migrant crisis, as large numbers of Syrian refugees migrated to Europe, Merkel famously declared, 'Wir schaffen das!' – 'We can do this!' A year later, many were less convinced. The vice chancellor admitted that Germany had underestimated the challenge and that 'there is an upper limit to a country's integration ability'. Although some of the refugees have since found employment, the effect of this sudden, unplanned influx on German culture, national norms, and institutions, as well as on the European Union more broadly, won't be known for decades.

No country explicitly uses an umbrella model, but the country that comes closest to doing so is Australia.

Lessons from Australia

Australia has adopted a strategy of sustainably managed migration that has mostly ensured cultural cohesion without undermining economic growth. In general, the competent civil servants of the country are highly pragmatic, borrowing different strategies wherever they find them.

It uses a points-based migration system with more points given for factors that are associated with more successful migration. For example, preference is given to applicants in the 25–32 age group. It's a sweet spot that typically means they already have some education or training and will make economic contributions over the next three to four decades of their working life. Preference is also given to those with strong English-language skills ensuring that they can fully participate in the broader community. Australia prioritizes those with skilled work experience and those with education in industries most in need of labor. These industries and criteria change based on changing needs and new data. It's similar to the way a company might use an evidence-based approach to recruitment.

Australia limits the number of migrants with incentives offered to move to regional areas in need of particular skills. Its borders are tightly controlled (it helps not having any land borders) and illegal immigration is disincentivized through offshore processing. This particular aspect of the policy package is controversial and by no means a perfect system. The long processing times do serve as a disincentive but are also arguably inhumane and can be a source of national and international outrage. They are in stark contrast to how Australia treats refugees once these non-selective migrants meet the refugee criteria.

Australia invests large amounts of money in integration. Refugees first go through a five-day pre-arrival orientation program – the Australian Cultural Orientation (AUSCO) Program. Everything is covered: what immigration will look like; what assistance they can expect when they arrive in Australia; an introduction to the Australian lifestyle and social and cultural norms; essential day-to-day life skills, such as how to find accommodation, get around with transport, deal with banks, and register for health care; and Australian laws around gender equality, religion, discrimination, their rights and responsibilities. It's a crash course in how to be an Aussie before they even arrive.

When refugees arrive in Australia they are provided with further support. My wife, Steph, was a volunteer with a refugee resettlement program. The gaps could be large due to culture and trauma, but resources were invested in closing those gaps. Assistance was offered in everything from teaching refugees what fruits and vegetables were available in Australia and how they might be cooked to match immigrant cuisines (often taught by previous migrants from the same culture or similar culture where this was not possible), to how to catch public transport or even how to make friends and appropriate and inappropriate ways to interact with people.

Steph recalls the case of a refugee who refused to speak to the female volunteer tasked with helping him. The refugee wanted to speak with a man and waited for a man to arrive. After a few hours he left. He returned the next day with the same result. On day three he begrudgingly accepted the assistance of the female volunteer.

The story illustrates the policy of the time of concerted efforts to welcome refugees, but also the clear message that certain norms are non-negotiable. Each country must identify what those norms are. This general approach is often criticized for being insufficiently sensitive to different cultural norms, but it is precisely what helps Australia maintain cohesion and national character.

Australia calls itself the lucky country – one of the last lands to be colonized, it has large untapped resource wealth. Aboriginals have lived in Australia for 40,000 years or more – indeed they are one of the longest continuous cultures on the planet. The modern nation of Australia was founded by British settlers carrying British culture, values, and institutions to this resource-rich and ecologically unique land. Economist and commentator Noah Smith once jokingly described Australia as a 'mining outpost with an exotic petting zoo'. Like Norway, Australia has for a long time carefully managed a large space of the

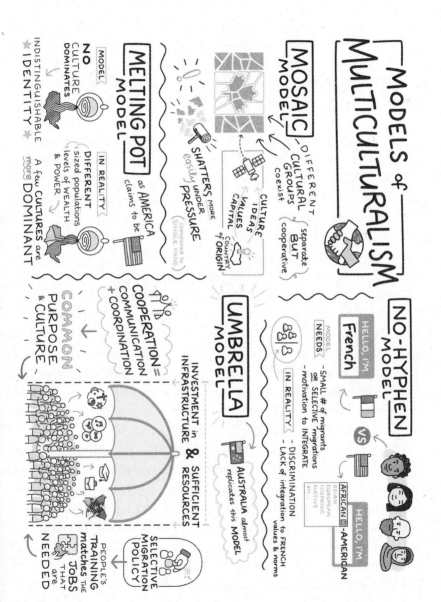

possible, but, unlike Norway, has also actively attempted to deal with the challenging paradox of diversity. Today Australia has the highest median household wealth of any large nation (over US$400,000).

Many other aspects of Australia also help in this success, including the persistence of culture through path dependence. Australia's culture, the descendants of the jailed and their jailers, is strong on rule of law, including both formal enforcement of laws and informal enforcement of norms. Road rules, such as speed limits and mobile phone usage, for example, are strictly enforced, with police officers known to hide in bushes to catch violators. During times of drought, it's not uncommon for neighbors to normatively enforce garden watering bans on one another. Australia also has many robust innovations in democracy that help bind the country and ensure that decisions represent more people.

Australia is not perfect. It has a shameful history of treatment of its indigenous citizens, with unresolved challenges persisting to this day. Corruption has also led to coalitions of mates swapping favors and supporting special interests, and large numbers of immigrants bypassing the formal system with the support of bribed politicians. As economists Cameron Murray and Paul Frijters document in their 2022 book *Rigged: How Networks of Powerful Mates Rip Off Everyday Australians*, the land of mates is unsurprisingly susceptible to cronyism, which many Australians have not woken up to but may do so when they realize how much money has effectively been stolen through the housing industry and from their superannuation funds.

Although Australia comes closest to the ideal umbrella model that we can derive from a theory of everyone, no country is currently well prepared for keeping their society together when our excess energy budget and the space of the possible shrinks. We don't yet know what currently wealthy nations will look like under conditions of both resource scarcity and potential social fractures. One context in which these conditions have been well studied is Africa, the continent with the lowest gross domestic product (GDP) per capita.

It's not obvious whether one can generalize the research from Africa to other contexts given the cultural, historical, institutional, geographic, and many other differences that have led to present-day differences in outcomes. Indeed, it's not clear how much we can even generalize one African context to another. But people are people and societies should be expected to display at least similar social dynamics under similar conditions. For this reason alone, it's worth knowing more about research from Africa, to anticipate the effects of diversity in resource-constrained conditions.

Diversity and low resources in Africa

The Scramble for Africa of the late nineteenth century saw the beginning of the European powers' colonization of the continent. Just as Namibia was separated from Angola and the Himba people were split in two by the decisions of men in faraway Lisbon, the borders of other African nations were drawn and redrawn with no concern for existing ethnic groups or tribal boundaries. Thus at the end of colonialism, Africans found themselves living in nations often with foreign institutions set up by the colonial powers and with tribes that may have been former allies or current enemies, kinfolk or completely separate ethno-linguistic groups. Borders would sometimes encompass entire groups alongside others or slice groups in two, who now lived in separate countries. One of the predictors of current conflict and failures to cooperate is the diversity that was created by arbitrary colonial boundaries – an effect with the same magnitude as the resource curse. Political violence is 57% higher in divided homelands, with higher rates of military incursion from co-ethnics of one ethnic group entering across the border. Take Zimbabwe, for example.

Zimbabwe, a country rich in gold and diamonds, is blessed with resources but also cursed by post-colonial diversity that it never had sufficient cooperation to resolve. The country has sixteen official languages, which represent the various ethnicities vying for power, often violently. The conflict is typically between the majority (80%) Shona and the second largest ethnic group (15%), Ndebele.

President Robert Mugabe was Shona. His role in the killing of over 20,000 mostly Ndebele dissidents is emblematic of the kind of intra-tribal cooperation and inter-tribal conflict preventing African countries from cooperating at a sufficiently high scale to use their resources to build their states and economies. Mugabe's failed state hit international news during the hyperinflation of 2007 and 2008, a viral 100-trillion-dollar bill an iconic reminder of the crazy heights the currency had reached.

Zimbabwe has for a long time had one of the highest corruption scores according to Transparency International. In 2022 it was around the same level as Iraq and both Congos. It's a stark contrast to its southern neighbor, Botswana.

Botswana is also blessed by treasures in the ground: diamonds. But instead of fighting over those diamonds, the Batswana used those diamonds to chart a path from one of the poorest nations in the world to the highest GDP per capita in real terms (adjusted for purchasing power) on the continent. The country's corruption level is the lowest on the African continent and similar to other developed, more ethnically homogeneous nations such as Israel and South Korea. Botswana had the ideal law-of-life package to resolve the paradox of diversity and avoid the resource curse.

Botswana is a small country and one of the most ethnically homogeneous countries in Africa, home to the majority Tswana people (referred to as Motswana if singular or Batswana if plural) who speak a common language and whose clans are closely related and intermarry (in contrast to nominally ethnically homogenous countries like Somalia with deep clan divisions). The Tswana also have an ancient democratic institution in the form of the *kgotla*, a public meeting held by the chief or headman, where decisions are made by discussion and voting. Such deliberative democracy was likely incentivized by the fact that it was easy for herders to leave if they didn't like the decisions of the big chief and the need to come together against outsiders. For example, in the 1852 Battle of Dimawe, against the white South African Boers, the Batswana tribes came together for a Batswana victory. The combination of ethnic homogeneity, proto-democratic institutions, and outside threat meant that when diamonds were discovered, the Batswana had sufficiently high levels of cooperation to reach an even higher level of cooperation and exploit their resources for the benefit of Botswana as a whole. Botswana was the Norway of Africa.

There are many factors that affect economic and human development, but as a general heuristic, a policy goal of reducing corruption is suppressing lower scales of cooperation, the ultimate source of conflicts of interests. Many countries, for example, are plagued by tribalism. Tribes often maintain tribal boundaries through endogamy – a preference for ingroup marriage or even prohibition against outgroup marriage. Often the preference is to marry even close kin, such as cousins, further reifying kin-group boundaries. Cousin marriage was once common throughout the world and is still common in South Asia, the Middle East, and much of Africa. Where cousin marriage is common, your uncle isn't just your mother's brother, he's connected to you through multiple relationships and this can be reinforced through cultural obligations toward these more bonded kin.

One strong piece of evidence for the effect of weakening these kin bonds is to be found in the Roman Catholic Church's prohibition against cousin marriage and other changes in marriage practices in Europe during the Middle Ages. In a 2019 study, Jonathan Schulz, Duman Bahrami-Rad, Jonathan P. Beauchamp, and Joseph Henrich have shown how these changes created the modern nuclear family of mom, dad, kids, aunts, uncles, and grandparents rather than all relatives being connected through multiple relationships. It was the creation of nuclear family trees rather than sprawling family webs that weakened the European tribes. The earlier the practices were implemented, the lower the corruption, the greater the impartiality, and the stronger the democratic institutions that exist today. By corollary, places that prioritize family are higher in nepotism.

After centuries of these changes, nepotism is less common (though still present) in the West. Instead, direct and indirect reciprocity – cronyism, lobbying, revolving doors – are more common sources of corruption. Just as undermining cousin marriage and other extended kin models reduced nepotism, undermining direct and indirect reciprocity reduces cronyism. Successful anti-corruption strategies include introducing long cooling-off periods before someone can move from policy to industry. Similarly, simply moving people around, so it's more difficult for them to form cliques and other reciprocal relationships that harm overall efficiency can also work. Botswana for example, regularly reassigns its civil servants, reducing their probability of working for their own tribe or region. These strategies are all part of a more general approach that reduces the ability to cooperate at a lower level.

Remember the law of cooperation – the optimal scale of a group is one where the group maximizes resources per person, that is, when resources are divided per person, a person gets more resources than they could get in a smaller or larger group. Sometimes this calculus can be achieved through alliances between groups. For example, a common alliance throughout history and even in present-day America is that between the holy and the wealthy – between religion and money. These groups have greater power together than by themselves. Similarly, the weakening of meritocratic institutions can be seen as an alliance between the wealthy and groups who traditionally perform more poorly in standardized testing, both of whom benefit from the removal of standardized admission tests such as the SAT.

And as per the law of cooperation, the general principle of weakening lower-order cooperative bonds to help strengthen higher cooperation still applies. As noted, there are many differences between countries beyond diversity and resources, but at least in terms of cooperation and corruption, it is an open question whether WEIRD institutions and norms are sufficient to sustain cooperation when the space of the possible shrinks. I suspect they are not. A new challenge has also emerged: the Internet.

The Internet has created new tribes and may have led us to a Second Enlightenment.

New tribes and the Second Enlightenment

In the irreverent comedy *Little Britain* an ongoing gag is Daffyd Thomas claiming that he's 'the only gay in the village': Daffyd is not the only gay in the village, and he's in denial of all the other gay characters he meets. But once upon a time the Daffyds of the village may very well have numbered in the low dozens.

Our worlds were once small. Most people married someone in their immediate vicinity, sometimes relatives. Geneticist Steve Jones praises the invention of the bicycle as one of the most important events in recent human evolutionary history. The average person could finally expand their dating pool beyond the people they grew up with to whole other villages within riding distance. But even with a bicycle, being a small minority meant meeting few others like yourself. You really might be one of very few LGBT people in your village, not enough to reach a critical mass. Similarly, if you were interested in uncommon hobbies, such as collecting porcelain plates or whittling wood, the club you could form would have been small. The long and the short of it is that in the past it was unlikely that you could form a community if you were a small minority or had uncommon interests. But thanks to population growth and the Internet, that's no longer true.

Today, no matter how peculiar, unusual, or uncommon your hobbies and interests are, you can form a community given the massive size of the world population and the way in which the Internet and social media allow you to connect with others who share your interests.

Are you into carrying sand in your pockets or just chatting about the many benefits of pocket sand?

No?

Well, over 40,000 people are, and they've formed a community in Reddit's /r/pocketsand subreddit. Pocket sand is a niche interest, I agree. It's not like the far more popular activity of stapling a slice of bread to a tree and posting pictures. Over 300,000 people over at r/BreadStapledToTrees are really into it. Go on, search 'bread stapled to trees' on Google images. Thanks to the Internet, you might discover that it's your thing.

I could go on.

This kind of hyper-connectivity has been a boon to groups and communities of all kinds. Those who suffer from rare medical conditions can find one another, discuss treatments and issues, and lobby for change. Tiny minorities with uncommon interests can not only find one another but also interest others in their niche obsession, and can thereby reach a critical mass. In essence, the Internet supports the creation of new cooperative cultural-groups – new human tribes.

The ease with which the Internet allows us to create new tribes is one of the reasons why the Internet is so disruptive. And that's both good and bad.

The Internet facilitates social change and broad sharing of information of all kinds. Arab Spring activists can coordinate, deep learning enthusiasts can find new applications, irredeemable racists can look for evidence for their racism, QAnon sympathizers can keep up with the latest top-secret releases. All find one another, form a culture, advocate, and cooperate. The Internet and social media enable even small groups to cooperate, grow, and compete with other cooperative groups and the rest of society as a whole. It's the same dynamics as corruption – one scale of cooperation undermining another. And so, for better or worse, these communication technologies are profoundly disruptive. But a dedicated group of like-minded individuals cooperating for change is how change has always happened. It's just that now it's easier for those groups to form, to get their voices heard. And there are simply more of those groups.

The Internet also amplifies existing real-world groups. Donald Trump grew his real-world followers thanks to Twitter until he was forced to move to Truth Social. The Russia–Ukraine war is perhaps the first that is being fought not only in the battlefield but also on social media timelines.

We can debate whether social media algorithms segregate us into separate social network feeds – echo chambers – radicalize us, or instead expose us to ideas outside our immediate bubble. It's an empirical question for which there is currently no general answer. Indeed, there may be no general answer to be found.

Algorithms are continually changing as they evolve to keep you engaged. Moreover, the specific format of social media can change dynamics.

Facebook groups can more easily become echo chambers because of how easily people with shared interests can come together and how easily people can be kicked out. TikTok's algorithms are opaque but it may have a larger influence due to its younger, more impressionable audience. Twitter's and Facebook's public posts, on the other hand, make people mad because they're exposed to ideas they strongly disagree with. That's a good thing.

But remember, it's not information that changes minds, it's people. Who you hear ideas from matters for whether you instinctively agree or vehemently disagree, often – let's be honest – without examining the evidence or even reading the article.

The ideas themselves are also part of a complicated ecosystem of viral posts and words spouted on traditional media – Fox News and MSNBC, *Washington Post* and *New York Times*, *Daily Mail* and *Guardian*. Articles and thoughts emerge here, are shared in various groups on social media, and may be fodder for the next mainstream media article.

Social media, I argue, is not the ultimate cause of polarization. Instead, it's the mirror to our society and the venue where we thrash out our differences. Which is also why polarization is not the same everywhere, even though almost everyone in the world is on social media. In some countries, platforms like Facebook effectively *are* the Internet. But the United States is particularly polarized relative to places with even more people on social media. Perhaps it's in part due to America's diversity, free speech culture, and size.

Small percentages are big numbers in a large population. This is true even without the Internet. For example, there are many fungal diseases that will almost certainly result in death for those who contract them, but with which we've never had to contend because most people are not immunocompromised, and even among those who are, most will never contract a deadly fungal disease. But with advancements in medical technology, there has been a rise in the number of people who survive illnesses that would have killed them even a decade ago but who now live with immune systems compromised by drugs taken after organ transplants, congenital conditions, HIV/AIDS, or cancer. Estimates suggest as high as 3%. That may be a small percentage, but in a population as large as the United States', it translates to around 10 million people. This number can continue to grow thanks to the marvels of modern medicine thankfully keeping more of us alive.

In the case of cultural traits, thanks to the global nature of the Internet even smaller percentages can find one another and grow into large percentages. Thus, perhaps paradoxically, the Internet gives us access to a larger base of common knowledge and simultaneously allows us to assort and affiliate with a subset of that knowledge based on common interests, common incentives, common values, or common goals. In other words, the Internet has empowered cultural evolution and cultural-group selection.

The global significance of this shift warrants a proportionately significant label: I call it the *Second Enlightenment*. Just as the First Enlightenment challenged our cultures and reshaped our societies, this Second Enlightenment is challenging many of our entrenched institutions, including democracy.

The First Enlightenment led to the French and American revolutions and to the Industrial Revolution. The Second Enlightenment can lead to social and technological revolutions, unlocking new energies and new forms of cooperation. It has the potential to lead to a true second Industrial Revolution.

Britain and Norway are a lesson in how divisions in a society affect how resources are allocated. Immigration and entrenched economic class differences have thus far been the major source of that division, but the Internet has opened a new space for new tribes and in turn has also created a new source of division. Thus, as with diversity, the Internet has the power to increase innovation and reach the next level of energy abundance, but that is contingent on discovering new ways to cooperate in a simultaneously more connected and diverse world. For this to happen, the institutional instruments with which we coordinate and govern each other need to evolve. Democracy needs to evolve.

Our democracies were built during a time when people interacted in geographically isolated communities. Bicycles weren't commonplace and Daffyd was one of a few dozen gay men in his village. Institutions such as the US electoral college may still have benefits, but many of the conditions they were created for and problems they were created to solve no longer exist. (Were the Founding Fathers alive today, they would no doubt develop a very different US democracy.) Indeed, democracy in its current form doesn't seem to scale well. We need governance for the twenty-first century.

8

Governance in the Twenty-first Century

The ideal organizational structure for human collectives in the twenty-first century is one that is able to adapt and evolve to the changing needs and challenges of society. As the world becomes more diverse, multifaceted, and highly connected, traditional hierarchical structures may no longer be effective in promoting cooperation, innovation, and progress. Instead, we need to explore new forms of governance that are able to facilitate the exchange of ideas and the emergence of new solutions to the problems we face. We now have some understanding of how our species got here, but where do we go next? How do we increase cooperation, maximize innovation, and ensure that our political organizations help rather than hinder our ability to break through to the next energy level? The answer is that we don't know the answer. But we do know how we can find out.

In 1992 American political scientist Francis Fukuyama boldly published *The End of History and the Last Man*. He declared that with Western liberal democracies we had achieved the final form of government. We had created the global optimum in governing ourselves and ensuring large, peaceful, innovative societies. At the time, this claim seemed right, perhaps even obvious. The only other competing form of governance – communism – had failed with the fall of the Soviet Union in 1991. Even Fukuyama admits that his book could never be written today, when the world witnesses the rise of China as a very different, rather novel political system that still seems to increase overall human welfare and economic growth. But China, too, at least historically, embodies some of the basic elements of efficiently evolving organizations and has the added benefit of being in catch-up mode – borrowing, recombining, and implementing innovations invented elsewhere.

If we define democracy as a government in which laws, policies, and leadership are directly or indirectly decided by the population then China's political system could generously be described as *democracy by assent* in contrast to WEIRD-style *democracy by selection*. In a democracy by selection, the will of the people is expressed by selecting people or parties that best represent their interests. In practice, this ability is constrained in some way by within-party politics; the will of swing voters and swing states often has a greater influence on policy; and overall democracy by selection places faith in the general populace's ability to select parties or people that are best suited to governing and representing their interests. But there is no reason to believe that people are particularly good at selecting who is best able to ensure economic growth and enough jobs, to keep crime low and war at bay, and to ensure that roads are drivable and the water is drinkable. What matters to most is that the government in power gets that job done, and there may be ways to ensure that happens other than a government of the people, by the people, for the people.

China's history is marked by long periods of stability punctuated by popular revolutions radically changing societies. The Chinese Communist Party (CCP) knows this history and lives in perpetual fear of that next popular revolution. To keep that revolution at bay, it controls the flow of information and suppresses, even violently, any groups or people with the potential to coordinate or pose a threat. But it also knows that it can spend less on censorship and suppression and continue in power by ensuring continued economic growth. As long as people's lives continue to improve, they'll care less about who's in charge and how they got there.

Under President Xi Jinping many of China's meritocratic selection mechanisms have been dismantled. In any case, even prior to Xi, China was by no means a free country. By no definition is it a liberal democracy. But the desire for freedom is also not a human universal. Desires for food, safety, and a better life for our children are. The CCP's legitimacy and assent by which it rules will be undermined when it can no longer meet these universal needs. As long as these conditions are met, many Chinese citizens will happily support the party through some mix of fear of speaking out, lack of alternative information, and fear of an alternative form of government, but also because they enjoy their rise in material wealth, and take pride in China's rising place in the world. But for the CCP to remain in power in a weakening economy would require far greater suppression or a common enemy, such as a war with Taiwan or the United States.

The recent rise of China can also in part be attributed to it being in catch-up mode, where the need for new innovations is lower. China can copy the West with some recombination for the Chinese context, and can implement policies that have worked elsewhere without dealing with opposing factions. But autocracies traditionally fail when it comes to innovating new solutions. China, though, has potentially solved this traditional autocratic challenge.

The power of liberal democracies, for example, is that by protecting individual rights and freedoms they allow for a diversity of ideas to flourish and recombine in a cooperative and freely thinking collective brain. When combined with mechanisms that allow for the best ideas to rise to the top – from meritocratic college admission to states as 'laboratories for democracy', as Justice Brandeis described the United States – this creates an environment that is conducive to social and technological innovation.

Instead of Brandeis's laboratories for democracy approach, China has pursued laboratories for economic growth in the form of special economic zones and promoting provincial leaders on the basis of economic performance at a local level. In principle, this has led to policy experimentation on an unprecedented scale, but in practice it is hampered by three forces. First, smaller coalitions within the CCP cooperate toward corruption, undermining the meritocratic selection. Second, local leaders' success at a local level doesn't always replicate at a national level, perhaps due to chance, effort, or local activities that artificially raise growth that can't be replicated nationally. And finally, the lack of diversity and freedom of speech and limitations on the free flow of information ultimately weaken China's collective brain and its ability to innovate. China is unlikely to offer a replicable model for economic growth or governance in the twenty-first century.

At the same time, liberal democracies don't seem to scale well to large, diverse, online socially networked populations, at least not without certain prerequisites and certainly not under conditions of limited resources. So what does governance in the twenty-first century look like? One starting point is actively pursuing new democratic innovations and new forms of democratic participation that have been tried in culturally similar countries. For example, Americans love democracy, but Australians might love it even more, continually innovating new democratic policies. Indeed, Australia is the only country with positive migration from America – as a percentage of the population, more Americans move to Australia than Australians move to America.

Australian democratic policies that have stuck around include secret ballots (popularized in Australia and now often called the Australian ballot). Another effective policy is compulsory voting for all citizens over eighteen years. As an added incentive, free or cheap food (BBQ sausage sizzles, also known as 'democracy sausages') are available across the many polling stations on election days.

Normatively, voting is seen as a duty to be incorporated into one's day. On voting days, people often pop into a polling place as another errand in their daily routine. It's not uncommon to see people voting in their pajamas, with nothing else planned that day, or in bikinis and board shorts, on their way to the beach.

Voting is in many ways an irrational act. It's a collective action problem where each vote is unlikely to influence any election but high turnout is essential to preventing small cooperative groups from dominating politics.

Compulsory voting solves this problem and also forces politicians to appeal to the large majority of moderates rather than fight over the votes of politically active extremes or well-coordinated special interest groups, as often happens in most other countries. To win an election, politicians and political parties have to have policies that most people want.

Voting is also preferential, such that people rank their candidates. This is another active area of innovation. Parties will often provide voters with a suggested ranking with other parties that best match their positions in case they lose.

From the 1980s to around 2016, if voters didn't want to rank candidates, they could just pick one party and if that party didn't get the majority, the party allocated its votes to the next party that best represented its position. This led to vote trading between parties and so at the moment, Australia is experimenting with forced ranking and with optional ranking where not all parties have to be ranked.

This continual process of innovation in governance is often data-driven, the goal being to balance ease of voting with best representing the will of the people. These measures ensure a vibrant multiparty system where people can vote for minor parties that represent the most obscure preferences they care about without wasting a vote.

Want weaker intellectual property laws? The Pirate Party has you covered. Want free parking? Vote for the No Parking Meters Party. Want weaker gun and fishing laws? The Shooters, Fishers, and Farmers Party has you covered. Want a candidate who has no positions but just votes on issues based on online direct voting? Try the Flux Party.

The preferential voting system ensures that the most preferred party is voted into office and the least preferred party is not. In contrast, consider what happens in Canada.

Canada also has a multiparty system, but if 30% of voters vote for one left-leaning party (e.g. the New Democratic Party: NDP) and another 30% vote for another left party (the Liberals), this split vote would allow the Conservative Party to win with 40% of votes, even though 60% of the country doesn't want a conservative government.

A first-past-the-post system of the kind found in Canada is also what America uses, which reinforces a two-party system and polarization of these two teams. As a result of not using preferential voting, policy changes can only occur within a party. In America, to change the left, the Democratic Party itself must be changed. To change the right, the Republican Party itself must be changed.

Running as an independent in America risks undermining the majority parties. Ralph Nader and Pat Buchanan were accused of undermining votes for Al Gore in the 2000 election leading to George W. Bush's election victory.

Americans may be horrified if they realize that their approach is much like China's CCP, where change can only happen within the party, but with two parties instead of one.

Existing democratic innovations are an adjacent possible, more easily achievable from where we are. Other visions are more radical. Indeed, many visions of the future of governance are utopian. Utopias are unconstrained by the realities of where we are today in terms of current political systems, current norms, and current incentives. As such, they are a fantasy or at a minimum require a nasty shock of war, disaster, or revolution to create a complete overhaul of the system as it is. The truth is, where we should go next is not obvious. If it were obvious, we'd have done it by now. But we can put some boundaries to this thorny question of what the future of governance looks like. For one, given the size and diversity of modern societies, long gone are the days when fifty-five people could sit in a room and make decisions from the top down that affect the other 2.5 million and the future of everyone, as was the case in the founding of America. Our populations are now too large, too diverse, too educated, and too connected for a radical overhaul from the top down.

Instead of trying to design the future of governance or democracy, we must use the principles of democracy itself to help it *evolve*. But that requires radical diversity combined with meritocratic selection.

When the Roman Empire fell, it eventually made way for new innovations in governance that led to the modern world. The vacuum it left created an environment where new ideas and new approaches could be attempted, the best outcompeting others through cultural evolution, cultural-group selection, and the laws of cooperation, energy, and innovation. But it was a messy, violent process that is best not revisited.

The general principle, however, is possible, and should be revisited. Rather than design *efficient* institutions, we can design more *efficiently evolving* institutions.

Indeed, the United States owes its robust democracy to this very approach, enabled by its federal structure and lack of strong central authority. Different states can try different strategies like separate countries trying different policies. If they fail, they fail at a state level. But if they work, they can be borrowed by other states and eventually bubble up to the federal level. Each state, in other words, can operate like a start-up.

The same model has been successfully used by militaries using a so-called team of teams approach. The same approach also helped Hinduism persist. In India, 'the country of a hundred nations and a hundred tongues, of a thousand religions and two million gods', as per Mark Twain, it is not uncommon to find pictures with a Hindu pantheon that includes Jesus. It's a model for diversity through syncretism held together by common beliefs such as karma and the *samsara* reincarnation cycle.

Other institutions that have stood the test of time have used a similar distributed evolutionary approach. The Roman Catholic Church has succeeded in maintaining a central structure for 2,000 years by allowing some degree of autonomy via various orders and regional rules.

Research on the history of protestant churches in America shows a common pattern of conservative rule-enforcing churches maintaining their borders but being outcompeted by more liberal churches. These liberal churches eventually grow too liberal and splinter or fail. To strike the balance between stability and change requires a mechanism for generating innovation.

This approach has been used to rejuvenate corporations. Since 2014 Microsoft CEO Satya Nadella has been revitalizing the company's innovative ability by turning the hierarchical, monolithic behemoth into what can best be described as an ecosystem of start-ups, thereby revitalizing the ageing tech giant and boosting its stock price.

In each of these cases the specifics differ but the general principle is that rather than centralize authority and administration, power moves back down to local rules, cantons, or cities who, like firms in a marketplace, freely cooperate and compete with one another with free movement of people, their labor, and ideas allowing some groups to grow at the expense of others. It's cultural evolution. It's cultural-group selection. It's an empowered collective brain. It's Silicon Valley.

How can we replicate this success for the future of governance? Many approaches speak to the devolution of power down to a local level, but this is a weak version of the ultimate solution. Such devolution rarely changes much more than how things are administered and implemented. Ultimately, institutions and all their rules remain the same with little ability to innovate. In any case, the local level remains geographically bound and solutions are rarely shared.

A radical option in the longer term is programmable politics, which we will discuss in a moment. But in the shorter term, an approach achievable as an adjacent possible from where we are today is that of start-up cities.

Start-up cities

Cities are increasingly where humans live. It has been estimated that 2007 was the first year in history when the world's urban population exceeded its rural population. Some cities, such as Tokyo, Shanghai, Delhi, Seoul, and New York, have become megacities with populations larger than those of many countries.

Singapore and Hong Kong are thriving, wealthy metropolises with well-educated populations, high income, and low corruption. Compared to both the United States and Canada, they have better educational outcomes, longer life expectancy, higher incomes, and lower corruption. Both are remarkable not only for their dominance on almost all metrics of human progress but also because they weren't always this way.

Both Singapore and even more so Hong Kong would once have been derogatorily described as backwater fishing villages. When Hong Kong, for example, was ceded to the British in 1841, it comprised a few coastal villages with a total population of under 8,000.

The secret to the success of both city states is the importation of culture, institutions, and capital from Britain. From here, they culturally evolved and recombined with local culture, implemented, and administered at a city-state level. Today, new measures of culture reveal that Hong Kong is culturally halfway between China and Britain. This same success can be difficult to administer at a large, national level. It's easier to manage a small company than a big one. It's easier to manage a city than a geographically large country.

China understood this. When Hong Kong was returned to China in 1997, it served as an engine of development, able to export culture, institutions, and capital to the mainland. China, rather than replicate Hong Kong throughout the country, used the autonomous city state as a model for special economic zones such as Shenzhen and Guanzhou, which were able to achieve similar trajectories of success. All these cities can be described as start-up cities.

Start-up cities bring the Silicon Valley ecosystem to democracy and governance. I remember when Google first offered free food on its campuses. It seemed like such a waste of funders' money. But with

Google's success, free food and other similar strategies that keep people happy at work were soon copied by other companies. Start-up cities take the same approach. Successful firms like Google spread their successful practices to other firms via explicit copying or Googlers moving to or starting new companies. Similarly, start-up cities can learn from one another, seed new cities, and help develop the areas around them. This model of economic development is now actively being used by China in an almost neocolonialist manner in dozens of projects primarily in Africa and Asia. Regions are often controlled through debt.

Start-up cities don't have to be neocolonialist and China should not be the only country using this approach unless we are all happy for China to be the only superpower spreading its influence in this manner. Especially because economic development isn't the whole story. Singapore, Hong Kong, and in particular China are placed lower on freedom indices. Lack of unfettered free speech may help with stability but it curtails the potential of the collective brain to generate radical new innovations, limiting it to incremental innovations.

Start-up cities don't have to be implemented in other countries or even by countries. All a start-up city requires is what any group requires by the laws of life: a large space of the possible based on the laws of energy and innovations in efficiency, strong cooperative mechanisms by the law of cooperation, and free-flowing information to adapt, evolve, innovate, cooperate, and share information in a collective brain connected to other collective brains.

Start-up cities require capital, culture, and institutions. An ecosystem of start-up cities competing and cooperating, much as companies do, can lead to new innovations in governance and faster decision-making. Indeed, start-up cities may be more legitimate when they are created as a collaboration rather than implemented from the top down. A bottom-up, collaborative approach is more in line with the spirit of start-up cities and what distinguishes them from more traditional charter cities or special economic zones.

Further, start-up cities don't have to be created *de novo*, and indeed shouldn't be. A point of failure of US foreign policy in the occupation of Iraq was the purging of Saddam Hussein's Ba'ath Party. It was a failure to understand the culture and need for civil servants with know-how and connections to ensure the real gears that keep society functioning are still there when a new leader emerges. Even terrorists and rebels realize they need to use existing infrastructure when they take over. Thus, start-up cities can take people carrying cultural traits and existing constitutions and institutions as their starting point. Institutions rest on invisible cultural pillars and thus both institutions and norms are essential to success.

When implemented in places where institutions and local norms are more culturally distant, it may be better to borrow from successes nearby which represent a more implementable adjacent possible. A viral tweet from screenwriter Debbie Moon captures how voters can move toward better adjacent possibilities:

> Voting isn't marriage, it's public transport. You're not waiting for 'the one' who's absolutely perfect: you're getting the bus, and if there isn't one to your destination, you don't not travel – you take the one going closest.

In the same manner, there may be no easy path from the Democratic Republic of Congo to Denmark, but the DRC may have something to learn from Rwanda.

A start-up city may be more ideal when it is created as a collaboration between cultures and institutions that bridge to others because of prior cultural history – such as Hong Kong as a bridge to China or Singapore as a bridge to the rest of South East Asia.

Politically, economically, and culturally, cities increasingly dominate and are expected to grow in dominance. Two-thirds of the human population is expected to be living in cities by 2050. By giving cities greater political control, enhancing and formalizing the way in which they can spawn and create alliances of city states, and creating an ecosystem of start-up cities, we can solve the age-old dilemma between the slow decision-making of democracies and faster decision-making of autocracies. We can more quickly escape suboptimal equilibria that we may be trapped in through path dependence. That trap may include liberal democracy itself, at least in its current form.

Today, many successful WEIRD nations have some form of

representative democracy, with wartime powers when decisions have to be made quickly – though it's often a grey area when these emergency powers should and shouldn't be used, creating a second-order speed of decision-making challenge (when to deploy the wartime powers). In contrast, autocracies like China can move a lot faster. But they can move a lot faster in any direction, even the wrong one. The Great Leap Forward was a leap toward death and disaster.

A structured, fast-fail, start-up city environment solves the challenge of evolvability by pooling risk at a lower level and bubbling up successful solutions to a higher level. It is an embodiment of the laws of cooperation and evolution. Setting up diversity and allowing people to vote with their labor and location. Not engineering new solutions to governance, but evolving them.

Before you say wait, Michael, this sounds like a classic tech bro solution – What kind of people would join these cities? What happens if a city fails? – let me say there may be many ways to found a new start-up city. A large company could move headquarters to a less-developed region and become more involved in governance. A group of people could form a community and change the laws of a town. A country could administer a region within another country. A consortium of various groups could come together to revitalize an ignored region. An existing large city could be allowed to expand and merge with nearby cities. The key is creating an environment that tests different laws, constitutions, and missions, competing for people and allowing the best ideas to spread.

Issues with who will join these cities and what happens if they fail are the same issues with business start-ups in the conventional sense that employees face when choosing between an established company and a new start-up. Initially, those early employees who join start-ups will tend to be young and childless. But then those start-ups that grow into large established companies naturally need to attract a wider range of the population. At first, a start-up may offer free food, table tennis tables, and massage credits, but as it grows it needs to also offer childcare and other support for parents. So too with start-up cities. Initial citizens will be younger, full of fire, and risk tolerant. But if the experiment works, the city will grow.

Innovations in democracy have been stagnant for too long. The institutions are cracking under the pressure of an educated, Internet-connected, social-media-consuming, diverse population. Start-up cities are an adjacent possible and achievable way forward. They are a recognition that it is hard to know in advance what works best, but we can arrive there by allowing a thousand flowers to bloom. It's the same principle that makes Silicon Valley so successful.

The secret to Silicon Valley's success is a popular topic with many hypotheses. Was it a founder effect with early microchip companies starting by chance and a community building around them? Was it Stanford and its spin-offs aggressively seeking military funding? Was it the synergy of Stanford, growing capital, growing talent, and companies feeding off each other? The easy mobility created by California banning non-compete agreements in the nineteenth century?

All these factors probably played a part in empowering the valley's collective brain. But what many forget is that Silicon Valley is less a bastion of success and more a graveyard of failure. And that's part of what makes it so innovative. In evolution, there is a delicate balance between diversity and selection; a trade-off between adaptation and adaptability. It's called evolvability. In my commercial work, I am sometimes asked what is the optimal culture of innovation? There are many ways that companies can suppress their innovation potential, but there isn't one single innovative culture. Instead, there are trade-offs in evolvability.

When countries, companies, or people decide to do something different, more often than not they fail. If it was so easy to improve on the average, more people would do it and the average would quickly improve. Instead, large leaps in innovation are rare, hard-won, and hard to predict. Diversity and deviation naturally lead to inequality as the few winners take the large prizes and the losers are left with less than they would have had if they'd followed the crowd. As populations and the marketplace grow in size so the gap grows wider.

On the other hand, without sufficient deviation and diversity there is nothing from which evolution can select. Diversity means deviation from the optimal strategy and there are different ways to do it. One strategy is well captured by Japan.

Japanese culture has created mechanisms such as constant small improvements (*kaizen*) and following the master for a long time before deviating (*shuhari*). These philosophies lead to ever greater refinement in many arts, but rarely large deviations. By following the successful scripts, more of the population can ensure they succeed by not failing, reducing inequality in outcomes and ensuring more of the population does well.

In many Asian societies, where competition for limited university places and jobs is high, obedience and persistence are valued over creativity and cutting your losses – focusing on your weaknesses rather than sharpening your strengths. A safe bet leads to better outcomes on average for more individuals but is unlikely to lead to large break-throughs for the population.

The other strategy, captured by the United States, is to focus on individual freedom or creativity. You be you. A go big or go home strategy over a more cautious risk aversion. Most of the time, at an individual level, the Asian strategy will lead to more success and less inequality of outcomes, but as a population, we may be better off under the American system.

The few successes – the Amazons and Apples – lead to a wealthier country as a whole, even if there is more inequality among its citizens. The question is how best to distribute that wealth. We'll discuss that in the next chapter.

Which of these is a 'better' strategy for your company, country, or start-up city? It depends on many factors.

Not all countries, corporations, start-up cities, or individuals for that matter, can afford to fail. For example, size matters. A large country or company is more able to try high pay-off, high-risk skunk-works than smaller countries or companies, who may be better off sticking to a successful script. Similarly, capital matters.

One of the largest predictors of being an entrepreneur is not over-confidence (believing you're better than others; knowing that among those who believe this, most will fail), nor confidence in confidence (confidence that you're right in thinking you're better than others – what often separates entrepreneurs from 'wantrepreneurs'), nor even good ideas, good education, and good connections, but *rich parents*. You have to be wealthy enough to have a safety net to handle the risk of failure and stretch your creativity. For the same reason, at a population level, both strong bankruptcy laws and social safety nets predict increases in entrepreneurship. People need to know that they can get back up if they fail.

For every Facebook, Google, Apple, and Netflix there are many many more Myspaces, Cuils, Veetles, and other failed companies you've never heard of or have long forgotten. The astounding successes are called unicorns for a reason. For every deified unicorn founder there are many more overconfident entrepreneurs of similar skill who would have been better off taking a salaried job, but we are all better off for the culture that encouraged them to take the risk. Successful institutions were never designed. What we're seeing are the winners left after the process of evolution. We focus on trying to understand the winners and forget all the losers that had similar traits. The religions that never became Christianity or Islam. The companies that never became Apple or Alphabet. The political organizations that never led to the Westminster parliamentary system, Chinese Communist Party, or United States.

Consider that perhaps the Founding Fathers of the United States may not have chanced upon the ideal form of government, but had key mutations that led them to a better form of government than those of the entrenched powers of old-world Europe. But America has become like old-world Europe. Thanks to the stickiness of path dependence, it is difficult today to make the kinds of changes to American governance needed in a rapidly changing world with urgent problems that require new solutions. At least without taking the traditional European approach of a revolution.

In the technological sphere, the start-up model has led to technological marvels. Start-up cities may do the same in the social and political sphere.

Start-up cities may choose their migration policies, they may be run as anarchies or under the multinational umbrella model of 'hiring' citizens. They may be collaborations with not-for-profits, existing governments, or tech companies. But whoever founds the start-up cities and whatever their initial form and established process of change, start-up cities are ultimately still geographically located. This contrasts with modern companies or universities, where people freely move and collaborate and work remotely with little concern for borders. This privileged position leads to great innovation and serves as a model for an entirely new form of political organization for the twenty-first and twenty-second centuries – a radical possibility I like to call *programmable politics*.

Programmable politics

It is doubtful that Elon Musk would have become the richest man in the world if he had remained in South Africa. Musk happened to have a mother born in Regina, Saskatchewan, allowing him to claim Canadian citizenship, which opened a world of opportunities for him to maximize the returns on his talent.

The world is filled with amazing people but many live their lives in nations with less than amazing governments and infrastructure. Our opportunities are often defined by our citizenship. A person with less potential in a well-run country may have a far better life and make far more contributions to our future than a person with much more potential in a poorly run country. To maximize the use of their talent, people move, but our ability to move is hampered by our citizenship.

Passports, for example, remain an often ignored source of inequality. Some passports allow you to move freely between most countries on earth, rarely thinking about visas, and even when you do, treating them as a mere formality. Other passports make travel an expensive, stressful, and sometimes impossible process. In academic conferences I've helped organize, we have worked hard to diversify the participants, but geographic diversity is almost always hampered by amazing researchers being unable to come and present their work in Europe or North America. Passport privilege is something many people with 'good' passports don't fully appreciate.

We live in a world of nation-states, but, for some, borders are effectively open, allowing us to freely and efficiently allocate our talent where it is most valuable. For others, borders are effectively closed no matter how talented we are. It is worth remembering that this world of nation-states is fairly recent, starting around the seventeenth century and becoming the norm around the nineteenth century.

A nation-state is a community of citizens who cooperate through institutions such as government in areas such as defense, roads and other infrastructure, and law-making and law enforcement. At the moment, each generation's starting point is based on the decisions of the past, even by those long dead. As philosopher G. K. Chesterton described it, tradition is the 'democracy of the dead'. But we now have new technologies that may eventually allow us to govern ourselves more flexibly in a way that empowers cultural evolution and cultural-group selection, allows for rapid innovations in politics and governance, removes the threats of corruption and lower-scale cooperation undermining higher scales of cooperation, and overcomes the challenges of sticky group membership and sticky dependent rules: *programmable politics*.

Technology has enhanced so many aspects of our lives. The Internet, social media, and other communication technologies connect us as never before. Grandparents can speak to grandchildren across the world. Each of us can forget so much, knowing that we can just look it up with the magic black box in our pockets. We are all able to connect to talented people from all walks of life eager to share their knowledge and skills with the world on so many platforms both paid and free.

Technology has also empowered cooperation. We can meet and date beyond friends of friends without relying on the inefficiency that is randomly meeting someone at a social gathering. Online reputation tracking has enabled indirect reciprocity to facilitate higher scales of cooperation, allowing us to share cars and houses and use the experience of others to more carefully pick restaurants, products, and services. But one area where technology has had a much more modest impact is governance and politics.

Digital interfaces – some better than others – have replaced people and paper forms in government and public services. And some attempts have also been made to introduce technologies like the Git version control system to enable editing and annotation of legislation, just as a programmer might submit a software patch or an academic collaborator might edit a paper. But fundamentally, democracy and voting, the way in which we decide our laws, the way in which we deliberate and discuss, and the way in which laws are implemented, remain as they were long before even our grandparents' generation.

We are a digital generation still governed by analog systems.

Programmable politics is a way to instantiate all the contracts, rules, and mechanisms for change that make up a nation-state, in a series of programmable polities. The technologies for creating these polities are rapidly emerging in communities such as decentralized autonomous organizations (DAOs). One important technology that may offer the critical breakthrough is blockchain.

You can't go too far online or even in the real world without bumping into someone talking about Bitcoin, Dogecoin, Ethereum, cryptocurrencies, or the many related technologies. The crypto space is a much hyped but nonetheless creative ecosystem of solutions and scams, iterated and feeding off each other with much excitement and capital. But as it stands, blockchain technologies remain a solution looking for a problem. That problem may be the future of the nation-state. Blockchain may be uniquely capable of implementing programmable politics, so it's worth understanding what this technology is and how it emerged.

When money went funny

In 2008 an unknown person or group published a white paper under the name of Satoshi Nakamoto. The title of the paper was 'Bitcoin: A Peer-to-Peer Electronic Cash System'.

It presented a solution to a long-standing challenge within the digital money community. Digital money sees money as fundamentally about tracking debits and credits similar to an account ledger: who gave you money? and who did you give it to? The problem was who could be trusted to keep this ledger.

Satoshi's solution was to trust no one. A trustless ledger to track money as claims on goods and services based on who pays who what amount and without the problem of double spending – spending digital currency twice. Bitcoin added a supply that goes up – new money could be created by solving a difficult computing problem – but which becomes successively harder to acquire over time. It's like mining a gold supply that's slowly running out. A clever trick was that solving this computing puzzle also confirmed transactions on the ledger.

There are many problems with Bitcoin. Even from the perspective

Bitcoin: A Peer-to-Peer Electronic Cash System

Satoshi Nakamoto
satoshin@gmx.com
www.bitcoin.org

Abstract. A purely peer-to-peer version of electronic cash would allow online payments to be sent directly from one party to another without going through a financial institution. Digital signatures provide part of the solution, but the main benefits are lost if a trusted third party is still required to prevent double-spending. We propose a solution to the double-spending problem using a peer-to-peer network. The network timestamps transactions by hashing them into an ongoing chain of hash-based proof-of-work, forming a record that cannot be changed without redoing the proof-of-work. The longest chain not only serves as proof of the sequence of events witnessed, but proof that it came from the largest pool of CPU power. As long as a majority of CPU power is controlled by nodes that are not cooperating to attack the network, they'll generate the longest chain and outpace attackers. The network itself requires minimal structure. Messages are broadcast on a best effort basis, and nodes can leave and rejoin the network at will, accepting the longest proof-of-work chain as proof of what happened while they were gone.

of our theory of everyone, Bitcoin has inflation uncorrelated to growth in the space of the possible, though the relationship to energy consumption for mining is interesting. Nonetheless it was a convincing proof of concept for digital money and a major step forward in removing the power of the state as the sole manager of money.

Who exactly Satoshi Nakamoto is remains a mystery. We don't even know if they are a single person or a group. Their idea was implemented in open-source software in 2009. Nakamoto mined the first Bitcoin block – known as the Bitcoin Genesis Block – with a reward of 50 Bitcoin. They continued mining blocks and then, in 2010, handed over the control of the open-source software, disappearing never to be heard from again.

I first heard about Bitcoin in 2011. It had recently reached parity with the US dollar. Back then you could get free Bitcoins from 'faucets'. I collected these free, seemingly worthless digital coins and left my computer mining coins when it wasn't being used. After a while I lost interest and, to the best of my knowledge, deleted my wallet when clearing up hard-drive space – a decision that still crosses my mind.

Bitcoin was interesting to me as an engineering problem, but Nakamoto's motives were even more interesting. When a block is mined, information can be stored in the block forever, each new block reliant on the previous chain. In the Genesis Block Nakamoto left the message, 'The Times 03/Jan/2009 Chancellor on brink of second bailout for banks' – a headline from Britain's *Times* newspaper reporting another bailout due to the global financial crisis of 2008.

In a post introducing their software, Nakamoto later explained themselves:

> The root problem with conventional currency is all the trust that's required to make it work. The central bank must be trusted not to debase the currency, but the history of fiat currencies is full of breaches of that trust. Banks must be trusted to hold our money and transfer it electronically, but they lend it out in waves of credit bubbles with barely a fraction in reserve.

They were referring to fractional reserve banking, the requirements of banks to have a minimum percentage of money held in reserve when lending. The gap between what they hold and what they lend is where new money is created in the conventional system.

Nakamoto wanted to bring what encryption did – allowing anyone to encrypt anything and thereby control privacy – to money. Money without a gatekeeper. But blockchain technologies allow for far more than currencies. They allow us to create DAOs where the constitution and laws are fully programmed and automatically instantiated.

A contract, for example, requires us to trust the other party to hold up their end of the deal – buying a house, paying your salary, paying out your insurance. If the other party reneges then it requires going to court, which requires trust in the government and judicial system to impartially step in and enforce the contract. In contrast, the programmable political solution is smart contracts. Here the contract is agreed by both parties, instantiated as code, and automatically run when pre-agreed conditions are met. The agreement can be changed as long as both parties agree, but it does not require trust in either the other party or in the impartiality and fairness of the enforcer. The agreement and its enforcement are one and the same, embodied in code.

Currencies are currently tied to central banks. Buying a currency and using it in a particular country is in some sense an investment in that country, subject to its laws and taxes. The relative buying and selling of different currencies affects the exchange rate and ultimately the value of a country's goods and services.

But imagine a world in which currencies are not tied to countries or controlled by central banks. Imagine a world in which different laws and taxes are programmatically embedded in different programmed polities. A world in which contracts are entered and automatically enforced. A world where you can participate in multiple polities and coalitions of polities, much as a lucky few have multiple 'good' citizenships. Your passport or citizenship will no longer restrict you.

In this world the equivalent of nation-states, states, regions, cities, and polities and alliances between them are effectively programmed into an ecosystem where they compete with one another. In one programmed polity one might allocate 1% of every transaction to hospitals. Another might fork off – creating an identical copy – but now including funding for schools, firefighters, and police forces who use the currency. Another might include basic income and only charge taxes for large transactions or allocate money to those under some threshold who use the currency. Another might scrap some of these and instead allocate funds to environmental causes and allow users to vote for which causes are supported. Still another might fork and implement *quadratic voting* – rather than voting for people or voting for all issues, people are allocated votes that they can store up for the things they care about most, as a way to express how much they care about something.

Programmable politics allows for forks, debates, and deliberation over issues large and small or simply over details such as proof of work or proof of stake or how transactions are verified. And programmable polities may work together just as NATO, the European Union, or US states work together. Cryptocurrencies may be exchangeable at fluctuating rates based on demand and usage, just as dollars, pounds, or euros are today.

Some of these ideas are based on real cryptocurrency and DAO projects, but the possibilities for programmable politics are endless. Yet all are complicated and, for many, irrelevant to their lived experience of buying groceries, getting paid, sending their kids to school, or going on vacation. To fulfil the vision of programmable politics requires infrastructure, like DAOs, smart contracts, digital constitutions, and payment platforms, all of which are being actively developed right now. Eventually paying with different cryptocurrencies may be as seamless as using your credit card to pay in foreign currencies; belonging to different programmed polities may be as seamless as buying items from different websites.

In experiments with different virtual political systems playing a public goods game, people will sometimes choose a game where there is no punishment – no peer punishment or institutional punishment. An anarchy of sorts. But players quickly realize that this institution is unable to sustain the public goods provision and so when they are allowed to migrate, they choose to move to punishment regimes. At least in WEIRD countries, an institutional punisher, where a person is selected to extract taxes and punish freeloaders, is preferred. In this world there may be very little actual punishment. The credible threat of punishment is sufficient to ensure people cooperate. This happens very quickly in a laboratory experimental setting. And indeed, the same cultural-group dynamics also happens in the real world as countries grow, shrink, succeed, and fail on the back of their economic policies and political systems, although this process is hampered by corrupt leaders holding their citizens hostage to lack of food or health care until foreign aid or a loan is provided. In the real world it is slow and change is difficult even when the 'right answer' is known.

In a world of programmable politics, policies and currencies compete, prices emerge, and exchange rates are set based on which currencies people choose to use. Taxes are automatically extracted without the need for a fallible tax officer or appointed Leviathan. Voting and identity can be stored within this ecosystem and shared between cooperating communities. We will belong to multiple overlapping and embedded communities that may freely coalesce into the equivalent of large nations as they arrive at and agree on the most effective set of rules and institutions, or freely fork into small communities experimenting with different approaches, as a start-up city might.

With the full history of programmable politics available, AI can also help us evaluate and learn from the network to decide what changes to make and where to go next.

Programmable politics will eventually remove the need for liberal democracies and nation-states as we know them, creating a fairer world. Governance in the twenty-first century will not be designed; it will evolve.

This vision creates a radical new way for the law of evolution to explore new ways to optimize the law of cooperation less hindered by path dependence. Ultimately, that cooperation will empower our collective brains, making the laws of energy and innovation more efficient. But to really understand this vision, we must first understand the relationship between politics, money, energy, goods, and services and inequality in all of these.

9

Shattering the Glass Ceiling

Inequality is a waste of human potential that harms us all. It prevents people from contributing all that they can to our collective brain. Entrenched inequality makes us less innovative and less able to break through to the next level of energy abundance. As evolutionary biologist Stephen Jay Gould so eloquently put it, 'I am, somehow, less interested in the weight and convolutions of Einstein's brain than in the near certainty that people of equal talent have lived and died in cotton fields and sweatshops.'

Inequality is particularly pernicious when it persists over generations, eventually leading to an almost impossible to eradicate, permanent elite. In this chapter we'll focus on exactly what the problem with inequality is and the changes we can make to permanently create a fairer world with abundance and opportunity for more people.

In our unequal world in which wealth is transmitted intergenerationally, the happenstance of birth puts a ceiling on what we can reasonably achieve. In such a world, ideals like the American Dream have become a fantasy. People live in entirely different worlds with vast differences in wealth, networks, access to education, and acquisition of the constellation of cultural traits that lead to success. We are running different mental software, see the world in entirely different ways, and have different experiences.

It's not just that some people start life on third base and never look back to notice that others are hustling with all they've got from first, but that some people are playing entirely different sports in entirely different leagues with entirely different trophies available to them. And because we live in different worlds, we don't see enough of the other side to appreciate just how large these differences are.

The average American has a median wealth of around $120,000

(about £100,000). The figure is $190,000 (£160,000) if you're white, $250,000 (£210,000) if you own a house or are over sixty-five years old; $300,000 (£250,000) if you have a college degree. And so to the average American who has a lot less than one million dollars, the difference between a multimillionaire and a billionaire can be difficult to appreciate.

The difference is staggering.

To see how large it really is, let's convert money to something we have a better sense of: time.

If dollars were seconds, the difference between a person with a net worth of $10 million and someone worth $1 billion is four months versus thirty-two years.

January to April versus 2020 to 2052.

Ten million dollars is roughly what it takes to be in the top 1% in America. But the top 1% often don't feel wealthy, because they have their own 1%, which requires a net worth of around $400 million. And the top 0.01% don't feel all that wealthy, because they have their own 1%, and so on upwards.

Elon Musk, Jeff Bezos, and possibly Vladimir Putin have a net worth that may be north of $200 billion. Convert that to time and we're talking about over three millennia.

From when the Israelites arrived in the Promised Land to the present day.

And that's wealth that's well tracked. The Saudi royal family are estimated to have a net worth of around $1.5 trillion. That's a family whose wealth represents roughly the entire GDP of wealthy countries like Australia, South Korea, or Canada. Or about 45,000 years in time – when Aboriginals first arrived in Australia to the present day. All of that wealth in just one family.

Even converting to time, these are unfathomable levels of wealth. Imagine that I offered you a job that paid $1 million every . . . single . . . day.

It would still take you over five *centuries* to reach $200 billion. Most people aren't earning $1 million a year, let alone $1 million a day.

Consider the proposed US minimum wage of $15 an hour. Imagine you were one of the first members of our species born around 250,000 years ago. Imagine that for some reason you were immortal. And let's

SHATTERING THE GLASS CEILING

say you were a hard-working immortal who didn't sleep and instead worked twenty-four hours a day, every day of every year, up to the present day. Today, in this unrealistic hypothetical, you would have around $33 billion, still only around *half* the net worth of Charles Koch of the Koch brothers. How long do you think it will take you to get to $200 billion? The answer is well over a *million* more years. More time than humans have been around, and you've been working your exhausting all-day, all-night job for $15 an hour.

If you're wealthy, it may be difficult to imagine what it means for $10 to be a lot of money. If you're poor, it may be difficult to imagine what it means for $200,000 to not be that much money.

Many people argue that such wealth differences shouldn't exist. That billionaires should be banned.

I disagree.

There is value in there being a strong relationship between a person's wealth and their contribution to society. With a larger population and the power of technology, it is possible for individuals to make that large a contribution.

Technological innovations multiplied across our large populations have allowed for the legitimate, innovation-based accumulation of vast wealth. If you can sell widgets at a profit of $10 each to every Austrian, you would make about $90 million. If you can sell those same widgets to every American, you would make $3.3 billion. But not everyone got rich selling widgets. *How* you become wealthy matters to human progress and reaching the next energy level.

Some billionaires may have earned their billions by actually expanding the space of the possible. When the acquisition of wealth comes from innovation in energy or efficiency, it represents an expansion of the actual pool of wealth – the space of the possible. This is wealth that is in some sense 'created'. And the billionaires rewarded for these innovations are taking a fraction of the new space of the possible that they have made. Their innovations improve life overall even if those put out of business or 'creatively destroyed', as economist Joseph Schumpeter described it, are less well off. Similarly, those who bet on the success of the successful innovator through investment in their companies also get a share of that wealth. This is *wealth creation*.

In contrast, when acquisition of wealth comes from what economists call rent-seeking – wealth acquired without a contribution to productivity – then it is simply taking a piece of someone else's pie without contributing to human activity. It is not expanding the space of the possible, but merely controlling some of that space and charging a rent for that control. And that kind of wealth harms our ability to innovate. Here, levels of wealth are not matched by contributions to society.

In contrast to wealth creation, this is *wealth appropriation*.

Wealth creation and wealth appropriation are often conflated, but they must not be confused. They have opposite effects on our society.

We think of money as something we are given based on the skilled use of our time for a particular job. And we are free to spend that money as we wish. But what we choose to consume affects what is produced. And people who have more money have more power over production. More power over how our energy budget is allocated in terms of goods produced and services rendered.

Imagine a person who has a magic money printer that produces as many counterfeit bills as they want. Let's call him Benjamin. The magic of Benjamin's printer is that those bills are indistinguishable from real bills. Benjamin can buy whatever he wants and what Benjamin loves are donuts. But Benjamin's not a bad person. In fact, he's altruistic. Thankful for the blessing of his magic money printer, he feels the urge to give money away to causes he feels matter the most: eczema research and hair loss solutions. It might seem as if Benjamin's philanthropy and injection of cash into the local economy is a good thing. It's not like he's stealing from other people working hard for their pay checks, right? In reality, he is.

Benjamin's money really represents control over what others produce and ultimately how we allocate our energy budget. Benjamin's love of donuts leads to a bespoke, gourmet donut industry with a lot of research into the best possible donuts at exorbitant prices. Maybe Benjamin also likes super-yachts and the best ocean-front real estate. Energy is devoted to producing and running these super-yachts and much of the best ocean-front real estate is now Benjamin's. And because Benjamin's philanthropy dwarfs all others', eczema and hair loss research end up as the best-funded research, affecting what our brightest minds end up working on. Benjamin's magic money printer has allowed him to steal opportunities from others working hard in their jobs, oblivious to how Benjamin acquired his wealth.

The point of this story is that consumption determines production and so who can consume more matters. It matters for whether we allocate our energy budget in ways that expand the space of the possible to make all of us better off and ensure our species' future, or whether we simply crowd out the ability for most people to buy a reasonable house, work reasonable hours, or go on nice vacations.

Wealth creation

The ability of Elon Musk, Jeff Bezos, Bill Gates, Larry Page, Sergey Brin, and other entrepreneurs to acquire vast wealth is often the result of the law of innovation in efficiency. Specifically, technological changes that have allowed Musk to build electric cars (Tesla) and put satellites and astronauts into space with cheaper rockets (SpaceX); Bezos to more efficiently host Web applications (Amazon Web Services) and hollow out high streets and shopping malls with vertically aligned, more efficient commerce that consumers seem to prefer (Amazon); Gates to create software with low marginal costs and high marginal profits, because once software is written, it costs much less to distribute (Microsoft); or Page and Brin to more efficiently give us access to the world's trove of knowledge (Google and Alphabet).

Such efficiency gains have, either directly or indirectly, made us all better off. And the wealth acquired by the above individuals and the investors and stakeholders in their companies is the result of a system that allows people who have made good bets in society – bets that have done more with less or created something new – to have greater control over our vast energy budgets.

They are given this control to make bigger bets on the assumption that they continue to make good bets. And if they don't, they lose money. While some may rail against capitalism, the alternatives, such as centralized planning, are far worse. Capitalism done well gives control over concentrated created wealth to those making good bets as innovators or investors. It leaves us all better off.

Concentration of wealth is required for any long-lasting, large-scale project – many architectural wonders, great works of art, and break-throughs in science have all been the result of concentrated wealth being put to good use. But in the past capitalism was even less fair than it is today, and those who wielded that wealth also wasted it. You see rent-seeking from workers or slaves, opulence for personal benefit, and wars of dominance to take from other groups in a zero-sum transaction.

The evolution of capitalism and the marketplace in its current form, where we can freely allocate our skilled labor and money to firms we choose at a cost the market is willing to pay, leads to concentrations of wealth. The system is far from perfectly meritocratic with a fair playing field and has many distortions in how much money people make for what they create, but even with these imperfections, the current system does a reasonable job of rewarding success and allowing the successful to make further bets.

Today it means Musk can leverage his past success at PayPal to make large bets on space exploration, energy technologies, or a social media company, and Gates can leverage his past success at Microsoft to make large bets on health care and development. And sometimes, especially when aligned with profit-seeking, this can be more efficient than governments as an alternative mechanism for concentrated collective action. At the very least, it exists as a complement.

But even wealth creation did not happen in a fair system. Many who would have chosen to invest in Microsoft or Tesla simply could not because they didn't have the capital to invest. They weren't born in the right country, at the right time, to the right families.

How you get rich matters.

Wealth appropriation

Wealth appropriation involves acquiring wealth without making a corresponding contribution to society. Here, control over our energy budgets is in the hands of those who don't make our societies better off but instead control a resource and charge for its use with little improvement or work on their part, using it entirely for their own benefit. Examples include literal landowning rent-seeking, multigen-erational wealth transfers, and corruption.

Britain is one of the last surviving European countries with vast wealth concentrations via wealth appropriation rather than wealth creation. Perhaps this is thanks to the efficient private school pipeline to powerful positions in British society. But whatever the reason, it cripples Britain's innovative capacity and forces it to rely on financial transfers, often from illegitimate sources such as money laundering and property purchases by Russian oligarchs. Because of its wealth inequality, Britain relies on these outside injections of capital rather than on innovation. Wealth inequality is notoriously hard to measure, but long-standing inequality in land offers a proxy.

Half the land in Britain is owned by just 25,000 people. Twenty-five thousand people is less than *half a percent* of the UK's population. So less than 1% of UK citizens own half of the UK. These are often multigenerational wealthy landowners, royals, and aristocrats, who might be starkly contrasted to the 20% of the UK population (in 2021, amounting to 13.4 million) who live below the poverty line, or even to what we might call an expanded productive working class, not in the traditional sense, but the working class of those who have to work for an income to live.

Browse Britain's most popular property website, Rightmove, and you will find listings such as a nine-bedroom, eight-bathroom mansion in Marble Arch, central London, the details of which I was asked to remove for legal reasons. The property is listed among those sold as leasehold. A leasehold property is a property where the owner doesn't actually fully own the site, and must instead renew their lease by paying a fee to the actual owner – the person who owns the freehold. The time horizon is usually very long, something like ninety-nine years, which means little to most Brits but a lot to those with centuries-old multigenerational wealth. In a leasehold, ownership is illusory.

In the case of this particular mansion, rather than a £10,000,000 fee charged every century, £100,000 is charged annually. What is this fee for? It's for nothing other than the fact that the owner controls the freehold. It is not an exchange of money for any added value. The freeholder does nothing other than freeload, collecting their regular fee. And as the advert states, 'The freehold is not for sale.' Why would it be?

How did the freeholder get this property in the first place?

In this case, the owner is Portman Estates. Portman Estates began in the sixteenth century, when Henry VIII – yes, the king with all those wives he didn't like – gave his Lord Chief Justice, William Portman, over 100 acres of land in central London. The Portmans held on to their property for four centuries – four hundred years – until inheritance taxes were raised during the Second World War.

Research shows that wealth is often resilient to shocks and revolutions, but inheritance taxes, when actually implemented, as they were at the end of the Second World War, can weaken multigenerational wealth transfers.

William Portman's descendant, the 7th Viscount Portman, was subject to high inheritance taxes (75%) when he died, in 1948. But wealth offers ways to shield wealth, which is why it's so difficult to tax the ultra wealthy.

A deal was struck with developers to hold much of the estate in a trust for the benefit of the family. Although much was lost, much was also retained. The benefit of the high estate tax, however, was that it forced the aristocracy and other wealthy landowners to transition toward more productive businesses – some wealth creation – in addition to wealth appropriation.

Not far from the Portman Estate is Grosvenor Estate. In 2016 Gerald Grosvenor, the 6th Duke of Westminster, died. By 2016 inheritance taxes were much weaker than they had been in 1948, when William Portman died. By 2016 methods for shielding wealth had also become much more effective. This allowed Gerald's twenty-five-year-old son, Hugh Grosvenor, 7th Duke of Westminster, godfather to Prince George, to access the £9.3 billion fortune held in trusts and businesses.

The Grosvenors had successfully transitioned to property development. Gerald was an introspective man, often publicly musing over the position in which he found himself as the inheritor of a vast fortune and business empire. A *Financial Times* reporter once asked him what advice he had for young entrepreneurs. His candid answer was: 'Make sure they have an ancestor who was a very close friend of William the Conqueror.' Grosvenor was referring to his Norman French ancestor Gilbert le Grosveneur, who caught a boat to England to fight alongside William the Conqueror in 1066. About a thousand years ago.

Contrary to the saying that the first generation makes the wealth, the second generation preserves it, and the third generation loses it, measures of social mobility in Britain using rare surnames reveals that it takes approximately ten generations, or another three centuries, for the descendants of the elite to become average, and that there's huge variation. Research reveals that wealth is highly robust to revolutions and wars, which may perhaps be seen as inter-elite competition – elites selectively amplifying and backing different middle- and lower-class revolutionaries, effectively trying to invest in the winners or sometimes both sides.

It's very difficult to completely lose large amounts of money. Money grows and compounds. Remember, 'Money makes money. And the money that money makes, makes money.' Some elites lose in the inter-elite competition, but many stay on, and in the meantime innovation, growth, human welfare, and potential are all harmed.

No one chooses their ancestors and we can't blame aristocrats and those born into old money for the position in which they find themselves, any more than we can blame those born into poverty. We can't blame those who invest in property as landlords if that's the system they find themselves in and that's where the returns are. In any case, it's not about people, it's about allocation of their resources. People aren't purely wealth creators or appropriators, they're typically a mix. So we shouldn't hate the player, we should hate the game. It's a systems-level problem and the solution too needs to be at a systems level to disincentivize or remove the ability to appropriate wealth. But norms matter too, so the first step is recognizing the difference between wealth creation and wealth appropriation.

There are many dubious ways for people to preserve their wealth, including tax avoidance and wealth preservation vehicles like trusts. But quite apart from the many clever ways the wealthy and their accountants have found to protect their money, being born into wealth also offers social connections – a powerful cooperative group of people you know. These connections persist and open opportunities available only to a few. This is another reason that wealth weathers even popular uprisings. For example, analyses of surnames in China starting from after the Communist Revolution of 1949 – a revolution that deliberately targeted elites – reveals that elite surnames are still over-represented among China's elite today. Similar findings can be seen across the world, even in countries seen as socialist and equal, like Sweden.

These factors are not simply the result of choices made by individuals, but reflect the entrenched systems of power in which they exist. Weak inheritance taxes inevitably lead to persistent class divisions, shuffled only by the occasional war, revolution, or shock, and then only imperfectly. This is the state of established societies like Britain as well as many developing countries, and will be the future of relatively new countries with weak or weakening inheritance taxes, such as Australia and the United States, if they don't take action. Both have weak or weakening mechanisms for minimizing intergenerational transfers of unearned wealth, and so too risk a slow and inexorable march toward oligarchy, the appropriation of wealth by a small elite, and a weakened ability to innovate.

Why is this a problem?

It is a problem because consumption determines production and, like Benjamin with his magic money printer, leads to misallocation of our energy budgets from fuel for private jets and super-yachts to control over the media and other means of shaping what most people think, what they believe, and how they act. It's not an accident that Rupert Murdoch, owner of large, influential media companies, including Fox News, is rarely mentioned in the news, or that new billionaires like Jeff Bezos have bought news media, in Bezos's case, the *Washington Post*.

The unchecked dominance of wealth appropriation over wealth creation marches us from meritocracy to mediocracy.

Rule by the best?

The etymology of the word 'aristocracy' is 'rule by the best', which may have been accurate when education and other opportunities were only available to a small section of society; when aristocrats were indeed the best educated and had the ability to make the best decisions when they were the only people with access to the knowledge and information that wealth brought with it. But it was not selection of

How YOU CHOOSE to ACQUIRE, INVEST, & SPEND $ MONEY BENEFITS OR HINDERS OUR FUTURE

START

WHERE IS YOUR WEALTH FROM?

HOW ARE YOU BUILDING WEALTH?

- INNOVATION
- SHAREHOLDING
- SALARIED JOB
- ENTREPRENEURSHIP & BUSINESS INVESTMENT

- PROPERTY LEASING
- MULTIGENERATIONAL WEALTH TRANSFERS
- LANDOWNING
- NEPOTISM, CRONYISM, BRIBERY, etc

THIS IS WEALTH CREATION

THINK INNOVATION & JOB CREATION

EXPANDING → THE SPACE of the POSSIBLE

through INNOVATION OR IMPROVED EFFICIENCY (doing more with less)

WHICH DIRECTLY — OR — INDIRECTLY IMPROVES our WEALTH & QUALITY of LIFE

THIS IS WEALTH APPROPRIATION

THINK MONOPOLY OR THE Mafia

CONTROLLING a RESOURCE (eg. land, natural resources, access to jobs & contracts) & CHARGING for its use with little improvement or work on your part

REDUCES INNOVATION & PRODUCTIVITY

IMPEDES the EXPANSION of the SPACE of the POSSIBLE

With GREAT WEALTH POWER comes GREATer RESPONSIBILITY

+ INFLUENCE over our ENERGY BUDGET

WHAT DO YOU CHOOSE?

CONTRIBUTES TO & GROWS SOCIETY'S ENERGY BUDGET

INEFFICIENT USE of SOCIETY'S ENERGY BUDGET

INCENTIVIZE THIS BY IMPLEMENTING

HENRY GEORGE'S LAND VALUE TAX

& OTHER TAXES ON UNPRODUCTIVE MONEY

the best based on potential. There were no doubt many non-aristocrats who would have achieved more with the same privilege. And sometimes wealth, power, and privilege mask mediocrity or less.

Donald Trump is an example of a trust fund inheritor losing vast amounts of inherited wealth, severely underperforming the average of the stock market, but still using that impossible-to-completely-lose inherited wealth and connections to forge a powerful and influential career. The Koch brothers and the Koch network of influence (dubbed the Kochtopus) are a similar story, having shaped American society through the strategic investment of money in an ecosystem of influence through universities, think tanks, and politicians.

As a society, we need to ask ourselves whether these inheritors of vast wealth are best placed to decide how we use our still-vast energy budgets? And what might happen if others had the same opportunities.

One of the major predictors of becoming an entrepreneur in America is having wealthy parents and receiving a large inheritance or cash transfer. A child born into the top 1% income bracket is ten times more likely to become an inventor than someone from a below-median-income family. Indeed, a top-achieving math student in the top quarter from a lower-income family is *less* likely to graduate than a low-achieving math student in the bottom quarter from a wealthy family.

Being born into wealth is (a) a social safety net that allows a young person to take risks, (b) offers access to high-quality education, information, and cultural knowledge, and (c) means having connections to turn ideas into reality.

We can see the effect of these multipliers even in the outlying, outsized success of Bezos, Musk, and Gates. But we must be careful: people's stories are complicated.

Bezos was born to teen parents who divorced not long afterwards. He was raised by his mother and Cuban immigrant stepfather. Some amount of financial security was provided by his grandparents, who owned a large Texas ranch, and his self-made stepfather, who eventually had a well-paying job at Exxon. By the time Jeff was thirty-one years old and ready to take the leap on a hunch that the potential for commerce on the Internet was under-realized, his parents had enough savings to invest $250,000 in Amazon. This was in 1995, so about half a million – the average price of an American house bought outright – in today's money. Where would Bezos and Amazon be without that safety net, capital, and education?

Musk's background is the subject of much speculation and limited information. His father seems to have been reasonably wealthy if not comfortable, allegedly owning shares in an emerald mine (though both Elon and his mother Maye dispute this). Musk attended expensive private schools in South Africa, so may have had some early advantages. His advantages later in life are less clear, estranged from his father, raised by a dedicated mother, and decidedly making his own way in the world. If advantages existed, they may have been through some in-theory possible level of safety net his family provided.

Gates, in contrast, was indisputably born to a wealthy family. His father was an attorney, his mother a teacher and businesswoman who famously sat on a board with the CEO of IBM, opening doors for her talented son. Gates went to an elite private school with access to computers at a time when few others did.

This is not in any way to denigrate the achievements of these three entrepreneurs. All are brilliant, driven, and hard-working, with many of the traits in the constellation of attributes it takes to even have a shot at such wild success. Yet while they may freely admit the role of their upbringing, it's easy to forget that so many with the same advantages achieved so much less. And it's just as easy to forget that so many with the same potential don't have the same advantages.

The great shame is not Bezos, Musk, and Gates, but all those with the same potential who didn't have that necessary capital and safety net, the necessary guidance and cultural input, and vital networks, connections, mentorship, and resources to turn ideas into reality. Houston's impoverished Hispanic community near where Bezos grew up; those without stable employment, many of whom are Black South Africans, near Musk's hometown of Pretoria; the unemployed of south-east Seattle, south of Gates's home in the north-east – *that* is what we must fix.

It is not that capitalism and commerce must be eradicated – there exists no better model for setting prices and maximizing human potential. But capitalism must be made more fair. Fairness is a laudable goal in its own right, but fairer capitalism is also more efficient. Inefficient allocation of capital and wealth leads to inefficient prices and inefficient progress.

Ultimately, to fix capitalism we must create a fairer, more level playing field for each generation. In practice, that fix might involve ensuring fair taxes during a lifetime, but even more important is ensuring that the control over our economies, energy, and future that wealth affords isn't passed on from generation to generation of a small elite. By preventing generational transmission, many of the problems we currently face in entrenched wealth differences by class or race will disappear permanently within a few generations. But unlike many forms of redistribution that are economically inefficient proximate solutions like the introduction of Australia's cane toads, there are ways to make capitalism fairer while also making it *more* economically efficient.

A fairer game

Slavery, wars of conquest, and explicit exploitation are all forms of wealth appropriation. Before we created what Buckminster Fuller called *energy slaves* – putting energy to work for us – it was common for some to put actual enslaved humans to work for them. Slavery coercively channeled the energy of the enslaved toward slave owners. Like other forms of wealth appropriation, and quite apart from the immorality of such an action, taking from others the value of their work and even their freedom does not expand the space of the possible. Innovations may accidentally emerge, but this is neither a fair nor efficient collective brain. Moreover, it incentivizes zero-sum competition because little new wealth is created.

Meritocracies create a culture of achievement, opportunity, and education that aids progress for more. The control of power by a hereditary elite creates a culture of networking in small cooperative groups to outcompete and control the many. In this more unequal and less socially mobile world, people increasingly rely on the wealth and resources of their parents and family rather than their own labor. And so for obvious reasons, in this world, inheritance taxes are often unpopular. It's a sad self-reinforcing loop. Yet inheritance taxes are an important way to break the loop, not because of the few thousand or even millions you may or may not get after the sale of your grandmother's house, but because they determine what happens to the vast accumulated wealth of people like Bezos, Musk, and Gates when they die. Are their children really best placed to have that much control over the vast portion of our energy budgets that their parents' wealth represents?

The answer is that if they were, they would similarly and independently achieve levels of wealth as their parents did without that inheritance. John Stuart Mill, arguing for lifetime limits on inheritance, described it well when he wrote:

> I see nothing objectionable in fixing a limit to what anyone may acquire by mere favor of others, without any exercise of his faculties, and in requiring that if he desires any further accession of fortune, he shall work for it.

Inheritance taxes instinctively go against our primal urge to protect, prepare, and provide for our children, to give them more than we had. They lead to fears of our children missing out on what we have worked so hard to earn or that we will deprive them of opportunities in a harsher world to come.

But, when we level the playing field, the children of almost everyone except the children of the very richest billionaires will be better off than they are now. And the children of those very rich will hardly be poorer. We all benefit by living in a society where more people can contribute to making us all wealthier. To understand why leveling the playing field is essential, imagine if we inherited debt in the same way we do wealth.

If debt were inherited over generations, it's easy to see how it would quickly leave some trapped in permanent multigenerational poverty, unable to escape no matter how hard they worked or how high an income they managed to secure. The interest on debt would compound over generations and eventually no amount of work would be enough for it to be paid off. Being trapped in debt is something many can understand. But the reverse – inherited wealth – has exactly the same effect on society.

A few people and their descendants, like Benjamin and his magic money printer, have a distortionary effect and control over our society and our energy budgets, reducing the space of the possible, our ability to innovate, and ultimately preventing us from cracking the next energy level that leaves everyone better off. In turn, we gradually enter a new feudalism.

In 2013 French economist Thomas Picketty published *Capital in the Twenty-first Century*. Over 696 pages, he carefully provided table after table and graph after graph of empirical evidence showing that inequality was a natural consequence of capitalism. Notably, when the rate of return on capital exceeds the rate of economic growth, those with capital (the wealthy) will grow richer at a faster rate than those earning an income. To everyone's surprise and no doubt his too, the book became a run-away success, a *New York Times* bestseller.

Through the lens of the laws of life, we can intuitively understand Picketty's discovery. When the space of the possible grows through energy or efficiency, new wealth is created. This is when the economy grows. When this space is growing, the relative share of those who already have wealth is smaller than if their wealth grows but the space stays the same size. It's like building more space in your house by building upward or downward. You can conceivably have a large house and so can your neighbor. But if we build an extension on limited shared land, we leave less space for our neighbors to also extend or even build their house.

Amazon and the washing machine have given you back more of your time. Microsoft and Google's software and hardware have led to more efficient working. The inventors of insulin and the cervical cancer vaccine have extended our lives. These are all efficiencies and those associated with them are deserving of some portion of the increased space of the possible that they helped create. So too when new energy technologies are developed.

If the space of the possible is fixed then the richer get even richer and take ever more of that limited space. The poorer have nowhere to go and become more and more squeezed. They scramble over scraps. Mobility falls and inequality grows.

But if the space is grown, either by energy or by innovations in efficiency, then the actual space has grown and whoever grew the space can keep some of this newly created space in the form of money. The rest is redistributed through either direct benefits such as taxes or philanthropy. Or indirectly, because the innovation can be used by others in their own innovations and in their own work.

All of this becomes more difficult in a world of diminishing EROI. This is what we are seeing today.

Inequality makes innovation less efficient because those with the talent to improve society don't have the opportunities to do so. In energy and economic terms, while the nominal value (number value) of money goes up, the real value (what you can do with the money) is decreasing because less true wealth is being created. The space is no longer growing as it once did and this in turn is moving us from a positive-sum win-win world to a zero-sum win-lose world with consequent effects on cooperation. And this in turn is creating vicious feedback loops.

We are not selecting the best in each generation. We are not ensuring the best people are at the helm. And, lacking the ability to generate new wealth, inheritors of vast fortunes instead spend their money on finding ways to entrench their positions in society, switching from wealth creation to more wealth appropriation – the reverse of what happened to Britain's aristocracy when estate taxes were raised.

The control of unearned intergenerational wealth transfers is *critical* to the health of a society and its continued ability to innovate. We must level the playing field to create a fair game for each generation. We must tax the dead for the sake of the living.

Taxes, taxes, taxes

The World Economic Forum in Davos, Switzerland, doesn't usually go viral, but it did in 2019 when a young Dutch historian, Rutger Bregman, got everyone talking about, of all things, taxes. The Dutch are known for their directness, and Bregman did not disappoint.

This is my first time at Davos, and I find it quite a bewildering experience, to be honest . . . 1,500 private jets have flown in here to hear Sir David Attenborough speak about, you know, how we're wrecking the planet . . . I hear people talking about the language of participation and justice and equality and transparency, but then almost no one raises the real issue of tax avoidance, right? And of the rich just not paying their fair share . . . Just stop talking about philanthropy, and start talking about taxes . . . just two days ago there was a billionaire in here, what's his name? Michael Dell. And he asked a question like, name me one country where a top marginal tax rate of 70% has actually worked? And, you know, I'm a historian – the United States, that's where it actually worked, in the 1950s during Republican President Eisenhower, you know, the war veteran. The top marginal tax rate in the US was 91% for people like Michael Dell. You know, the top estate tax for people like Michael Dell was more than 70% . . . this is not rocket science . . . We can invite Bono once more. But, come on, we've got to be talking about taxes. That's it. Taxes, taxes, taxes. All the rest is bullshit in my opinion.

It was an unscripted, somewhat rambling rant, but it hit a chord and quickly went viral. They say nothing in life is certain except death and taxes. These days, only death is certain.

Bregman is simplifying the challenge. It is notoriously difficult to tax the ultra-wealthy because, as we have already discussed, the ultra-wealthy have clever accountants and more money to invest in tax avoidance. Even the high taxes in the middle of the twentieth century were avoided or weakened in a variety of ways. The wealthy found ways to evade them through loopholes and elaborate structuring. For example, private foundations were a compromise whereby the wealthy could avoid inheritance taxes in exchange for giving away a small percentage of their wealth to the public good. Many philanthropic foundations exist under this model. These foundations have done a lot of good and advanced society in a variety of ways. But they also represent a powerful ability to direct our current energy budgets. Consumption determines production. The priorities of foundations and their vast wealth affect the priorities of scientists, researchers, and politicians, directly or indirectly.

Foundations are in theory limited in how much lobbying they can engage in, but in practice, some lobby in all but name. For example, the selective amplification of academics and think tanks exists in an ecosystem of influence over government and public institutions, including the IRS. This power has been used to change the laws themselves, for example, weakening inheritance taxes (rebranded as 'death taxes' in the US after focus groups revealed that even average people were more likely to reject 'death taxes' than 'inheritance taxes') to further reduce tax obligations, further increasing inequality.

Trusts exist to evade personal taxes. In a famous exchange with a heckler, US presidential candidate Mitt Romney famously said, 'corporations are people, my friend'. Indeed, corporations are 'legal persons' and, through trusts, people can become a kind of corporation, effectively becoming immortal, their wishes and wealth existing across multiple lifetimes. But if trusts or other corporations are people then we can constrain them and oblige them to pay into the social good as we do with flesh-and-blood people.

The solution to this challenge is not straightforward. It's not clear what to tax, how much to tax it, and how to administer the tax.

Taxes must be economically efficient so that growth continues. They must not undermine innovation in self-defeating distortions. They must not cause people to work less. And it's difficult to set a tax rate knowing that differential tax rates across countries mean that some locations will exist as tax havens. Just as some people free-ride in a public goods game, some countries encourage wealth flight to their shores. Tax rates are another way that countries compete in a process of cultural-group selection.

And because of these challenges, there is often no countervailing force against tax avoidance.

The Panama Papers, the Paradise Papers, the Pandora Papers, and even non-alliterative evidence of tax avoidance and other unfairness are leaked. A few people lose their jobs, but many don't and the system doesn't change. It's a challenging situation because it's also unclear what we should be advocating for. But we can establish some goals, evaluate some options, and lay out how to get there. Let's start with the goal.

A fair system is one where we ensure equality of opportunity such that the brightest of each generation bubble to the top. We want to give the young people in each generation who can push forward our species the knowledge, connections, and capital to do so. But we also want to ensure that everyone else has the knowledge, connections, and capital to contribute to the collective brain, including in allocating their investments toward the things they think matter to them, to their children, and are likely to expand the space of the possible.

A fair meritocracy can only function if the playing field is level.

But when wealth begets wealth, a powerful few can collude and corrupt our institutions. We end up with a double whammy. Innovation and economic growth slow as what's left in real terms for those with less becomes a smaller and smaller percentage of production. And wealth, legitimate or ill-gotten, created or appropriated, persists beyond a lifetime.

The natural starting point is to have billionaires pay more taxes and have larger inheritance taxes putting a cap on the total one can receive over a lifetime. The trouble with both these approaches is manifold.

First, loopholes exist. As mentioned, tax codes differ in different places, leading to wealth flight as people move their money to tax havens. As each leak reveals, almost anyone with substantial wealth, including those in charge of the tax system itself, are sheltering money in tax havens. Don't hate the player, hate the game – though it doesn't hurt to shame players. It's hard to audit and catch these cases and then recover funds.

Second, inheritance taxes must be set in a way that doesn't undermine one of the incentives to produce later in life – the desire to leave more for your children.

Third, while many from Bill Gates to Warren Buffett and others agree that wealth taxes and inheritance taxes are good and have signed up to campaigns like the Giving Pledge – contributing the majority of their wealth to philanthropy – this philanthropy, even when no longer under the direct control of the wealth creator, passes a disproportionate control over our current energy and production into the hands of unelected, unaccountable people who may or may not be efficient allocators. And even if we could tax these philanthropic foundations, it's not clear that our governments would use these new funds in ways that aid human potential and economic growth over, say, more drones, missiles, or contracts for cronies.

Inheritance taxes are critical to human progress, but they are hard

SHATTERING THE GLASS CEILING

to achieve and implement. Of course, this does not mean we shouldn't try, but in the meantime, there are other options on the table that avoid many of these challenges. Remember, the overall goal here is to tax unproductive money; to reduce wealth appropriation and rent-seeking. One solution is what Norway did in taxing its oil-rich land in the North Sea.

Taxing unproductive money

Monopoly is a ruthless game that tears families apart. There's nothing more annoying than rolling the dice and making your way across the board knowing you're going to get caught by rents you cannot pay to landowners who take too much away. Just as the board game tears families apart, the real-world equivalent tears societies apart. And unlike the board game, where the conditions are reset every time you play, these inequalities persist and grow across generations. Imagine how awful Monopoly would be if you had to start each game from where you left off last time. But that's real life. In real life the game doesn't get reset. This core issue of persistence across generations is not fixed by a basic income every time you pass Go.

Monopoly was actually designed to teach people about the unfairness of the unchecked capitalist dynamics we've been discussing. More than just a fun game, it was intended to be a tool for teaching how unchecked capitalism inevitably leads to persistent control by a few. Its original formulation was the Landlord's Game, and it was designed by Lizzie Magie, an activist known for her unconventional methods of conveying the logic of injustice.

Magie once bought an advert to sell herself to a potential husband as a 'young woman American slave' – 'intelligent, educated, refined; true; honest, just, poetical, philosophical; broad-minded and big-souled, and womanly above all things'. It was her way of highlighting the unequal, slave-like position of women in her nineteenth-century society.

The Landlord's Game was an introduction to the work of political economist Henry George – namely his emphasis on a fairer, more efficient capitalism through land value taxes. This wasn't socialism.

In fact, Karl Marx was not a fan, because Henry George's proposal remained firmly within a capitalist model. George was proposing a better capitalism. One that removed a key distortion in what can be owned. George's solution still applies today.

Land value tax is what everyone agrees is a better tax system but is not sure how to introduce without risking a revolution. The basic idea is that, unlike everything else we can own – shares, businesses, patents, or art – land is not something anyone creates; it is never wealth creation. It is entirely wealth appropriation, and without land development it does not expand the space of the possible. Land value taxes therefore impose a tax on the value of the land, excluding what's built and developed on it. That is, three adjacent blocks of land, one with a house, another with a block of flats, and another undeveloped, will have the same tax obligation. This incentivizes development, improvement, and reallocation of land for its most productive use. It also disincentivizes holding on to undeveloped land for speculation.

Land value taxes cause people to reallocate money to growth and property development over idle land speculation, leading to more efficient use of land, more available land, and more affordable housing. In general, it incentivizes the flow of capital away from unproductive rent-seeking or wealth appropriation toward more productive wealth creation. Unlike, say, raising income tax, which can disincentivize people from working harder to earn more lest it move them to the next tax bracket or cause them to spend money on avoiding tax. Both reduced productivity and tax avoidance reduce overall tax revenue. Land taxes do not distort incentives for production. Land taxes also have the advantage of being inescapable, because moving your block of land to the Cayman Islands is not an option. They incentivize people to reallocate land they control to more productive uses or sell to those who can use it more productively. In this way, land taxes are economically efficient and non-distortionary, encourage the flow of money, and, unlike other wealth taxes, have no danger of wealth flight because you can't move land. In any case, being a simpler tax, land value taxes are in general harder to evade. Indeed, some proponents have argued that a land value tax would do away with the need for *any other taxes*.

Imagine a world with no income tax, no sales tax, and no capital gains tax. That may sound like a fantasy, but land value taxes have the support of a range of economists across the political spectrum. Friedrich Hayek, the 1974 Nobel Prize winner, was a proponent in his various writings (indeed, Henry George is said to have sparked Hayek's interest in economics). Milton Friedman, the 1976 Nobel Prize winner, described it as 'the least bad tax'. Joseph Stiglitz, the 2001 Nobel Prize winner, formulated the Henry George theorem, which lays out the conditions when a land value tax can finance 100% of public expenditure without the need for any other taxes. And 2008 Nobel Prize winner Paul Krugman argued that the theory was sound for financing city growth, though, with more of the economy in the stock market and other assets, it may not generate enough to replace all other taxes; it would, however, make housing affordable, increase development, and do away with the kind of rent-seeking we discussed earlier using London as an example. Whether land value taxes can replace all taxes for government spending remains an ongoing debate, but even conservative estimates suggest that it could do away with income and sales tax.

In the United States, even though less wealth is held in land, a land value tax of 6% would pay for Social Security, Medicare and Medicaid, the entire Defense budget and more. In countries such as Australia and Britain, where more wealth is held in land, it could effectively replace all other taxes and solve the crisis of lack of affordable housing. And because a significant portion of loans are made for purchasing property, more new money would be created for wealth creation rather than wealth appropriation, reducing inflation.

The two biggest challenges land value taxes face are how to transition from the current system to a land value tax system and how to value the land minus improvements, such as buildings. Both have solutions.

Transitioning to land value taxes

The transition problem exists because in the current system in the US, UK, and many other developed countries, the middle class own land. But what many middle-class landowners don't realize is how little land they own. That the amount of land that even the richest own, let alone the middle class, is tiny in comparison to that of the ultra wealthy. We previously discussed the UK land ownership gap, but even in the United States, the top 10% of wealthiest Americans own about two-thirds of all privately owned land. What the middle class should know is that they would be much wealthier under a land value tax system. One transition pathway would be to reduce and eventually eliminate income, sales, and capital gains tax while gradually and proportionally increasing land value taxes to compensate for this. This pathway would also offer time and incentives for shifting from wealth appropriation to wealth creation. Many people's property taxes would go down because what is taxed is only the land, not the value of what's on it.

Psychologically, land value taxes also seem to undermine the very notion of property rights – indeed, the word 'property' is synonymous with land. But there is also a moral case against land ownership. Land is unlike everything else we can currently own. Land is more like water or air. We did not create it. So beyond the economic case for a land value tax, there is a moral case that no one should own land indefinitely without paying a tax to the common good for their control over our common land. Because of the unique status of land ownership compared to other assets, wealth often persists thanks to the persistent intergenerational transfer of land, which was often acquired generations ago, and often through dubious means, such as conquest, slavery, theft, and other forms of exploitation. Since those early means of acquiring land are no longer acceptable, newcomers have little access to large tracts of land that are effectively permanently controlled by a few.

As a result, in both the UK and US, research suggests, the gap between the wealthy and the poor will persist indefinitely. In the US, this gap is also correlated with race as a result of the history of slavery. To resolve this large and persistent Black–White wealth and asset ownership gap, reparations are sometimes proposed. Reparations, however, have low levels of support among Americans as a whole, and suffer from practical and ethical challenges in identifying the beneficiaries of slavery, the victims of slavery, and exactly how much they benefited and suffered.

Land value taxes have the added benefit of sidestepping many of the challenges of targeted reparations while still leveling the playing field against past injustices that are currently band-aided by inefficient redistribution and affirmative action efforts – proximate solutions. Land value taxes are a systems-level solution that removes the path dependence of history and removes the practical and ethical challenges of wealth transfers on the basis of ancestry, identity, or skin color. Indeed, the abolishment of slavery, one of the greatest moral victories for our species, also offers guidance as to how we might transition toward land as the basis of taxation.

Today, while slavery or effective slavery still exists in parts of the world, including developed countries like the United States and United Kingdom, it is not only illegal but unthinkable. But slavery was once common and uncontroversial. The idea that you could own another person was taken for granted by every major civilization. Indeed, the holy books of major world religions, including the Bible and Quran, do not admonish against owning slaves but rather describe how slaves should be treated.

How that transition happened is a guide for other moral transitions.

In the United States, it took a civil war to remove an entrenched slave-owning class, but abolishing slavery there and elsewhere also required economic solutions. Slave owners (though notably not slaves themselves, leading to the case for reparations) were often compensated for the loss of their 'property'. Britain's 1833 Slavery Abolition Act, for example, which freed all slaves across the British Empire, cost 40% of the Treasury's annual budget and required a loan that wasn't paid off until 2015. But this compensation reduced the need for revolution or other violence.

Similarly, the least disruptive method of transition to land value taxes is some form of compensation to owners of land that encourages a transition to wealth creation, such as tax breaks that incentivize a transition toward investment in business, entrepreneurship, the stock market, and other, more productive investments. These could include tax offsets for property development and the reduction of income and sales tax proportional to the increase in land tax. There are also lessons and models from patent law. Just as there are time limits on ownership of intellectual property so one could introduce time limits on land, after which there are high taxes, such as at the point of death or land transfer. The focus here is land, so trusts, businesses, and other similar vehicles offer no protection and of course land can't be moved offshore. Such solutions help transition us from a bad system to a better one, force companies profiting from natural resources to pay more based on the value of what they're extracting from the land, and have the added benefit of incentivizing landowners to innovate and develop to pay the taxes needed to retain their control over land. No more buying empty land and holding it in the hope its value will go up.

With house prices so high on the political agenda, there may be popular support for a politician or party running on this platform to lower or eliminate income, sales, and capital gains taxes. This shift would leave almost everyone with a lower tax burden. The final piece of the puzzle is how to value land minus what's built on it, but there are many solutions to this problem.

Valuing land

The valuation problem has various solutions. Land value taxes differ from a property tax in that only the land is being taxed, not improvements such as buildings, which is why many people's property taxes will decrease. Land intrinsically has value based on what's under it (e.g. natural resources) or near it. Buildings, such as houses, add value to the land, but the value of the land and the value of the buildings can be separated. For example, two plots with identical buildings, one closer to the city and one further away, will have different overall values because of the land value. Similarly, two identical adjacent plots of land, one with a house and another empty, will have a different overall value based on the value of the house. In both cases, only the land is the target of taxes. So in the first case, the property closer to the city would have a higher tax obligation despite having the same buildings. In the second case both adjacent properties would have identical tax obligations despite only one having a house. There are various solutions for how to tax only the land without taxing what's on the land.

These solutions calculate land value based on, for example, the value added by improvements or using the rental value of similar properties in different locations as a guide. Of the many methods of land valuation, one of my favorite solutions, because of its simplicity, is self-valuation. People tell you what their property is worth. Surely, you may counter, people would self-value at the lowest reasonable amount and pay less tax. Here's the catch. You have to be willing to sell the property at the price you say the property is worth. For example, the government may wish to buy these properties to create a parallel public housing system similar to Singapore's Housing and Development Board, where any citizen is eligible to own a public property, but only the one they live in. It may also discourage speculative holding of undeveloped land, freeing it for more productive purposes. Self-valuation is a system that can be used even when the property is not rented. It encourages higher rather than lower valuations, increasing tax revenues.

Indeed, the self-valuation approach has a long history. Denmark's seventeenth-century King Christian IV is famous for the Sound Taxes whereby ships' captains were allowed to self-declare the value of their cargo. No inspections were carried out, but the crown reserved the right to buy the cargo at the declared price.

'Sound' here refers to the strait between Denmark and Sweden, but this solution is also sound in the sense of a 'good' tax. Solutions such as self-valuation exist as a class of emergent solutions to fairness and are often preferred because they're less susceptible to corruption. Another example is the 'You cut, I choose' procedure – the person who cuts the cake must take the last piece, incentivizing fair division and efficient allocation. A start-up city or programmable politics would be an ideal place to implement and test this solution.

Land value taxes are far simpler and more efficient than the current tax systems they would replace. Even so, the goal here is not to go through all the details but instead show that a better system exists, even though it is not currently in the Overton window of political discussion. Indeed, both Henry George and land value taxes are memory-holed for many and there are large incentives to keep them from being widely discussed. Unlike other aspects of inter-elite competition, including the case of slavery, where not all elites owned slaves, most elites own land and would not want their land taxed lest they should lose it. Those who hold vast swathes of land, often the most valuable prime real estate, hope that people's fear of losing their relatively small property portfolios, family farms, or family home will prevent land taxes from being discussed or implemented. And so, the majority of society, not realizing the level of inequality, ends up fighting over scraps of space. For land value taxes to succeed, people will need to be shown how much more they would have under these taxes. They will need to be shown that housing will actually be more affordable for all. And beyond land, all but the ultra wealthy will enjoy reduced taxes overall.

Land value taxes are in a class of taxes on unproductive money. This is not a particularly socialist position. As mentioned, Marx was actually against land value taxes, which he regarded as entrenching capitalism. Land taxes will have the result of redistributing wealth that was appropriated, but without disincentivizing productive uses of money. But redistribution is not the goal, just a by-product. The goal is simply to create a more economically efficient, fairer capitalist system that incentivizes productive uses of our energy and resources every single generation.

By taxing unproductive money, we create a better capitalism that doesn't kill the golden goose by disrupting incentives in the way that communism or extreme socialism does. And it widens the group of people who are incentivized to work and bring their talents to the benefit of all.

Norway's tax on its North Sea resources is an example of how unearned wealth can be made to work for all. When unproductive money, rent-seeking, and wealth appropriation are taxed, it reduces the need for other forms of social welfare reallocation, limiting them to cases of bad luck in life rather than bad luck in where you were born and who you were born to. It's a way to ensure a fairer capitalism.

Philosopher John Rawls asked us to imagine a veil that shields you from knowing the circumstances of your birth. You don't know if you'll be male or female, bright or average, dark or fair, hot or not, rich or poor, or what country you'll call home. Like the 'You cut, I choose' solution, a fair society is one designed from behind a Rawlsian veil. We can achieve Rawlsian capitalism and a fairer playing field by taxing literal fields. Greed can be good when incentives and scales of cooperation are aligned and the world is positive sum. The invisible hand works best when the right rules are enforced.

There is an urgency in getting to this point quickly. Baby Boomers are about to retire and their massive inequality is about to get passed on to the next generation. This isn't about those with even a few million in housing portfolios but those with hundreds of millions and billions that dwarf everyone else, such as the 25,000 people who own half of Britain. We must act now as part of the package that leads us to a better solution. By encouraging innovation, retirees too will have access to more affordable and higher-quality goods and services. The past should not prevent us from reaching a better future.

WTF happened in 1971?

In shattering the glass ceiling, we must consider the mechanism by which wealth is allocated and reallocated through taxes, but we also need to consider the total size of the pie – the space of the possible – which is a function of both energy and technological innovations in efficiency. The decoupling of total wealth and wealth distribution is most obvious when we look at what has happened to our societies since the 1970s.

A popular website asks a simple question: WTF happened in 1971? A series of graphs show rising inequality, lower wages for the middle class, increased costs of living, later marriage, falling birth rates, and general reduction in quality of life, all starting around 1971. The website implies that the answer is 'Nixon shock'.

On August 15, 1971 President Richard Nixon effectively ended the Bretton Woods system by moving the United States off the gold standard. This gave the Federal Reserve the flexibility to control the money supply to adjust for stronger and weaker economic growth, avoiding booms and busts. Control over the money supply meant that we now had to trust central banks like the Federal Reserve to not devalue the currency. It also meant that one could now get wealthy by being closer not only to sources of energy and innovations in efficiency but also by being close to the money supply. The finance sector takes the best minds away from innovation and efficiency gains and pays them what seem like extraordinary salaries with multi-million-dollar bonuses. But these salaries are tiny compared to the billions being extracted. Being closer to the financial sector gets you closer to this new source of capturing money, capturing a larger space within the space of the possible decoupled from innovation and economic growth. And that is because of the monetary equivalent of share dilution – capturing the representation of wealth rather than wealth itself. Your money is like your share in the economy. When new money is created and distributed without growing the space of the possible – that is without creating new wealth through energy or efficiency – it devalues your shares. We call this inflation.

As money enters our system even when productivity stagnates, we may wonder why inflation hasn't gone up more than it has already. But it depends on how you measure inflation. Inflation is typically measured by a basket of goods and services that households typically consume – food, clothing, transport, utilities. By this metric, inflation is typically low. But inflation can be artificially kept low as long as the price of this basket of goods stays low. If inflation were measured by rising asset prices – housing, the stock market, even Bitcoin, then inflation is actually in the double digits and as high as you might expect given the creation of money and low productivity. But higher inflation is now measurable even by the basket of goods and we should see corresponding rises in each country based on their access to excess energy. Inflation will rise if new money is created while EROI and energy availability fall.

Around 1973 the Great Stagnation in innovation started – there was less room to expand innovative new efficiencies, but Nixon is probably not entirely to blame for WTF happened in 1971. Instead, the real departure was the end of cheap oil shrinking the space of the possible that started in 1973.

In 1973 the recently created Organization of Arab Petroleum Exporting Countries (OPEC), led by Saudi Arabia, embargoed oil to the United States, United Kingdom, Canada, and other nations that had supported Israel during the Arab–Israeli conflict or Yom Kippur War. Overnight, the price of oil quadrupled from $3 per barrel to $12 per barrel. This led to the recession of 1973–5. In 1979 oil production fell during the Iranian Revolution. This led to the recession of 1980–3.

We can now clearly see why the energy ceiling fell, the space of the possible shrank, and how production and productivity and all that we do are ultimately contingent on continued access to plentiful excess energy. In financial terms, that means cheap energy, particularly cheap oil.

Thus the solution isn't just in the money supply or financial instruments, which can increase inequality and harm our ability to innovate. It's not just about setting the right neutral interest rates or increasing consumer confidence. It is ultimately all about how much excess energy we are able to produce, a function of our energy sources' EROI and availability. We feel this directly in the price of energy as a proxy, though it is an imperfect proxy due to subsidies and other distortions. Ultimately, our future depends on expanding the space of the possible.

The next step is perhaps continuing to work on battery technologies and slowly expanding solar where it can be used. But what we really need is to transition to the nuclear age – a revolution stillborn for fears of problems with early reactors long since resolved in modern nuclear power technologies. Some countries already realize this.

China, for example, is entering the nuclear age all by itself, with 228 nuclear reactors in development in 2022, as I write. This will help at least China and others who choose this path reach a next level of energy abundance, increasing their wealth, power, and influence, as cheap coal did for Britain. And that wealth, power, and influence may lead to the next energy level – fusion.

But to get to fusion or even to solve the renewable battery and EROI challenges, we need to trigger a creative explosion.

10

Triggering a Creative Explosion

On April 26, 1956 a truck driver from North Carolina changed our lives forever. Thanks to Malcolm McLean and his standardized shipping container, products could be manufactured in the places where they were cheapest to make and then shipped anywhere around the globe. This innovation in efficiency irrevocably changed the world economy. Geography was no longer a barrier as items crisscrossed the world in a new globalized economy where consumers were spoilt for choice.

Before the historic 1956 voyage of McLean's fifty-eight containers from Newark, New Jersey to Houston, Texas, goods were sent by trucks to docks and then loaded onto ships. But first the goods that had just arrived by ship had to be unloaded. Ever seen old pictures of dock workers rolling barrels and moving huge sacks with ropes, pulleys, and planks? That's what it was like before McLean. The process was expensive, inefficient, and time-consuming – goods often sat on docksides for weeks. The high cost and long wait times disincentivized long-distance trade. And there was no way you would ship anything perishable.

McLean was an entrepreneur who started with one used truck and eventually owned a fleet. He took more of a Mark Zuckerberg Facebook-style, 'move fast and break things' approach. Ship and iterate, in this case, literally. Initially, he simply drove his trucks onto ships. That was an improvement from loading and unloading goods, but he quickly realized the waste of space the engine and wheels represented. So McLean started loading just the trailers. These trailers were incrementally improved to make them more easily movable by crane, which also made them easier to transport by train. Like many innovations, from trains to transistors, it was a massive net gain. But, as always, this creativity destroyed occupations. Dock workers, for example, became redundant.

Thanks to the standardized container, goods could be packed into a box, which could be put on a truck, then a train, then a ship, then back on a train, truck, and into a warehouse, from where they could be distributed. Of course, it took another fifteen to twenty years before a process of refinement, logistical innovation, and revised regulations would lead to one-click orders that arrive the next day.

The shipping container led to the rapid growth of Asian manufacturing and the decline of US manufacturing. Out-of-season fruits and vegetables were now cheaply available all year round. Television sets could be shipped from China to the United States for just two dollars per device; iPads to Europe for just five cents. It was such an efficient way to transport goods that the carbon footprint of sun-loving tomatoes could be lower if grown in hotter places and then shipped than if locally grown under a heated greenhouse. Eating local isn't always better for the environment.

The price of goods also fell because manufacturers could take advantage of arbitrage opportunities, finding the cheapest labor anywhere on the planet and then cheaply shipping the product to where it would sell at the highest price. In the West, more people could afford what were previously luxury products. In the East, exports raised wages and grew the economies of what became known as the Asian Tigers – Singapore, Hong Kong, South Korea, Taiwan, later China and others. Standards of living rose everywhere.

The story of the standardized 8-foot by 8.5-foot by 10-foot, 20-foot, or 40-foot corrugated Corten, all-weather steel shipping container is a story of wealth creation and the expansion of the space of the possible by energy savings under the law of innovation. And it follows many of the innovation lessons we learned about in Chapter 4.

It was an adjacent possible – more people were driving cars, creating congestion on roads. Trucking became less efficient. It was a simultaneous discovery – McLean wasn't alone in his idea of putting trucks on to ships. The basic concept had been developed by the military, and others were also trying to develop a standard approach, calculating the most efficient size and logistics for the boxes.

Today, 80–95% of manufactured goods are shipped with shipping containers. The next waves of innovation in shipping may be electric self-driving ships, loaded and unloaded by autonomous robot cranes. It's an easier self-driving problem than cars on roads – there are fewer lanes and pedestrians in the open water. But the results will be incremental relative to the invention of the shipping container itself.

Innovation drives economic growth by expanding the space of the possible. It is by the law of evolution and our collective brains that we innovate. Our innovative efforts have focused on efficiencies such as the shipping container, but we're running out of ways to become more efficient. Remember, higher income beats frugal spending; higher revenue beats lower overheads. Without new energy technologies, improvements in efficiency become smaller and smaller and harder to find, leading to what economist Tyler Cowen calls the *Great Stagnation*.

The Great Stagnation

Our story is one of slow but steady progress – the background sound of everyday life – punctuated by leaps in human capacity driven by key new technological innovations that unlock new energy. This in turn increases cooperation. And in turn allows us to innovate improvements in efficiency by the law of evolution. From hunter-gatherers to farmers to industrialists to technologists.

Some innovations open up new possibilities, creating a scramble for low-hanging fruit. The Internet led to a burst of activity in websites and e-commerce; the iPhone led to a proliferation of apps; artificial fertilizers vastly improved agriculture in the Green Revolution; the shipping container revolutionized trade; lithium batteries allowed for electric vehicles with a reasonable range; CRISPR has led to massive advancements in gene editing such as GMO foods increasing the efficiency of agricultural production; machine learning and AI may supplement or even replace human cognition.

The real leaps, however, happen when those innovations raise the energy ceiling. And so while it may seem like the Internet is as large a leap as the Industrial Revolution, it is not. Instead, the Internet has led to a Second Enlightenment which we hope will lead to a true Second Industrial Revolution – an energy revolution.

The Internet and its spaces, such as Facebook and Twitter, are the modern coffee shops of our Second Enlightenment, allowing scientists, engineers, politicians, entrepreneurs, and everyone else to share tweets, podcasts, blogs, newsletters, TikTok videos, and other modern pamphlets. This new Enlightenment includes us all. While we lament the rise of fake news and apparent polarization, all the Internet has done is speed up the processes that always took place, of misinformation and correction.

The fake news of the past – that carrots improve eyesight, Napoleon was short, or that the Great Wall of China can be seen from space – persisted for decades. But with the Internet, misinformation spreads and so too do the corrections. In the xkcd comic *Duty Calls*, Randall Munroe immortalized this urgency we feel to correct people on the Internet. The protagonist is asked, 'Are you coming to bed?' They reply, 'I can't. This is important. Someone is <u>wrong</u> on the Internet.'

The urgency with which netizens seem to want to correct people has a name: Cunningham's Law, after the inventor of the Wiki, Ward Cunningham. Wikipedia, arguably one of the greatest contributions to the modern world that exceeds whatever the Library of Alexandria might have been, is powered by this inexorable desire to correct others.

Cunningham's Law states that the best way to get the right answer on the Internet is not to ask a question but to post the wrong answer. A software engineering friend once sheepishly admitted that when they ask questions on StackOverflow – a Q&A platform for programmers – they use a puppet account to post a wrong answer to their question, which typically brings in a brigade of programmers eager to correct the injustice.

'Someone is wrong on the Internet! I must correct them!'

Just asking the question elicits far fewer responses.

And so just as the Enlightenment laid the foundation for the Industrial Revolution and then empowered it, so the foundation for the next true energy revolution is being laid by the flurry of ideas flowing around the Internet, that are debated and dismissed; the fifteen minutes of fame and falls to infamy; the calls for cancellation and the reading beyond our bubbles. And with that revolution will come the unlocking of the next level of human potential.

The next energy revolution will be fusion. Perhaps a mix of harnessing the fusion reactor in the sky – our sun – through solar power, and control over fusion, if we ever achieve it. With greater energy control, we can begin accessing the vast trove of resources relatively close to us. For example, mining asteroids for rare metals using robots and then gradually figuring out how to do it ever more efficiently. *The Jetsons* promised us flying cars. *Back to the Future* promised us hoverboards. In contrast to other science-fiction tropes that are now commonplace – cellphones, video conferencing, robotic surgery – flying cars, hoverboards, moon bases, and interplanetary travel have been stifled by the stillborn nuclear age and transition to the next energy level – fusion. With fusion, all these and more once again enter the realm of the possible.

As we improve fusion, its EROI increases and the virtually limitless availability of hydrogen will lead to an unimaginably large space of possibilities. It is impossible to predict the heights our grandchildren will reach, but it will be staggering. Our lives and all that we have achieved on the back of fossil fuels will seem as primitive to our descendants as the Middle Ages are to us.

But to get there, we need to increase our scale of cooperation, incentivize wealth creation over wealth appropriation, and trigger a creative explosion.

The fuse to fusion

Cultural evolution creates punctuated innovation – lots of incremental innovation with some occasional leaps via serendipity and recombination. Some of these innovations open up new spaces that lead to an innovation gold rush. Computing and information technology has seen such a gold rush in recent times: various innovations big and small in the microchip, Internet, lithium batteries, communication, machine learning, and most recently artificial intelligence. How do we trigger that same kind of creative explosion in other spheres of life? In Chapter 4 we focused on COMPASS's seven secrets of innovation and how they can be used at an individual level and at a corporate level to vastly improve our collective brain and the speed of cultural evolution. We can apply similar insights from our theory of everyone to practical policies at a societal level. Here are some examples.

Unfettered free speech

Language, speech, talking to one another. These are social synapses that are firing in our collective brain. Restricting that firing reduces trust in one another and cripples our capacity for innovation. It is collective brain damage.

In his vigorous defense of liberty, philosopher John Stuart Mill reminded us 'that it is important to give the freest scope possible to uncustomary things, in order that it may in time appear which of these are fit to be converted into customs'. Mill was describing what we would now call cultural evolution – the evaluation of different ideas and the spread of those that work best. But cultural evolution is most efficient when ideas are allowed to flow freely. Limitations on free speech are like blockages in the pipes of progress, preventing us from seeing the world as it is. And if we can't see the world as it is, we can't figure out how to fix it.

We used to believe the Earth was the center of the universe. You got into trouble for suggesting otherwise. We used to believe that it was natural that some people should be able to own others. Wars were fought to end slavery. For the uncustomary to become customary, we must be exposed to alternatives so that we can evaluate them.

To solve problems, we must first understand them. Take the gender wage gap for example – the gap between the average earnings of all women versus all men, often quoted as 70 cents to the dollar. If we stubbornly insist that the main explanation for this gap is bias, discrimination, and/or sexism, we are guilty of either not being specific enough about what constitutes and causes bias, sexism, and discrimination to be able to act, or, worse still, we may jump to solutions that may not work or, like proximate patches, create new problems. The large industry of HR implicit bias training or attempts to debias hiring committees are examples of failed policies that lack evidence. A 2019 review of 492 implicit bias studies consisting of 87,418 participants found no evidence that implicit bias training programs brought about behavioral change. Other reviews suggest diversity training can even backfire, increasing bias.

An alternative approach is a free exchange of hypotheses with critical tests and remediations based on evidence. Seventy cents to the dollar is a statistic that looks at the median wage of all men and all women across all occupations – the unadjusted wage gap. It compares high-paid majority male CEOs with low-paid majority female social workers. What leads men and women to these different careers with different tasks, wages, and lifestyles may be a result of limited choice, boys and girls receiving different encouragement to pursue different careers, differences in long-standing evolved preferences for types of tasks, inequalities in male and female participation in child-rearing, lack of societal support for balancing child-rearing and careers, or many other possibilities. When related factors such as occupation, experience, or hours worked are controlled – the adjusted wage gap – the size of the gap shrinks or disappears. But these controls, regardless of what they are, are precisely the policy levers and potential solutions we need to address the wage gap.

Easy answers or certainty over what the answer must be – or worse still, sanctioning unfavorable hypotheses – hinders our ability to discover the truth and act on it.

The world is a complicated place and doesn't always conform to what we think or hope the answers are. We can only arrive at the truth in a diverse environment of different backgrounds, considering all hypotheses and ideas – both those we like and those we don't. Cultural evolution needs the fuel of diversity and free speech to create the variation for transmission and selection. This kind of exploration needs a diversity of people with different experiences to come together in a safe space that enables unfettered free speech.

This is how science works best.

We are all biased. We all think we're on the factually and morally correct side of any debate. None of us is immune and none of us can even see the full extent of our bias – what's called the *bias blindspot*. I'd like to think that scientists are more likely to change their minds in the face of new evidence, but I know that this is rare even among scientists trying their hardest to be unbiased.

Science doesn't work because we're enlightened humans who see past our incentives and our life experience. It doesn't work because we readily change our minds in the face of new evidence. No, science works because we commit to a method of discovery, there is agreement on what counts as evidence, and, most importantly, we are incentivized to show others that they're wrong. It's a collective act that slowly converges on the truth. But our findings can only be trusted if we are free to find the opposite to whatever current political sentiments suggest is the right answer.

In 2020 researchers Bedoor AlShebli, Kinga Makovi, and Talal Rahwan published a paper in *Nature Communications*, a journal within the prestigious *Nature* family of academic journals. The paper was titled 'The Association between Early Career Informal Mentorship in Academic Collaborations and Junior Author Performance'. Using the Microsoft Academic Graph database of scientific papers, citation networks, and information about authors, they found that the work of female scientists with female mentors is associated with lower scientific impact than female scientists who have male mentors. They argued that this finding raises 'the possibility that opposite-gender mentorship may actually increase the impact of women who pursue a scientific career. These findings add a new perspective to the policy debate on how to best elevate the status of women in science.'

Understandably, the paper was met with a swift negative response on social media. There is a real potential that these findings could lead to female researchers receiving fewer applications from talented female students, a feedback loop that could further perpetuate whatever might be driving the finding (assuming the finding replicates). The negative response eventually led to a retraction. With no further context, this might have been the story of a bad paper with careless consideration for the implications of the findings. But it is instead the story of how bias can creep into the scientific record, distorting truth and affecting our ability to understand the world.

In 2018, just two years earlier, *Nature Communications* had published another paper, this time by Bedoor AlShebli, Talal Rahwan, and Wei Lee Woon. This was titled 'The Preeminence of Ethnic Diversity in Scientific Collaboration'. Using the same Microsoft Academic Graph database, the authors had used a similar method, finding that 'ethnic diversity had the strongest correlation with scientific impact. To further isolate the effects of ethnic diversity, we used randomized baseline models and again found a clear link between diversity and impact.' They concluded that 'recruiters should always strive to encourage and promote ethnic diversity'.

This paper, published by two of the same authors in the same journal with the same dataset and similar methods, was met with praise and remains, unretracted, in the scientific record.

As with any paper, both studies have their strengths and flaws, and supporters and critics can point to these as evidence for why one, both, or neither deserved to be retracted or never published.

The main critique of work like this often boils down to perceived harmful effects. A finding from the same dataset that female mentors are bad for female scientists is harmful, but a finding that ethnic diversity is good for impact is beneficial. This harm and benefit may very well be true, but there is a long-term cost to this short-term thinking. Selective condemnation based not on accuracy but harm supports what Plato called a noble lie – something false that is nonetheless maintained and promoted because it has perceived positive effects on society. Noble lies, especially when naked in their bias, squander scientific credibility. Whatever the truth, if people don't believe that for any finding an opposite finding could have also been published, then science becomes nothing more than untrustworthy propaganda, mirroring and supporting what we already believe.

The solution to misinformation is more information. The answer to the infamous fire in a theater analogy is that when someone falsely shouts 'Fire!', we need other voices to shout 'No there isn't!' And we need to exploit indirect reciprocity, tracking reputations like the boy who cried wolf, turning false fire alarmists into untrustworthy sources who lack credibility in other domains of life.

Many point out that there are problems associated with a policy of free speech, but these problems are not resolved by restricting speech. For example, some fear that power differences mean that speech is never truly free because some voices will be louder than others. But this problem is only made *worse* by restricting free speech. Powerful voices that can shout loudly in an environment of free speech can be balanced by whispers that grow into roars. But in an environment where speech is restricted, those same powerful voices can ensure alternative softer voices never speak at all.

This is not a left or right issue. There are real legislative proposals restricting speech in both directions. In US states governed by Democrats laws have been proposed to regulate social media companies when it comes to hate speech and extremism. In US states governed by Republicans laws have been proposed to prevent social media companies from removing hate speech. Scientists are disincentivized from saying things that would upset donors or their funders or cause them to lose their livelihoods. Some have attempted to pass laws that explicitly restrict what scientists can say. This is a bipartisan issue that affects all of us. None of us likes hearing things we disagree with, things that undermine our livelihoods or challenge features of the world that benefit us. If you have enough information to audit people's incentives, you'll often find that they advocate for things that directly or indirectly improve their material well-being, reputation, or status.

Unfettered free speech is historically unusual. That's why it was explicitly enshrined in the First Amendment of the US Constitution. It was an unusual and important cultural innovation in a world where laws prosecuted blasphemy, offensive speech, or insulting the monarch. Many of these laws are still on the books in the old worlds of Asia and even Europe. In 2009 an Austrian woman was fined for insulting the Prophet Mohammad, a ruling upheld in 2011 by the European Court of Human Rights. In 2022, after the death of Queen Elizabeth II, British anti-monarchist protesters with Not My King placards were threatened with arrest. British police regularly arrest people over online posts and comments deemed racist, offensive, or hate speech – not protected speech in the UK. Where the freedom to speak is uncertain, it has a chilling effect on people's willingness to express their honest opinions and the free exchange of the ideas that lead to innovation and creativity.

Of course, the US First Amendment only restricts government, but the broader principle of free and open exchange – free speech – protecting even and especially speech we would rather see banned, is essential to science and progress. The First Amendment has bled over into a culture of free expression that requires protection.

The importance of a safe space to speak one's mind and where we can reveal how we think the world really works is emphasized by research on psychological safety.

In a famous analysis conducted by Google, Project Aristotle, psychological safety – the freedom to express what you really think and resilience to hearing ideas you don't like – emerged as a key prerequisite for successful teams, especially diverse teams. Diverse minds can only combine into a brilliant collective brain when they trust one another. When people trust each other, they feel more comfortable saying what they really think. And that's critical to being shown why they're wrong, showing others why they're wrong, or discovering a new truth at the intersection of ideas and beliefs previously isolated in the heads of different people.

We can create a culture of psychological safety by encouraging in interpersonal interactions what's called the robustness principle (Postel's Law) in software engineering: be conservative in what you do, be liberal in what you accept from others. In programming this refers to, for example, outputting well-formatted files but robustly accepting poorly formatted files from other programs. Don't create files with missing sections or stray semicolons, but be able to read them if they have these mistakes. In communication it means being generous to others' intentions – 'steel-manning' rather than 'straw-manning' their position – but personally making an effort to communicate clearly. A steel-man argument attempts to make the best version of someone else's argument, perhaps even better than they made themselves. Steel-manning is the strategy for people who see arguments as a means of together arriving at the truth rather than sparring to win for the sake of winning. Straw-manning is the opposite, taking the weakest version of an argument and arguing against it with the goal of winning rather than learning.

It's hard to know in advance what will and what won't cause harm in the long term, and so the bar for what we should ban must be incredibly high and err on the side of not banning or suppressing speech. This is particularly important when it comes to freedom of speech in science, because reality is complicated and science is never settled. If we can't trust that the scientists are speaking freely then how can we trust science? Take, for example, group differences and the role of genes that we discussed in Chapter 3. Based on my reading of the evidence, I come down on the side of culture being the primary cause of these differences, but you should only believe me if those who believe otherwise are equally able to express the most defensible version of their position. If the only voice that could be heard in polite circles was mine then how could you know that there was an alternative argument and hear the evidence for it?

No one has a right or even a reasonable expectation to not be offended. We make progress not by bludgeoning our opponents into believing they are bigots and bad people, but by finding out what people think and why they think it. What evidence they have and what critical evidence they would need to change their mind. The reasons why we disagree with someone have to be based on logic and evidence, not group membership or prejudices. To err is human, but being wrong should not end people's livelihoods or their ability to say something else. Though it may change the credibility of what they say, it should not change their ability to say it. It would be like preventing an entrepreneur from starting another company because their previous company had failed. So many innovations would be lost in such a closed culture. Free speech is critical to the process of discovery and the triggering of a creative explosion, especially for social innovation. The vibrancy of the United States – its robust debates, fights, and protests, and even racist and anti-racist polarizing discussions – is a product of the freedom to speak freely. But the goal should not be to win the argument, it should be to arrive at the truth as a society. The norms of how we argue are critical to this process.

The world is a complicated place. Yesterday's obvious truths are today's falsehoods and vice versa. As scientists we try our best to steel-man an opposition's arguments. The goal is not to win the argument, but to arrive at the truth. And that means seeing our opponent not as an opponent but as a fellow truth seeker who deserves whatever assistance we may render in developing the best version of their argument no matter how wrong we think it is.

That's why censorship by government institutional misinformation tribunals or making Mark Zuckerberg, Elon Musk, or whomever they designate arbiter over what is true and false is not a solution to the real problem of misinformation. Instead, like many proximate, band-aid solutions, it creates new problems. Misinformation is a real challenge, but banning or suppressing speech is not the answer.

The person in charge of deciding the truth may not always be the person you want. Remember the opposition may one day be in power and a person with different politics may one day run the company. Laws must be made in a Rawlsian manner, prepared for a time when you and yours are no longer in charge. We must be ruled by principles, not people. For tackling misinformation, we want emergent, systems-level solutions such as offering more, not less, context and information.

There is value in minimizing the ability to create bots and fake accounts that poison our social network streams by flooding them with content that is not representative of what actual people think in an attempt to exploit our social learning biases. But real people should not be restricted in what they have to say.

A lack of free speech cripples the collective brain. It deprives it of information, the sparks of ideas moving from person to person that could light the next fire that radically changes our lives. It is cultural evolution and cultural-group selection that sorts it out through differences in scientific discovery.

None of this abrogates our responsibility as individuals to consider the impact of dual-use discoveries or harms we care about. Good science and good policy require careful consideration of the broader impact of what we discover and how it can be contextualized for mass consumption. In engineering, these are referred to as dual-use technologies. As we discussed previously, nuclear technology gets you power plants and bombs. Rockets can launch satellites and warheads. Care is required in these areas of research, even if they are the path toward energy abundance.

The rigorous defense of speech – even speech you vehemently disagree with – is perhaps the most important policy for triggering a creative explosion. At an interpersonal level, laws and cultural norms that protect free speech and provide psychological safety ensure that ideas can flow. At an institutional level, the same is achieved by ensuring people have more access to the ideas and discoveries of others. These institutions and technologies then become knowledge and skill boosters.

Knowledge and skill boosters

Creating greater connections between people and greater access to diverse cultural packages, ways of thinking, analogies, metaphors, skills, and knowledge lead to greater educational and income outcomes. Interpersonally, this often requires talking to people you disagree with, opening yourself up to a wide array of weak links with different ideas. Your best ideas and new jobs come not from those you regularly interact with – your strong links who share your preferences, consume the same information, and think like you – but those you know peripherally who belong to other, outside networks. Tighter, more closed communities are the least likely to see new solutions.

Beyond interpersonal dynamics, access to high-quality education, libraries, and the Internet have all been shown to boost graduation rates, income, and lifetime outcomes. One study found an up to 300% return in spending on annual Internet access by a school district in terms of incomes for former students, mediated by their improved academic performance. Those are fantastic returns. Obviously, like providing micro-nutrients to calorie-deficient communities, the impact of school Internet access will be larger among communities with lower household Internet access.

Before the booster that is the Internet, people found information through public libraries. When Andrew Carnegie, the richest man in the world at the turn of the twentieth century, created free libraries across the United States, cities with those libraries had 7 to 11% higher patent rates than comparable cities without Carnegie libraries. It's the kind of philanthropic spending that concentrated wealth allows and that governments don't always provide. Instead, in a beneficial exchange, Carnegie can serve as an innovator through the philanthropic sector while governments can then select the successful ideas and scale them in a way that few philanthropists can. The private sector often excels at innovation that the public sector can then more efficiently deploy at scale.

The brains of a cultural species do well when they have access to high-quality information and skills transmission. Indeed, this is part of why Silicon Valley is so innovative.

Optimal intellectual property law and non-competes

The story of Prometheus stealing fire from the gods is a story of intellectual property (IP) law. After Prometheus stole fire from the gods, the gods still had fire, but now humans did too. And with that IP theft, we cooked on camp fires, smelted steel, blasted rockets, and made unimaginable progress. The ability to recombine IP is critical to innovation.

Japanese brands were once synonymous with poor quality. The car industry is a good example. Early Japanese cars were poor-quality knock-offs of European and American models. Some were hybrids of the different designs that Japanese car manufacturers could access. But eventually, copying turned to recombining. The careful, incrementally improved culture that led to modern bonsai was recombined with Western culture to create new processes such as kaizen, the Toyota Way, and other ways to continuously and incrementally improve technology. The result was Toyota, Nissan, Mazda, and other Japanese car companies becoming the reliable brands they are today.

Chinese brands, for example Huawei and Haier, are now going through a similar process.

Patent laws give inventors exclusive rights over their invention for some set period of time. The relationship between patents and innovation is one of balance. Too strong patent laws stifle recombination by restricting others from building on the invention. Too weak patent laws create weak incentives to innovate without the ability to economically benefit from the risks taken to reach the invention. In fast-moving industries, policies that err on the side of opening up discoveries encourage more innovation while still allowing inventors to enjoy a first-mover advantage.

In the tech sector, the restrictions of patents are overcome through cross-licencing. Companies are mutually dependent on each other's intellectual property (IP). Faced with the possibility of a cold war of patents and mutually assured destruction via patent litigation, the industry has adopted a policy of allowing the free use of others' patents within reason.

In Silicon Valley, the flow of IP, ideas, skills, and culture is also enabled because California doesn't enforce non-compete agreements.

Job hopping is common and one way to acquire IP and talent is to simply hire the engineers responsible for creating a particular product. This means skills and cultural traits can recombine across start-ups, empowering recombination in the Valley's collective brain, giving the West Coast a distinct advantage over the East when it comes to innovation and giving the United States an advantage over other countries.

It also helps that merit and performance are highly competitive and equally highly rewarded in Silicon Valley. These incentivized demonstrations of merit have side effects such as the empowerment of the open source software movement. People are eager to demonstrate what they can do, even for free, because reputation is heavily rewarded in the long run. The power of the meritocracy and rewarding ability has a long history in successful enterprises and empires.

Meritocracy and rewarding ability

Evidence for the way in which meritocratic systems outcompete those based on group membership or who you know is manifold.

China's imperial examinations led to a scholar class, selecting the best regardless of background to administer vast empires that lasted for centuries. Genghis Khan is famous for implementing a meritocracy that selected the best and brightest for the most important positions regardless of tribe or family. The Great Khan's meritocracy incentivized followers to give their best in bloody conquests. They knew that they were judged by their ability and that there was no preordained limit to what they could achieve. In the United States, there is a long history of successful concerted meritocratic efforts, including the Manhattan Project to develop a nuclear bomb and the Apollo program to put a man on the Moon.

Today, that meritocracy is under threat as more objective measures are replaced by measures that are easier to game. Standardized tests such as the SAT and GRE are being dropped in favor of more subjective, less comparable, and easier-to-game measures such as teacher grades, letters of reference, or work experience open to only a few. These measures are more susceptible to outside influence and create the uneven playing field and subsequent inequality we discussed in the previous chapter. Some privileged students may have more access to test preparation than others, but an objective test gives everyone a shot and has been a leading source of opportunity for talented poorer kids with huge potential to prove themselves. References, on the other hand, require connections. Work experience requires time, opportunity, and funding, all scarce resources for the less affluent. This move to avoid objective measures of ability is a coalition of the privileged who want to protect that privilege and those who want more flexibility to create better representation. Some of this is nobly motivated, but it ends up as an unfortunate alliance that will harm our innovative capacity and harm those it's trying to help. This eventually hurts us all.

Moreover, these policies lead to increased intergroup division and stigmatize minority candidates who may have comparable or greater ability than their peers in other groups. If the bar seems lower for one group than another, any graduates or hires in that group could be perceived to have less ability even when this is not the case. This leads to the problem of rational racists and Bayesian bigots that I'll discuss in Chapter 12. When some win prestigious places at the expense of others on the basis of uncontrollable factors and not skills or cognition, they turn an individual competition into a zero-sum intergroup competition. They create Enron effects.

Avoiding Enron effects

In the year 2000 a large energy company – Enron Corporation – was flying high with a market capitalization of over $100 billion and a share price that had risen from $20 to over $80. By the end of 2001 it was pennies a share and the company went bankrupt. What became known as the Enron Scandal was widespread fraud and illegal accounting practices hiding Enron's failures and true value. As a result of the cooperative fraud, Enron's collapse also took down its accounting firm, Arthur Andersen, turning the Big Five into the Big Four (leaving Deloitte, EY, KPMG, and PWC). Tens of thousands lost their jobs, investments, and retirement funds. The CEO, Jeffrey Skilling, was sentenced to twenty-four years in prison, but was released after twelve.

Skilling was a huge fan of evolution. His favorite book was Richard Dawkin's *The Selfish Gene*. Skilling attempted to implement evolutionary principles in his business practices and corporate culture. But too little knowledge is a dangerous thing. Skilling understood that at the heart of evolution is selection and competition but forgot that there is also diversity and cooperation. Skilling's policies created a zero-sum environment. One in which another's failures were predictive of others' success. We can call this the Enron effect.

'Rank and yank' was one of Enron's more famous policies. Indeed, it was celebrated in business books and media at the time. Performance reviews involved ranking employees on a bell curve. The top 5% were considered superior and received the largest rewards. The bottom 15% were fired. This led to a toxic culture of destructive competition where people lied, deceived, formed alliances, and undermined one another to avoid losing their job. Remember, it's easier to be nice when there's lots to go around. Skilling didn't invent rank and yank, but where it's found, it creates zero-sum conditions conducive to destructive rather than productive competition. People work hard to harm one another rather than work hard to be better.

Though it may seem counterintuitive, helping other countries, even former enemies, avoid Enron effects by improving their development is good for the world as a whole. The Marshall Plan, providing aid to help rebuild Western Europe, including former enemy Germany, contributed to the long-standing peace in Europe following the Second World War.

We are clever because our culture is clever and we are kind because there is plenty to go around. Only conditions whereby rewards are shared and the success of others contributes to our success can keep us together. Otherwise, ever-present diversity causes us to come apart.

Today, we all face Enron effects thanks to rising inequality, slowed growth, and economic decline triggering zero-sum biases and incentivizing destructive competition between people and groups. This in turn is weakening one of the solutions to the paradox of diversity: structured diversity.

Structuring diversity

Microsoft, the United States, and the Roman Catholic Church all have one thing in common. It has led to their persistence and historical success. All have incorporated structural diversity into their organization to boost knowledge and spread solutions. They offer a middle-path solution to the paradox of diversity.

There are different ways to distribute diversity. As we discussed, at one extreme, one could solve the paradox of diversity by each person having a greater diversity of experiences. For example, workers and researchers with an interdisciplinary educational background have within them a diversity of skills and knowledge. Similarly, people who have a more culturally diverse background or have lived in many places around the world, particularly when younger (often called 'third culture kids'), have within them a greater diversity of cultural experience. The most famous third culture kid is probably Barack Obama, who had parents and step-parents from different cultures and who grew up in Indonesia and Hawai'i.

At the other extreme of personal diversity, diversity could be completely diffuse but with a common communication protocol. Examples include interdisciplinary teams of people with different training and educational backgrounds or multinational teams working on a common challenge.

A more attainable strategy is clustered cultural diversity whereby separate parts all work together toward a common goal. These are the administrative units and states, the departments and divisions, the work functions and specializations whose futures are inexorably tied.

At the core of this puzzle is what is called *evolvability* or *adaptability* in biology. The most innovative groups are diverse. So too are the least innovative groups. The difference is in how much they trust one another and share enough to work together. Because resolving this paradox is difficult, many companies and some countries opt for homogeneity. They may adopt a policy of superficial diversity of sex or skin color, but ultimately they hire according to the strategy of 'more like me' because it is easier. They do enough to tick the necessary boxes and avoid embarrassing imbalances in outward-facing situations and effectively practice a form of tokenism. But when we grab the bull by both horns and resolve the paradox through proper structuring, the rewards are massive.

One of Nadella's successful strategies at Microsoft was to turn it into an ecosystem of start-ups. It is the start-up city approach applied to an organization. There is a trade-off between centralization and diffusion. Nadella's change from centralized to diffuse at Microsoft was a way for it to innovate. The United States, too, benefits from its federal structure and lack of strong central authority – each state as a laboratory for innovations in democracy. And the Roman Catholic Church has found the balance between centralization and dissent through regional variation – bishops in each country have a certain degree of autonomy over their parishes and the same is true within the different religious orders.

Of course, not all countries or corporations can take this approach. It requires a certain size to engage in high-risk research and development skunkworks, to take massive business risks or allow for internal division. Smaller companies and countries exist as part of the units taking risks but can benefit from copying and learning successful strategies.

Our species needs the next Industrial Revolution – a true energy revolution. For that, we need a creative explosion. Policies that ensure that information is freely transmitted and that successful outcomes are rewarded without overly punishing failure lest we harm diversity and risk-taking give us the best chance of triggering that creative explosion. One place where it is essential we get it right is where the Second Enlightenment was born: the Internet.

I I

Improving the Internet

The Internet has embedded itself in every aspect of our lives. It's how we work, how we play, how we socialize, and even how we date. But the Internet is so much more than our favorite websites and apps. It represents the collective output and knowledge of our species in a way that deeply weaves us together. It interacts with every law of life. The Internet accelerates cultural evolution, creates new cooperative groups, and supports innovation. It will be critical to creating the connections in our collective brain that will lead us to the next generation of fission and fusion and even to spreading the knowledge necessary to disseminate and support these critical technologies. The Internet is also changing us.

Much has been written about what the Internet and social media are doing to us. Panicked polemics point to our reduced attention span, inability to focus, increasing tendency to multitask, and reduced memory for all the things we can now look up in seconds. Social media is said to be destroying our democracies, worsening mental health, and fomenting division and polarization. All of these things may in fact be true, but the most powerful disruptive effects of the Internet and social media derive simply from the fact that they connect us as never before, storing and giving us access to our trove of cultural knowledge.

Our culture is slowly catching up to figure out the best ways to deal with this hyperconnectivity. Older people, for example, are more susceptible to misinformation. How they learned to learn did not prepare them for a world where the signals of authority and authenticity are so easy to fake. Younger people are less susceptible to what, to them, seem like obvious fakes, but have yet to update their social learning strategies to account for people presenting their successes and best selves through posed posts, filters, and selective information or collapsing worth to likes and follower counts. This may be fueling fears of missing out (FOMO) and feelings of inadequacy leading to mental health issues, particularly among young women.

Social media is also disrupting us in ways that may be beneficial in the longer term. We are more exposed to ideas we disagree with. We get angry and frustrated as we doomscroll on 'hell sites', but our exposure to even ideas we find abhorrent forces us to deal with the existence of other people who think differently to us. It forces us to hone our own arguments and allows observers to judge who has the best arguments and evidence. We are also more exposed to the inequality that is reflective of broader changes in society. The previous generation of privilege and money learned to not advertise that wealth gap. And if someone flashed the cash, they were often looked down upon as being nouveau riche 'new money' in societies where inherited wealth was considered superior. But the next generation are online socially connected to a larger sector of society and often advertise vastly different lifestyles, making the wealth gap more visible. If you want an insight into the lives of children of the ultra-wealthy, watch Jamie Johnson of Johnson & Johnson's documentary *Born Rich*, featuring a twenty-two-year-old Ivanka Trump.

The Internet also enables us to find those like ourselves and form ad hoc groups targeted at a cause, from Anonymous hacking Russians to open source software projects with contributors who have never met each other in real life, and to academics and entrepreneurs discovering each other's work and finding ways to productively collaborate regardless of geographic distance.

These trends will only continue. For example, the astonishing success of machine translation may melt the linguistic barriers between us, opening up the full Internet, which currently remains somewhat segregated by languages. Machine translation has already increased the efficiency of international trade, increasing exports by around 10%. This will continue to increase, further shrinking our world as the shipping container once did.

The Internet has also sped up cultural evolution. But much of this has happened by accident and happenstance. Through trial and error and A/B testing, small changes – mute lists and blocks, stars changed to hearts, a larger range of emotional expression, revealing or hiding likes or followers, more or less characters to express ourselves – have large multiplied affects across the globe, but the engineers who implemented these changes may have had little understanding of what those changes were likely to do or even what exactly they have done. Because changes on the Internet and social media have such a large instantaneous impact on all users, it is essential that we use our theory of everyone, particularly cultural evolution for evaluation and improvement of these platforms. The low-hanging fruit is engineering social media to better match how people have evolved to socially learn.

Improving social media through what we know about social learning

The people we interact with unconsciously shape our opinions and what we think most people think. We deploy our social learning strategies instinctively without being aware of the way they write our brain's software. We are a product of what we consume – who and what we watch, read, listen, and speak to.

It's easy to develop a view of the world that doesn't represent most people if we spend a lot of time on the likes of Facebook, Instagram, Twitter, YouTube, TikTok, Reddit, Quora, or 4chan. But the opinions and opinionated posts we see aren't even representative of most of the people on social media platforms, let alone most people in the world. These platforms self-select for people with strong opinions and a strong desire to share those opinions. The data bears out something close to the 90-9-1 rule: 90% of people are lurkers reading the comments made by 9% of people on the content produced by 1% of people.

This isn't unique to social media. We all live in bubbles. We consume media written by journalists and have lives shaped by politicians and political advisors who may watch the same TV, follow the same people on Twitter, live in the same high-income or middle-class communities, send their kids to the same schools, and go to the same social engagements. Through conversations over cocktails and chats at school pick-ups, our views, too, are filtered. And then, as we share those views, they are filtered into others' perceptions and work. The best we can do to burst the bubble is to expand our networks as much as possible and try to interact with more people who are not like ourselves. That can be difficult to do in real-life physical interactions, which are shaped by neighborhoods and by who we most often bump into. But social media and the Internet can be used to intentionally burst that bubble.

Bursting the bubble

The Internet and social media have enabled us to access knowledge and conversations we never had access to before. You can join groups and listen in on conversations among feminists and men's rights activists, Wahabi Muslims and Orthodox Jews, white supremacists and Effective Altruists. You can hear how insiders talk to their ingroup in semi-public conversations on Reddit, Telegram, WhatsApp, Substack, Discord, forums, or other social media. But many don't.

Instead, we are often passive recipients of newsfeeds produced by algorithms whose only goal is to keep you engaged. But it doesn't have to be designed that way. Social media can be deliberately designed to expand collective brains while keeping people engaged. It can also be engineered to better exploit what we naturally pay attention to.

Prestige and other pay-off biases

Many social media platforms show you how many followers a person has – an indication of popularity. Popularity is a signal that our social learning psychology pays attention to, but a social media follower count is a very weak indicator compared to the data we attend to in the real world.

In the real world, we care not only about how many people know someone or follow them, but who those people are. Prestige comes not just from amassing followers but by amassing followers who themselves have many followers. In the language of network theory, we pay attention not just to direct centrality (number of connections) but to eigenvector centrality (connections of connections of connections onwards).

To put it simply, a high-school sports star may have many friends, but if they're all other high-school students, she will be less prestigious than someone with fewer friends but among which are former presidents, celebrities, and successful business people.

High eigenvector centrality has been empirically demonstrated to be a source of influence within communities and a more accurate marker of prestige than follower count. If you've ever looked at who someone's followers are, you are implicitly looking for eigenvector centrality. In fact, Google's success over previous search engines was thanks to its use of eigenvector centrality calculated by PageRank, which ranked web pages based on who linked to them and who linked to the linkers. But this information is nowhere to be found on social media.

Other missing biases include more honest indicators of expertise and success. Social media clout is easily faked. In fact, reputation itself – an important cue for both social learning and cooperation – is often poorly tracked.

Improving reputation

I mentioned previously the way in which online reputations can help enhance cooperation for ride sharing or room bookings, but reputation is often not leveraged in other online information.

When we read any article or see a post, we are rarely presented with more context about the author. Who they are, their expertise, what else they've written, how many times they've been right or wrong, what other people think of them and who those other people are. That information exists but is not aggregated, and so one of the most powerful characteristics of a person that we care about – what other people think of them – their reputation – remains untracked and not leveraged.

Instead of individual reputations, most of us use what we might call *securitized reputations*. Rather than the reputations of individual journalists, most people rely on the reputation of the outlet – *New York Times*, *Washington Post*, or other journal, magazine, or newspaper. Journalists who have large enough personal reputations moving to personal newsletter platforms such as Substack are one example of moving away from securitized reputations. But here, too, reputational information is often missing, limiting our ability to find valuable new sources of information.

Reputational information includes past work, expertise, what others think of the author, funding and conflicts of interest, and what people got right and wrong in the past, and which people think what about the author. But in the same way that statements about conflicts of interest can be useful, so too can information about incentives. What we might call *incentive audits* are useful, particularly when the stakes are high.

Incentive audits

An incentive audit is to self-reported conflicts of interest what tax audits are to self-reported tax returns. They make information that may be relevant to a person's beliefs more salient. For example, who they are friends with, their personal circumstances, who own the companies they work with or for, and their sources of income. If you scratch the surface of many people's beliefs, you'll find self-serving incentives. As the old adage goes, 'It is difficult to get a man to understand something when his salary depends upon his not under-standing it.' You can probably guess who will support which positions if you know enough about the opinion holders.

For example, it should be no surprise that people with more educa-tion are more likely to believe in a meritocracy, the upper end of the income distribution are more likely to argue for lower taxes, or that people will signal beliefs that gain their ingroup approval.

Sometimes signaled beliefs are costless to us personally; they simply signal our affiliation or that we are good people as judged by our ingroup. The accuracy of evolution as an explanation for human origins, whether we should defund or defend the police in other people's communities, or whether transwomen athletes should be allowed to compete in women's sports are often topics that people hold strong opinions on, despite them having little bearing on their own lives. We can advocate for and reinforce group affiliation without having a stake in the game or it impacting on our everyday existence. We can continue to enjoy the benefits of evolutionary approaches to medicine, living in a safe community, and not participating in elite sports. One way to reduce the effects is to make signals costly – for the signals to have consequences – revealing and making salient relevant information and potential biases and encouraging a culture that normalizes such disclosures.

Incentive audits are also worth applying to ourselves. It takes humility to admit the limits of our knowledge and that we are incentivized to seek out certain knowledge, be strategically ignorant about what doesn't serve our interests, and when there are competing beliefs, believe what serves us best – and that's not always truth. Remember, some beliefs are acquired through direct experience, but most are indirectly acquired from others, where the causal relationships are complicated or unclear. In an uncertain world of mixed evidence, why not believe what would serve you best if true?

These self-serving biases can sometimes harm us. For example, they may lead us to worry about less important but more controllable decisions. People often worry about the health effects of parabens in shampoo or pesticides in non-organic vegetables, which are decisions that are easier to exercise at the supermarket than actually getting more exercise at the gym or just eating more vegetables, organic or not.

It is also easier to hold self-serving beliefs when the information is socially acquired with a choice of what to believe and when causality is less clear. These are also cases where we tend to copy most faithfully, not knowing which bits of our beliefs are necessary for causality. Religion is an extreme example of beliefs and behaviors that are more indirectly transmitted than directly discovered. As such, religions often have more ritual behavior, with emphasis on exact copying.

For example, in the early 2000s some Catholic priests, rather than baptizing children with the conventional 'I baptize you in the name of the Father and of the Son and of the Holy Spirit', began using the more gender-neutral 'We baptize you in the name of the Creator, Sustainer, and Liberator of Life'. Some priests, wanting to hedge their bets, added, 'who is also Father, Son, and Spirit'. That may seem like a small change to non-Catholics, perhaps even a preferable one, but it led to great uncertainty about whether children were actually baptized or not. With the destination of souls at stake, the controversy went all the way to Rome and a decision was made that the incorrect wording rendered the baptism invalid. Hundreds of children had to be re-baptized. Although this tendency toward exact copying is common among more devout religious people, this psychology is not unique to religion.

The separation of the natural and the supernatural, the secular and the sacred is a rather recent WEIRD idea. For much of the globe and for most of history, the world was just the world, not a mixture of natural and supernatural. The real psychological separation is between beliefs and behaviors that are more directly or indirectly acquired. Non-religious people also hold indirectly acquired, evidence-free beliefs stated with certainty. Take the US Declaration of Independence: 'We hold these truths to be self-evident, that all men are created equal'. This idea of equality before the law or before God is not self-evident nor derivable from individual characteristics. It is a bold statement in a world of people such as Usain Bolt, who can run faster than the typical residential speed limit for cars (25 mph), and Simone Biles, who has four gymnastic skills named after her. At best, it is a belief that might be defended as being useful or creating less suffering. It is also not a belief that is practiced as it is written, not even at the time it was penned. The author himself, Thomas Jefferson, owned over 600 slaves. Other American presidents who owned slaves, including while in office, were George Washington (also over 600), James Madison (over 100), James Monroe (~75), Andrew Jackson (~200), John Tyler (29), James K. Polk (56), and Zachary Taylor (~300). John Adams and his son John Quincy Adams are notable for being the only two presidents in the first dozen *not* to have had any slaves. And yet, the self-evident nature of human rights and equality or the assertion that this is a better belief than the alternative are often asserted as obvious or objective truths. It is indeed a useful belief that can in turn create progress through coordination around the belief, accusations of hypocrisy, and runaway cultural evolution, as people compete over a more authentic and inclusive implementation of equality.

Self-similarity and more information

By now, it should be no mystery why some fall for what others call fake news or conspiracy theories. Those who hold these derided beliefs are simply people who have learned what they know from people they think are trustworthy but we think are not. That is to say, none of us is actually prioritizing truth, especially when that truth harms our interests. In any case, few people have the ability to verify first hand the many beliefs they hold about the world, from whether germs and not spirits cause illness to whether the world is round or flat, let alone whether equality as a moral principle is useful. Instead, whether or not we hold true beliefs is not a function of our ability to distinguish truth from falsehood; it is a function of whom we trust. And that's true of scientists as much as it is of anyone else.

Scientists trust and internalize scientific results based on the expertise or success of the person who ran the study, or the prestige of the journal in which it was published, or if the findings are considered good or bad. This is part of why it took decades for the psychological and behavioral sciences to realize that half the literature was untrustworthy. And before you assume this is restricted to these disciplines, after the highly publicized replication failures in psychology, other disciplines also ran replication tests. This should terrify you: more than half of the top cancer studies have failed to replicate.

To trust the science is to trust the people and the process. Rarely are beliefs verified through direct experience or personal replication, but rather their plausibility is checked against everything else we know about the world – or, at least, what we think we know about the world. Many of these assumptions are indirectly acquired from childhood onwards based on the culture in which we happen to be born, the subjects we happened to study, and the people we happen to encounter. And when scientists hold scientific beliefs about subjects outside their own field, then we have no direct experience of even how the theories were developed or the data was collected. Instead, a scientist's belief is founded on a generalized trust in 'science' as a process and scientists as honest practitioners of that process.

As cultural learners, we have evolved to focus on *who* is making the argument more than the argument itself. Among left-leaning

Americans, Trump's attempts to close borders during the COVID-19 pandemic were far less well received than Biden's use of the same strategy. The same argument or behavior may be defended when done by those on our team but condemned when done by those on the opposite team.

Those who design popular social media platforms and AI tools have incredible power to shape the software of our minds and therefore our societies. Small changes made by a few people guided by A/B tests and optimizing algorithms deployed to millions can help or harm our ability to cooperate and to innovate. These decisions currently do not reflect a theory of everyone and the ways in which we evaluate who to learn from and what to learn. But by marrying an understanding of cultural evolution and engineering design, we can use the tremendous power of the Internet and social media to enhance the brilliance of our collective brain. The Internet gave birth to the Second Enlightenment, but the future of this Enlightenment and a true Second energy-based Industrial Revolution depends on getting our information architecture right.

With small changes, such as more deliberately exposing people to alternative views and communities who hold opposite beliefs; by providing contextual cues of trustworthiness, honesty, and reputation that we use in the real world; by offering audited incentives that may shape a person's genuinely held beliefs; and by cues of a person's identity and relevance to oneself, we can help trigger a creative explosion and become brighter.

12

Becoming Brighter

Humans have become cleverer. In the last few million years a lot of those upgrades were at the hardware level. Our brains grew bigger, literally tripling in size. Our default psychological approach switched from relying on our own abilities to relying on socially acquired information from others. Each of us became a cultural neuron wired into a collective brain. That collective brain was more brilliant than any one of us and in turn it made each neuron – each cultural brain – brighter.

Around 200,000 years ago that hardware-driven intelligence slowed down and if anything, it has begun to reverse. Brain sizes became smaller in the last 10,000 years or so. But a smaller brain didn't stop us from becoming brighter because hardware isn't the only thing that matters. We continued on a path to becoming more and more clever thanks to our shared culture. Our software improved.

Indeed, improvements in our software were probably the reason why our brains shrank. We didn't need all that energetically expensive hardware because we were distributing our thinking to collective computation.

As our individual brains got smaller, our collective brains got larger and more sophisticated. And as the software rather than the hardware became cleverer, the best bits of software could be shared throughout the social network. We could all upgrade our ability to think by acquiring the latest knowledge and learning new ways of thinking – numbers, hypotheticals, more formal logic, and much more. Each of us solved a little sliver of the world's problems and through selective social learning, shared and recombined the solutions into a highly effective cultural package that could then be transmitted to our children. We deferred and distributed the innovation computation from individual brains to the collective so that even the least able among

344

us could benefit from adaptive behaviors, advanced technologies, and life-improving health care in which we had played only a peripheral part or even no part in creating. This is how our collective brains made each of us cleverer even while our individual brains shrank – through software.

Much of our thinking that we now take for granted is not universal nor the way we thought since the beginning of our species, but instead comprises transmitted products of our culture that have now reached fixation – become ubiquitous – in many populations. Such products seem like human universals because everyone we know shares them. They are not. The ability to count indefinitely beyond fingers or body parts; to read, write, store, and learn ideas through text; the tendency to reason abstractly with syllogisms and enthymemes and approximations of formal logic – all were tools for thinking that were culturally created and then transmitted. All have made us brighter and more capable of performing the many tasks needed in the modern world. These processes have enabled the modern technologies we have created. As Friedrich Hayek put it, 'it's culture which has made us intelligent, not intelligence which has made culture.'

Since the advent of IQ tests, we have been able to measure these changes. The Flynn effect reveals that IQ has been rising, and the rise is fastest in recently developing countries. This may have been partly due to increases in nutrition and decreases in pollution, parasites, pathogens, and other health insults that damage our brains. These nutritional and health improvements also involve cultural and institutional software improving the social and physical environment using everything from increased agricultural output, better nutritional knowledge, and less polluting vehicles to mosquito nets and medicine. But a lot has been driven by purely cultural software updates downloaded directly into our cognition. One of the main means by which this download takes place is school.

As we discussed, schools are not a human universal. In many hunter-gatherer societies there's not even a lot of active teaching going on. The kids are left to hang around and watch the adults hunt, cook, build, and deliberate. Pastoralists have more to learn and do a little bit more deliberate instruction. When there's more to learn, we might spend more time teaching. But as the 2020 pandemic taught many parents, teaching is a difficult and time-consuming activity that prevents you from doing other things. And that's why we have specialist teachers. An investment in teaching children pays off at a population level when there's a lot to learn. The more complex and complicated the cultural package, the greater the pay-off.

Today, people freely find their comparative advantage by performing well on general tests and then picking a major or career that suits their interests and skills. But prior to the Industrial Revolution, most people learned their skills through an apprenticeship. They were restricted to whom they knew or simply learned the family trade from a parent or relative. It was only the very upper echelons of society that had tutors offering a personalized education.

Alexander the Great could be greater than those around him because he was one of the few who had access to Aristotle. Aristotle was a student of Plato and Plato was a student of Socrates. We know less about the influences that created Socrates, but he was no doubt a product of the Sophists with whom he argued and other cultural groups in the ancient Greek intellectual environment of the first millennia BCE. Ancient Greek culture was itself a product of cultural transmission and competition with earlier groups in North Africa, the Middle East, and Asia. Remember, different groups have dominated at different times, and who was 'civilized', 'barbarian', or 'savage' has gone around and around.

When the Industrial Revolution was full steam ahead, our society needed skilled factory workers, which motivated the expansion of compulsory formal education. This Victorian model of the school was itself a factory; a factory for producing good factory workers. Many of us no longer work in factories, but we have inherited the basic school design through a process of path dependence. Remember, when a child is born, they must catch up on the last several thousand years of human history. We now use schools as an efficient way to help them do that, delivering a base-level cultural package from which they can acquire everything else they need in the modern world. Today, as the world becomes more complex, educational innovations try harder and harder to pack in more, earlier, and more easily.

And so we are hungry for educational improvements. But they're hard to prove and harder to implement. We have begun to teach some skills and topics earlier; subjects have changed; we've dropped some things and included others. Many curricula now include computing as a fourth pillar to the traditional three Rs: reading, writing, arithmetic, and now algorithms.

This is where we are today.

As a result of path dependence there are many inefficiencies in our current system, especially in the overall structure of schooling. We tweak at the margins because massive change is difficult. But this inability to educationally innovate may be causing a Great Stagnation in improving our intelligence.

The Flynn effect has slowed down in the developed world and, in some cases, begun to reverse. But drawing on our theory of everyone we can see that there are various changes, both large and small, that may continue to make us cleverer, putting our foot down and accelerating the Flynn effect.

Accelerating the Flynn effect

In 1996, five years after gaining independence from the USSR, Estonia founded the Tiigrihüpe – Tiger Leap Foundation – to rethink and redevelop their education curriculum. With a virtually blank slate, they looked for best practices from around the world. Estonians are a people who value education, and they recognized the increasing role that technology would play in our lives. The goal was to leap ahead of other countries by creating the most technically savvy educational curriculum in the world, thereby creating the most technically savvy people in the world. The program to transform Estonia into E-stonia was launched by Toomas Hendrik Ilves, Jaak Aaviksoo, and Lennart Georg Meri with the recognition that the future performance of Estonia depended on the future performance of its people.

Ilves was Swedish-born and American-raised – his parents escaped Estonia after the Soviet occupation. He had a rich and diverse life experience which no doubt prepared his mind and made him an ideal magpie. He graduated as valedictorian from Leonia High School in New Jersey and then majored in psychology at Columbia, later completing a masters degree at the University of Pennsylvania. Ilves had a diverse career, working in education, research, the arts, and journalism in various places from Vancouver to Munich. In 1993, after the fall of communism, he served as Estonia's Ambassador to the United States before returning to Estonia in 1996 to become Minister of Foreign Affairs, and in 2006 became the country's third president since the fall of communism.

The other two figures were also magpies. Aaviksoo was Minister for Education, holding a PhD in physics. The president at the time, Meri was a peripatetic filmmaker and writer who had been educated at nine different schools across Europe.

These magpies with prepared minds (COMPASS Secret 3) put together an amazing team that included people like computer scientist Linnar Viik. They invested in three key pillars: computers and the Internet as pathways to knowledge, basic teacher training, and native-language electronic courseware. Whether they knew it or not, they were creating a more effective cultural evolutionary environment and cleverer collective brain. They had filled the car with gas and put their foot on the accelerator.

In 1991 only half of Estonia had access to a telephone. By 2001 all schools were Internet-connected and all students had access to computers. This connected Estonia to the rest of the world's collective brain. The curriculum was changed, encouraging and teaching students how to learn through the Internet. Teachers were trained in technology and were encouraged to seek out the best ideas from around the world and from each other. The platform SchoolLife was launched to create a teacher collective brain where teachers could share ideas, resources, and course materials.

Tiigrihüpe recognized the importance of collective brain thinking, sociality, high transmission fidelity, and sharing. Even today, the model embodies the full COMPASS approach and has continued to improve, offering opportunities for radical revolutions not only in education but governance, health, and every other aspect of Estonian society, unrestricted by the constraints of path dependence, and built on the back of high-quality technology training for both teachers and students. When the first generation of Tiger Leap Kids entered university and then the public and private sector, Estonia was transformed forever.

Estonia continues investing in its people through a cutting-edge education system that has maintained their students' position as the top non-Asian country in math, science, and reading. In 2012 it was the first country to start teaching programming and algorithms in elementary school to six-year-olds. In 2013 it was the first country to implement a radical approach to math education, spearheaded by Conrad Wolfram, younger brother of prodigy mathematician and physicist Stephen Wolfram of Wolfram Alpha and Mathematica.

In a 2010 TED talk titled 'Teaching Kids Real Math with Computers', a frustrated younger Wolfram explained that 'people confuse, in my view, the order of the invention of the tools with the order in which they should use them for teaching'. His point was that we teach mathematics in the same order it was invented and this way is not necessarily the most efficient or effective. In biology there is a concept called recapitulation theory which says that as organisms grow they go through their evolutionary history; ontogeny recapitulating phylogeny. By learning mathematics in the order it was invented, children go through a kind of cultural ontogeny recapitulating history. Start with the Greeks and go from there. But there's no reason that this should be the most efficient or effective way to teach mathematics. And indeed, as a result of this approach, we never get to twenty-first-century advancements in probability and statistics, which are probably more valuable than calculating angles in triangles or memorizing rules for figuring out the length of a hypotenuse. Trigonometry, if it needs to be covered at all, doesn't need to be taught before algebra. And algebra doesn't need to be taught before calculus.

The difficulty many students face in calculus, for example, is in the mechanics. Remembering and algorithmically following the chain rule or quotient rule and knowing when to use them.

You may have forgotten these and perhaps you never really understood why they worked. Both facts probably had zero impact on your life – even if you're an engineer. And that's because calculus – and indeed mathematics in general – is not about the mechanics, it's about the thinking. What does it mean to take a derivative? What does it mean to calculate an integral? When and why are these useful? You can learn this intuitively, developing an intuition for mathematical reasoning, without knowing anything about chain or quotient rules, which are techniques invented for a world without computers.

Math isn't about adding and subtracting or remembering rules for calculating partial derivatives. It is about logic and reasoning, only sometimes with numbers. And in the real world, the mechanics of mathematics are done on computers, not on paper. We can introduce concepts such as derivatives and integrals in elementary school alongside programming, leaving the computation to the computer, and introducing mechanics later.

Conrad's thinking was just the latest approach that Estonia had sought to becoming brighter. It's no wonder that Estonian students leaped forward in their PISA scores across all subjects to become the top-performing non-Asian country in the world!

Estonia's success is a lesson to us all. It goes beyond the use of technology or any specific content. The secret is a product of an innovation mindset (distilled in the COMPASS approach), high cultural and social value being placed on education, and cooperative commitment to their people and future. These in turn led to Estonia plugging its population into the world's collective brain and, like magpies, borrowing and recombining a broader educational, cultural, and institutional infrastructure.

Any country or even individuals can do this.

From Shanghai to Sydney to San Francisco there is much that could be and should be shared and cross-pollinated. Not just content and curricula but also attitudes toward education itself. As just one example, take cross-cultural differences in attitudes toward mathematics.

In the West you'll often hear people saying something along the lines of 'I was never really good at math.' Rarely will you hear 'I was never really good at reading.' Western attitudes toward numeracy and literacy betray hidden assumptions not present in, for example, much of Asia.

It's true that not everyone needs to interpret Tolstoy, but if you can't read then you have to trust others to interpret a world of information for you, just as an illiterate person did in the past. Similarly, it's true that not everyone needs to transform tensors, but if you can't do simple math then you need someone to interpret personal finances, probabilities, and health decisions for you.

Education systems in many WEIRD countries have let down many students. Students who show an early aptitude in math do fine, but those who do not are not taught that they are just as capable at learning mathematical skills as their peers. Many WEIRD cultures have had to compensate for a lack of individual skill by legislating simplicity at a societal level – simplify investments, mortgages, and the presentation of important statistics. Despite these societal crutches, people fail to understand concepts such as exponential growth, contributing to the failure to save at an earlier age for their retirement, susceptibility to poor credit card usage, inability to compare loan and investment decisions, and suboptimal decisions in everything from health insurance to home loans.

We know that this is a correctable problem because the asymmetry between numeracy and literacy is not a cultural universal.

In many cultures math is not seen as an inherent trait that only some are good at. It's seen more like reading – a skill that requires practice. And in these cultures children perform better at mathematics, leading to stereotypes such as 'Asians are good at math.' In reality, Asians are good at realizing that math is a skill that can be learned and developed with the right instruction and attitude. For me, this point was starkly made during the 2020 pandemic when schools were closed.

Drawing on the experience of my friend from engineering school, Clinton Freeman, who had been home-schooling his daughters for some time, we took the opportunity during the lockdown to bring together different curricula around the world and test how our children reacted. In math, it was astonishing the progress our kids made using the Singaporean and Shanghai curricula (which have a lot of cross-pollination with each other). Simple things like drawing on relationships across multiple areas of mathematics or multiple approaches to solving the same problem encouraged generalization; explicit logical reasoning and clear and precise explanations for why and how an approach worked and hypotheticals about what would happen if something changed encouraged thinking about proofs; and the early use of conventional mathematical language, such as learning that letters could be numbers and introducing simple pre-algebra, removed barriers for later learning. As a result of little changes such as these, China and Singapore are able to teach algebra in elementary school, a subject reserved for secondary school in the United Kingdom. It is perhaps no surprise that China and Singapore top the PISA tables. I was good at math in school but was astonished as I watched my then six-year-old competently solve for the unknown variable 'x' in an equation and very quickly advance to rearranging multivariable equations to solve for different values. All thanks to small differences in how they were taught.

The general point is that there are low-hanging intellectual arbitrage opportunities (COMPASS Secret 3) for becoming brighter if we're willing to go off the beaten path (COMPASS Secret 2) of decisions made in the past. It requires sufficient school funding, investment in retraining teachers, and incentives to do things differently, but also a realization that it's not about math education nor overly competitive pushy educational curricula but rather the flexibility of the human mind and recognizing what we could be capable of if we try to find out.

Our psychology, in other words, is highly hackable. We are not blank slates but our minds are highly flexible. Formal education is the primary means by which we transmit our cultural package to the next generation. The things we assume people are bad at are often the things we don't prioritize, don't teach, or don't do enough research in figuring out how to teach better. For example, we assume people are highly susceptible to logical fallacies – straw-man arguments, *ad hominem* attacks, appeals to authority, or confusing correlation with causation – but formal logic, reason, and fallacies are also things we rarely formerly teach children at a young age in the English-speaking world. For a species so dependent on its software, logical fallacies need not be a permanent foible. Just as traditional societies were able to learn how to count, we can learn how to reason. And as the Flynn effect makes clear, we have always been capable of more than we currently do. Children from the 1940s learned far less than twenty-first-century children but would have done just as well if presented with a modern curriculum.

There are many degrees of freedom for change, but it requires a shift in attitude, a willingness to experiment, and political will, perhaps in a start-up city.

Consider even the format of the school day, which need not be 9 a.m. to 3 p.m. or similar partial days that force parents to leave work to pick up their children. This system may be a legacy of a time when many children would go home to help on a family farm and when many mothers stayed at home and could pick up their children in the middle of the working day. An alternative arrangement would be one that matched a typical adult working day, with lots of breaks, which reduced as children grew older. Sports, extracurricular activities, homework, and weekend work could be built into this time. Extracurricular activities would become curricular activities and homework would just be schoolwork. Expensive childcare could instead be reallocated toward reducing the cost of these extra hours. And as Finland demonstrates, paying teachers a salary comparable to other high-prestige careers and offering opportunities for development increases the prestige of the career, and attracts the best minds to take up one of the most noble and valuable of professions, charged with transmitting the light of knowledge to the next generation.

The structure of courses and assessments, too, is open to innovation. In 2017 the then president of the Royal Society – the world's oldest scientific society and Britain's top scientific society – Venki Ramakrishnan, condemned Britain's A levels in which students take just three to four final subjects assessed by subject-based national exams as 'no longer fit for purpose'. A levels are an educational model that is too narrow. Asking students at age sixteen what they want to specialize in leaves huge holes in their knowledge of the world that are rarely filled.

From a collective brain perspective and from my own experience moving through multiple educational systems and now teaching students who have gone through every major curriculum in the world, studying only three to four subjects at an advanced level leaves students ignorant of so much, preventing them from drawing the valuable connections needed for intellectual arbitrage and other means of innovation in the modern world. A broader and more balanced curriculum is necessary to better connect the collective brain and solve the paradox of diversity created by over-specialization. This skill perhaps is even more valuable in the age of AI.

In addition to these macro-level changes, within schools there are many micro-level opportunities for innovation. We train specialist teachers because the gap between children and adults is so large that extensive training is required to bridge it. 'Hey grown-ups, this is how kids learn.'

In stark contrast, in many traditional societies there is less adult–child instruction. Instead, children learn from slightly older children. Five-year-olds learn from six- and seven-year-olds, seven-year-olds learn from eight- and nine-year-olds, and so on. This smaller age gap offers a more gradual learning gradient, making it easier to understand what is being taught. This gradient has been lost under our current system. But that doesn't mean we need to go back to these other models, only that we can learn from them.

We typically teach children in age-group cohorts, which is assumed to represent similar skill levels and is therefore more efficient. But as any teacher knows, that assumption of similar skill levels is false. Skills vary dramatically between children and in different subjects. Inevitably, some struggle to keep up and others are held back from their full potential and this may differ from subject to subject. Modern technology allows us to create a more personalized education or even cohorts based on skill and maturity rather than age. Students then learn how to learn, an essential skill in a quickly changing world.

There are various other approaches to rethinking education. Famous among these is Elon Musk's Astranova school, formerly Ad Astra. The school is based on pillars such as first principle learning and real-world relevance. The idea is captured by contrasting two ways to teach children about an engine. One approach is to start by teaching them about tools like a screwdriver and wrench and how they're used. Eventually students learn how these tools help take an engine apart. It's an approach that focuses on learning the constituent parts and building them up to the final product. But along the way, many students fail to realize the relevance of what they're learning, which affects their motivation. They ask, 'What's the point of learning this?'

An alternative approach is to give students access to the tools and an engine and get them to take it apart and put it back together, learning the principles as they undertake the task – starting at the beginning and at the end and meeting in the middle with a practical and relevant focus throughout. Just as it was easier to derive the principles of hydraulics after we had steam engines, it's easier to learn anything by doing it and seeing its most real-world relevance. This approach is one of many and may not even be the best approach, but such innovations are necessary to step off the beaten path and step on the accelerator of the Flynn effect.

Schools are trapped in suboptimal local equilibria and seemingly unprepared for the demands of our current world. Many parents recognize this and so to better prepare their progeny, they compensate for these inadequacies by supplementing what public systems, and even private systems, offer. Not all parents have the resources, skills, or time to do that. The increasing irrelevance of public education to the acquisition of everyday life skills further reinforces inequality and group differences, which is costly to us all.

Not all the kids are all right

Talent is equally distributed; opportunity is not. That's not strictly true.

Our genes vary sufficiently that in a perfectly equal world in which everyone had the same resources, the same parenting environment, the same educational opportunities, the same cultural input, and the same access to unpolluted air, nutritious food, and clean water, there would still be inequality of outcomes. But while there may be prodigies and genetic geniuses – John von Neumanns and Terry Taos – there are many more with unrealized potential. Indeed, those we call geniuses on the basis of their contributions may have simply been 'bright enough' but in the right place in their collective brain to be the nexus of ideas. They stood out not because of raw talent but because of greater personal opportunity in a world of more unequal opportunity than today.

Sometimes it feels like there were more geniuses in the past, and people wonder where all these great minds have gone. One possibility is that it's not so much that there were more geniuses in the past, but that there were just fewer people with the necessary education, access to books, networks of knowledge, and resources. This is an answer to today's missing geniuses. Newton was the peak of a molehill at a time when rates of literacy, let alone higher learning, were incredibly low and there was much low-hanging fruit to be discovered. It is statistically unlikely, given rates of social mobility at the time – at a time when so few had access to education – that Newton happened to be the person with the most potential in England.

Similarly, Einstein and von Neumann both sat on a slightly larger but still very small hill. Today, it is almost statistically certain that there are many more people of equal potential and genius as Newton, Einstein, and von Neumann working in tech, top universities, and top finance firms. Or put it another way, take the average Big Tech engineer, physics professor, or finance quantitative analyst back to the seventeenth century without any knowledge of today and put them in Newton's situation, and they will probably re-derive Newton's laws and perhaps do much more.

Geniuses as we label them are not just genetic geniuses but are a product of their cultural software written by their position in a collective brain. Given the increase in population size and advancements in education since Newton's time, there are far too many Newton-level geniuses for any to be particularly noticed or make the history books in quite the same way as when the competition was lower. But given rates of poverty and the world population, it is also almost certain that there are even more Newtons, Einsteins, and von Neumanns who work far more modest jobs than tech, academia, or finance, simply for lack of opportunity. Born in another place, even today, Einstein may have lived out his days as a quiet clerk. Many could-be Einsteins still do.

And so, talent may not be equally distributed, but opportunity *definitely* isn't. We are a long way from an equal world. The potential for talent is far more equally distributed than is the opportunity to nurture that talent and have it benefit us all. As we discussed in Chapters 7 and 9, so much human potential is lost to the vast inequalities entrenched in our systems by the fractures between us and the unearned intergenerational transmission of wealth.

Evidence for this can be seen in the social mobility data. Social mobility indices track the correlation between a child and their parents' socioeconomic status. That is, to what degree your wealth, income, educational and other lifetime outcomes are determined by those of your parents. Obviously, some portion of that correlation can be attributed to genes, but the cross-national differences in this mobility are revealing.

It should be no surprise that the highest social mobility can be found in the Nordic countries, Denmark and Norway consistently topping the list. As we learned earlier when comparing Norway to the United Kingdom, being a poorer Norwegian doesn't prevent you from accessing high quality education, good food, and a safe and pollution-free neighborhood. Poorer and richer Norwegians mix more freely, allowing for the flow of ideas and culture, keeping a unified cultural-group and reducing inequality. And as a result, genes become a better predictor of outcomes in Norway than in Britain and a better predictor in Britain than in the United States. Mobility is higher and therefore so is heritability.

This contradicts ideas that the rich stay rich due to better genes or greater talent. As we learned, genes are a stronger predictor of cognitive ability among the wealthy in the United States, although not the poor, but a better predictor across the population in places like Norway.

It may be counterintuitive, but a more equal society is one in which genes play a *greater* role in success.

I hope these cases and policies inspire change around the world and offer a playbook for how to get there. All of them are approaches derived from a theory of everyone to accelerate the Flynn effect at a population level to give everyone the best opportunities in life to maximize their potential – not just for their own sake, but for the sake of our collective future. The goal is to maximize the probability of our children being healthy, wealthy, attractive, successful, and happy humans.

Maximizing human potential

Countries that invest more in education have greater intergenerational mobility. Even within a country like the United States, states that invest more in education have greater intergenerational mobility than those that invest less. The message is clear: more educational opportunities can help a society discover and nurture the next generation needed to take us to the next energy level.

Unfortunately, intergenerational mobility has been falling; wealth and other outcomes are becoming more entrenched. And as the energy ceiling falls, for the first time in a long time, by many metrics, children are leading worse lives than their parents.

An American born in the 1940s had a greater than 90% probability of being better off than their parents, a result of both economic growth and intergenerational mobility. A child born in the 1960s had around a 60% probability. The American Dream died in the 1980s with the oil crisis, when the probability of a child born in that decade earning more than their parents became a coin toss: fifty-fifty.

There remains tremendous variation in intergenerational mobility across the United States and across the world.

One commonly used intergenerational mobility metric is intergenerational elasticity of income (IGE), the percentage of a person's income that can be predicted based on how much their parents made. It's around 50% in both the United States and United Kingdom, which means 50% of a child's income can be explained by their parents' income and 50% by other factors, such as education, luck, or hard work. In contrast, IGE is less than 20% in more equal Canada, Finland, Norway, and Denmark, which means just 20% of a child's income is explained by their parents' income and 80% is explained by other factors. Financially struggling parents are far less likely to have an impact on their children's future finances in Canada or Denmark than in the United States or United Kingdom.

Drawing on our theory of everyone, we can consider what it takes to give everyone the best opportunity. I'll frame these as policies and as individual choices, but recognize that there are many barriers to implementing such policies and that most people don't have the ability to make such choices. Poorer people aren't always making poor choices, they simply have poor choices to choose from. Nonetheless, knowing the general direction helps us to steer away from where we don't want to go and toward where we do.

At a population level, it's almost obvious, but simply investing in education has a large return. A cleverer country is a better country for everyone. As John Green, author of *The Fault in Our Stars* and host of the successful YouTube series *CrashCourse*, explained,

> Public education does not exist for the benefit of students or the benefit of their parents. It exists for the benefit of the social order. We have discovered as a species that it is useful to have an educated population. You do not need to be a student or have a child who is a student to benefit from public education. Every second of every day of your life, you benefit from public education. So let me explain why I like to pay taxes for schools, even though I don't personally have a kid in school: It's because I don't like living in a country with a bunch of stupid people.

But it's not just financial investment in schools, it's how that money is used, the context and culture children find themselves in, the other values that are encouraged such as hard work and persistence, and the possible aspirational futures children can see through the lens of their evolved social learning psychology, sensitive to the success and pathways of potential models available to them.

Groups differ in their outcomes as measured by IQ, educational attainment, income, wealth, health, and lifespan, among other metrics. A constellation of cultural traits is required in order to succeed. If any are missing, you're less likely to get to the top: an excellent education, ambition, willingness to work hard, connections, resources, ability and willingness to take risks, and so on. And so given imperfect transmission and different cultural traits between individuals, families, and entire groups, it's not surprising to see group differences. That's an understatement. It would be astonishing if all groups had the same outcomes. But the size of the gap between groups is within our control.

The opportunities for tackling group differences upstream – childhood environments and opportunities – are much greater than where we tend to focus downstream – university places and jobs.

Many communities, even across the United Kingdom and the United States, suffer from ongoing brain hardware assaults from pollution, disease, insufficient nutrition, exposure to smoking, and/or toxins such as lead. Several reports, for example, reveal that many UK residents suffer from lead poisoning caused by lead leaking from old pipes into their water supply. I tested the water at our house and discovered unsafe levels – and our home is not in an area known for lead poisoning. Areas known to be even higher in lead poisoning, such as Glasgow, correlate with lower school performance and higher juvenile delinquency, among other associated traits.

Lead abatement alone may have large effects, if we measured and invested in it. Research is ongoing. A study published in 2022 in the *Proceedings of the National Academy of Sciences* suggested that leaded petrol – common in the twentieth century – reduced the IQ of half the American population.

Upstream group differences can also be found in the culture of families, how children are raised, the stability of parent relationships, and quality of schools. All are amenable to policy levers that lead to better lives.

By the end of high school, these differences can have remarkable outcomes. Estimates from the United States suggest that in 2016 no more than 2,200 Blacks and 4,900 Latinos scored above 700 on the math SAT. The maximum score was 800. In contrast, at least 48,000 Whites and 52,800 Asians scored over 700. Looking at even higher percentiles, say over 750, the number drops to 1,000 Blacks and 2,400 Latinos, compared to 16,000 Whites and 29,570 Asians across the entire United States. So although 51% of test takers are White, 21% are Latino, 14% are Black, and 14% are Asian, among those who score over 750, 60% are Asian, 33% are White, 5% are Latino, and 2% are Black.

The same forces that cause these differences propagate to universities, colleges, and workplaces. As with many public health interventions, getting in early can have larger effects, but rather than deal with these difficult multifaceted challenges upstream, we often take an easier but more defeatist approach. This well-intentioned strategy of affirmative action, quotas, removing standardized testing, or many anti-racist policies is perhaps implicitly racist insofar as it creates low expectations for some groups on the assumption that the problem is unsolvable upstream.

Affirmative action and anti-racist policies attempt to tackle very real challenges. Indeed, they can be useful in, for example, offering role models that children can aspire to or simply generating awareness of the deep inequities. They are also based on correct assumptions that these are very real issues that exist at a cultural and structural level. But the forces that lead to these large differences in school performance are not fixed by downstream patches. And because the ultimate root cause of the group differences remains unaddressed, we are left with an industrious ecosystem of proximate further fixes, policies, approaches, and advocacy that creates skewed incentives and unintended, sometimes self-defeating consequences. All because we never tackled the upstream, ultimate root causes.

As an example, consider research from Duke University looking at the trajectory of Black and White students. Compared to White students, Black students are far more likely to switch from more difficult to easier majors. This difference is *entirely* predicted by upstream high-school preparation, *before* they arrived on campus. In turn, these decisions and reduction in Black representation in more difficult majors have downstream effects on employability and lifetime income that are not resolved by admission to Duke.

To put it simply, policies that admit less-prepared students do not solve a performance gap that needed to be solved before students arrived at university. The gap needed to be closed by helping high schools better prepare their students. In turn, differences in high-school performance needed to be solved by helping elementary schools prepare their students for high-school. In turn, the gap in elementary schooling needed to be solved by helping mothers and fathers stay healthy and produce healthy offspring and provide a more nurturing environment.

The downstream effects of upstream differences can also be seen in the experience of migrants, where immigration policy fails to consider factors that affect the success of migrants. For example, in the United Kingdom, European migrants make a greater fiscal contribution than do non-European migrants, at least in the first generation. Data from France, Germany, and Canada show the same pattern – highly skilled, more culturally close migrants make the greatest economic contribution in terms of income and subsequent taxes. They are a larger net gain for the economy.

More highly educated migrants tend to have higher incomes and therefore contribute greater economic value, paying for themselves, so to speak – especially when those skills are needed and missing in the economy. Economics is not everything of course, but it is essential to supporting a social welfare state that takes care of those with disabilities, who have fallen on hard times, or are less well off for a variety of reasons. Social welfare states are supported by the tax base, so the more productive workforce supports the less productive members of a society. As a result, we need to carefully consider the welfare implications of people with different skill sets and how they contribute to the productivity of the tax base. If not, we need to consider how we will support them, or support them in becoming more productive.

But not everything is on the migrant side. Even with the same education and qualifications, migrants from culturally distant countries have lower incomes. This may be simply due to discrimination, but discrimination is not simple.

One compelling discrimination audit research design involves sending out identical CVs – same degree, university, work experience, and so on – changing only a name. For example, in a Western context, some employers might get a CV from say Abdul Mohammad while others might get an otherwise identical CV from Adam McKinsey. The Mohammads get fewer callbacks despite having the same education, same experience, and indeed otherwise identical CVs right down to formatting. Studies adjusting the names – stereotypical male versus female, minority versus Western, and so on, offer compelling evidence of discrimination on the basis of sex and ethnicity.

But evidence for discrimination isn't large enough to explain the entire income difference for culturally distant migrants or different ethnic groups. Moreover, discrimination isn't an ultimate explanation. We need to explain *why* some people discriminate more than others and why there is more discrimination for some groups than others even when they share many characteristics. Remember the income differences by immigrant country of origin that we discussed in Chapter 3? Consistent with these patterns, research reveals that South Asians are over-represented in executive leadership positions compared to East Asians, but also reveals that South Asians face *more* discrimination than East Asians. But, as this disaggregated data reveals, this too is a broad generalization.

The luminary South Asian CEOs of many American companies, particularly tech companies such as Microsoft, Alphabet, IBM, Twitter, and OnlyFans, are not South Asian broadly speaking; they are of Indian ancestry. But this *too* is too broad a generalization. These CEOs aren't just Indian, they are overwhelmingly upper-class Brahmins, who represent less than 5% of India. As you can see, the measurement of culture and its application is complicated without a theory of everyone. Broad-brush assumptions or statements about discrimination fail to understand the degree of discrimination, at whom it is aimed, and its ultimate causes. Simplistic accounts such as 'people are racist' do not explain why some people are more racist or more so toward some groups and even subgroups than others. A fuller explanation requires understanding what cues people are using to discriminate, and why.

Some of this may be due to perceived differences in the quality of education in different parts of the world. Migrants moving to Sweden from Switzerland may be more culturally close and have a better educational experience than those from more culturally distant Syria. But even with the same educational qualifications, Switzerland may have better schools and training than Syria. The effects of war and economic differences alone would create such educational differences. Both training and the cultural challenges of coordination and communication are known to result in differences in work performance.

Resolving challenges downstream by focusing on the tendency to discriminate, but not understanding what causes differences in discrimination, not only foments conflict between existing societal fractures but even risks rationalizing racism, sexism, or other forms of prejudice.

Let's revisit the topic of variations in SAT scores among different racial groups, a subject we touched on earlier. These variations can persist into university life, influencing academic performance. Universities aiming to boost the representation of minority groups might consider accepting lower scores from these groups compared to others. Consequently, race or gender could be seen as indicators of differing admission standards or academic performance. Now, imagine an employer reviewing identical resumes, but without access to the candidates' grades – only their degrees (since grades are typically not included on resumes). If this employer's goal is to select the highest performing candidate, they might make assumptions about the academic performance of a minority candidate, based on the knowledge that universities sometimes accept lower SAT scores from under-represented groups. While it's important to acknowledge that these candidates may have faced significant challenges, these challenges were encountered earlier in their journey and could potentially impact their performance in the workplace. An employer focused on maximizing performance might be concerned about this. However, in a scenario where affirmative action is not in play, the same employer might view the minority candidate more favorably, recognizing that their journey to this point may have been more challenging. If admission standards are consistent across all groups, it can be assumed that all candidates possess equal skills. In fact, under-represented minorities may even be seen as having an edge, considering the obstacles they've likely overcome.

By solving problems downstream rather than upstream, we risk creating rational racists and Bayesian bigots, who behave in a way that a computer just trying to optimize outcomes would also behave. And then, in order to reduce the effects of this discrimination, we need further fixes for the further problems that in turn further distort the goal of an efficient and fair system. We are a long way from an efficient system, but that doesn't mean we should embrace inefficiencies rather than try to improve them.

Problems flow downstream and when we try to fix them downstream instead of upstream, at the source, we create distortions that undermine values such as meritocracy or free speech. The trouble is that some people use these legitimate criticisms of policies such as affirmative action as an excuse to do nothing. But it is not sufficient to say that a model is broken – we need to develop a better model. We need to do as much or more work to fix these problems at their upstream source.

Many popular anti-racist approaches, though well intentioned, undermine our commitments to free speech and meritocratic promotion, demonize sections of the population deemed to have privilege, and poison intergroup relations. But the alternative of doing nothing is also not an option.

When someone is injured, we must staunch the bleeding and also treat the symptoms. To truly heal our society we must not only invest in bandages and symptom relief but also remove the underlying causes of the injuries. A theory of everyone demands a systems-level ultimate policy approach.

Harvard's Opportunity Insights initiative, led by economist Raj Chetty, is an example of a research program that brings the latest and most powerful data analytic, AI, and machine-learning tools to bear on the largest datasets with the goal of developing scalable solutions that close these gaps upstream. Although progress has been made in many small projects, large-scale initiatives like those pioneered by Opportunity Insights are rarer. Opportunity Insights confirms the effects of wealth inequality, which may be reduced by policies such as land value taxes, as discussed in Chapter 9. It also confirms the important role of cultural transmission and knitting together our collective brain. For example, children with poorer parents who have an opportunity to grow up in neighborhoods where there is more interaction between the rich and poor have considerably better outcomes. Though not measured, presumably children of wealthier parents also have an opportunity to develop a better understanding of the structural and material challenges faced by other, less-privileged members of society.

Insights such as these have led to scalable initiatives like the Creating Moves to Opportunity (CMTO) project and the Charlotte Opportunity Initiative Collaboration.

Every year, the United States spends approximately $20 billion on rental assistance through the Housing Choice Voucher Program. CMTO helps redeploy these funds in ways that increase the educational and life outcome returns on the money. For example, Opportunity Insights helps low-income families identify neighborhoods with opportunities to better meet their needs and then coaches and helps them apply for houses that bring them to these neighborhoods. Children who are able to move to these neighborhoods are more likely to attend college, attend higher-quality colleges, and are less likely to become single parents. The effects are stronger the longer children are able to live in neighborhoods with more opportunity.

Alongside mobility approaches like CMTO are initiatives that use data-driven approaches to support local communities in improving themselves. Opportunity Insights identified Charlotte, North Carolina, as having the lowest rate of upward social mobility of the fifty largest metropolitan areas in the United States. The Charlotte Opportunity Initiative Collaboration works with local leaders to make targeted investments that increase the likelihood of children from low-income families rising up the social ladder. Support includes helping local leaders make direct investments in health and education, from ensuring adequate health care and increasing access to high-quality pre-schools through to supporting college applications. The collaboration also helps desegregate neighborhoods to increase cultural transmission, using similar approaches to CMTO and through re-evaluating housing policies.

Opportunity Insights is just one example of where the latest data science and big data are being used to tackle the challenges of inequality in educational, income, and overall life outcomes. The approach is a way to help us become brighter. It is more difficult than any preferential admission policy, but it tackles the hard problem of creating a fairer world rather than miring us deeper in the mess created by the legacy of past injustices. It is imperative that we tackle these challenges as we enter a world of AI and machine-learning systems. Our computational companions are quickly entering every aspect of our lives and, despite no evolved xenophobic tendencies, nonetheless instantiate the kind of rational racism and Bayesian bigotry discussed before, simply by picking up on patterns in our unfair world. But these AI and machine-learning systems also have the capacity to make us brighter by joining genes, culture, and individual experience as a fourth line of information.

Artificial intelligence and machine learning: A fourth line of information

Marvin Minsky was a god in the world of AI. No history of AI can be written without mentioning his contributions. The dominant paradigm for building artificial intelligence from the 1950s to the 1990s was the symbolic approach. Human intelligence was assumed to lie in our ability to reason with a rich body of information. Therefore, our ability to program logic and apply it to rich representations of knowledge, such as in silico semantic networks, led to wild optimism about the creation of human-like intelligent machines. In an interview with *Life* magazine in 1970, Minsky optimistically declared that 'In from three to eight years we will have a machine with the general intelligence of an average human being.' But he was wrong. In fact, the assumption of intelligence as logic applied to knowledge was wrong. The secret to human intelligence was not just in our logic and knowledge representations. At the very least, these were difficult to program directly.

An alternative approach, notably associated with Geoffrey Hinton, was connectionist. Rather than directly programming logic and knowledge, Hinton, who had a degree in experimental psychology, sought to represent the very brain itself in silico – an artificial neural network. Initially, success was limited. Minsky famously stood up after a talk by a young computer scientist, who had just presented a neural network approach to AI, and asked: 'How can an intelligent young man like you waste your time with something like this? . . . This is an idea with no future'. What was missing from the connectionist approach was sufficiently large networks and sufficiently large training datasets.

In the early twenty-first century neural networks proved the value of the connectionist approach thanks to more powerful compute and vast troves of data that enabled the training of deeper and larger neural networks. Hinton was vindicated. The neural network approach has continued to offer surprising successes, from machines beating humans at board games such as chess or Go to writing essays and creating art from a description.

The neural network approach has continued to take inspiration from human analogues. The focus has been on neuroscience and brain architecture. AI researchers now recognize that this approach is improvable or even insufficient.

The theory of everyone reveals that human intelligence isn't just a function of substantial neural hardware but of socially acquired software. That is a paradigm shift. Human intelligence is not simply a result of our brains but of the sophisticated social learning strategies we use to acquire information from large, cooperative, collective networks of other humans.

Our intelligence is a product of 'pre-training' by millions of years of genetically evolved hardware and thousands of years of culturally evolved software, adjusted over a lifetime of experience. This body of research offers a new paradigm for AI. Within this, the focus isn't just on the neural hardware replicating a single human brain but on the way our many brains evolved for cultural learning as a collective brain, making each of our brains cleverer. We shaped our machines, our culture, and our technology, but they in turn shape us and our children. The addition of machine intelligences to our collective brain has the potential to truly supplement our cognition in ways the existence of computers has only scratched the surface.

We have used energy to power machines that supplement our muscles. We have used energy to power machines that calculate for us and connect us. The next step is to use energy to power machines that truly supplement our minds.

Remember that our three main lines of information – genetic, cultural, and individual learning – are all reinforcement learning with different limits and lags. Ultimately, all reward what works and punish what doesn't over different timescales with different information sets. Machine learning is the missing line of information, parsing the world's data, with the capacity to do it in a personalized way. It is the combination of cultural and individual learning.

To paraphrase a common joke in the AI community, trial and error, small changes to your software, is bad coding practice. But do it fast enough and it's machine learning. Machine learning is individual learning on steroids.

Machine-learning algorithms can parse our large cultural corpus to discover patterns that cultural evolution may miss. Moreover, they can help us make better decisions. If you want to know what can make you happier, wealthier, or more attractive, you are often forced to rely on data and evidence about the average person: what makes the average person happier, wealthier, or more attractive.

But none of us is an average person. The average person is, ironically, *rare*. Machine learning, however, by combining your data with the data of all people like you, can help you see what makes people like *you* happier, wealthier, and more attractive. It can also make you cleverer.

First, simply by providing powerful ways to crunch data and see new patterns. And second, by humans building on machine-powered discoveries.

In 1997 AI Deep Blue beat then world chess champion Garry Kasparov. Soon after the victory of machine over man, the *New York Times* ran a story with experts weighing in that winning at chess might be achievable by a machine, but AI was unlikely to beat human players at Go because the subtleties and space of possible moves is orders of magnitude larger.

In 2017 AI AlphaGo defeated the best Go player in the world, Ke Jie. What's interesting is that since then human players have improved by learning from how AlphaGo plays, discovering new moves and playing styles. Just as the steam engine helped us discover the laws of thermodynamics, AI is teaching us about how we think and improving our cultural software.

Machines now write stories. They write profound poetry. They draw pictures from descriptions. They write functional computer code. In this era of astonishing successes with no current limit in sight, people endlessly speculate about what the future holds. Computers doing science and engineering, improving themselves and leading to unprecedented quick advancement, perhaps helping us crack fusion and enter the next level of energy abundance. Computers achieving Artificial General Intelligence, brighter than any human. Perhaps even digital people, able to do more than we ever could, embodied completely in software living on any planet that has energy and conditions to run computer servers.

More progress is needed to know the true limits of what machines can achieve and our role in all of this. The tides of progress can only be held back for so long. Even if one country tries to protect human jobs at the expense of Great Stagnation, other countries will push forward toward a creative explosion, just as China is developing hundreds of nuclear reactors, speeding up their progress toward energy abundance.

We have a new fourth line of informational inheritance – artificial intelligence. We may not yet know the limits of what AI is capable of, but we do know that AI has the capacity to make each of us cleverer.

AI empowers the law of innovation and may help us crack the next level of energy. Biologist Carl Bergstrom described DALL-E 2, an AI which can create new images based purely on a description, as tapping into our collective unconsciousness. DALL-E 2, like text-generating AIs such as GPT (generative pretrained transformer), is perhaps the most sophisticated magpie we can conceive of creating, searching a latent space of possibilities and creating new art and writing never before seen. It is recombinatorial creativity in a very human sense. The same principle could be used in scientific discovery, which is based on similar principles. We can expect to see advancements ranging from protein folding to gene editing to advancements in our theory of everyone.

AI also empowers the law of cooperation, making us more cooperative by helping coordinate behavior. Today, people will often use memes or pop-cultural references to communicate the right metaphors and emotions. Similarly, in the near future, commonly used AI will help us find common ground to coordinate and communicate. As people begin to interact more with these artificial agents, their influence will grow. In turn, if these agents offer similar advice on appropriate responses to a particular context or appropriate behaviors in general, they will effectively become mediators, culturally diffusing these norms. In turn, this will help people coordinate and communicate through a more shared culture.

But AI also has the capacity to further exacerbate inequalities and fractures created by our current social systems. It is an innovation in efficiency that lets fewer people do more with less leading to lower scales of cooperation entrenching a few with power over many. In some domains, nations such as China and Russia have an advantage. AI is dependent not just on powerful computers but on large troves of data. Large language models such as GPT benefit from large corpuses of digitized English text. In other domains such as facial recognition and medical diagnoses, China and Russia can advance more quickly thanks to weaker privacy protection laws. China isn't quite as worried about using medical data or footage from CCTVs. Indeed, China has become a large exporter of AI facial recognition technologies. The advancement of AI-enhanced surveillance drives upgrades to high-resolution CCTV cameras in China and the rise of desk-based 'smart cops' replacing police on the streets.

There will soon be AI workers in our economy working alongside humans. In some cases they will replace humans, in others they will cooperate and enhance human abilities. The ability to work with agents, such as the new skill of 'prompt engineering' – learning the best approach and phrasing to ask an AI to perform a task – is quickly becoming more valuable.

The creation of these agents also exacerbates our need for energy. Unlike your phone, laptop, gaming rig, or even home server, which use very little energy, a lot of energy is needed for the computing power needed to train AI agents. These energy requirements grow every year because there are more agents trained for different tasks and because these agents train on larger datasets and become more sophisticated with more parameters. Of course, once trained, these agents can do more work per watt than the equivalent number of humans. Their control enhances the power of those few who control them. Every company in the world will be changed by AI as profoundly as they were changed by the Internet itself.

This may very well be the most important century in human history, but if our future is to be in the stars then we must resolve the cultural, institutional, and social challenges that threaten humanity. We must use the laws of life and the theory of everyone to pick the brighter future.

Conclusion

In *An Inconvenient Truth*, Al Gore uses the analogy of a frog placed in tepid water whose temperature is slowly raised to illustrate our disregard for the warning signs of a changing climate. It's a clever analogy that's almost certainly wrong. Animals have powerful instincts to survive in precarious and danger-filled environments. As soon as the frog becomes uncomfortable, it will leap out. In the same way, humans are not oblivious to the changes happening around them. We can feel in our bones that something is wrong. Like the frog, we are trying to make the leap.

The problem is not that we don't notice the warning signs, it's that we don't know what to do about it. As excess energy falls, life becomes harder. As the challenges posed by climate change, such as large influxes of people, weaken our institutions, our societies fracture and it becomes difficult for us to realize the role of energy, innovation, cooperation, and evolution, and how together these laws have created us and our civilizations. The resulting anger and frustration can make our societies ungovernable, and we may lose faith in the fairness of our systems and in each other.

As energy return on investment and energy availability continue to fall to precipitous levels, our civilizations are quickly losing the excess energy necessary to overcome the danger we find ourselves in. This may sound dire, but history has shown us that every major civilization has been crushed by a falling energy ceiling – as their space of the possible shrank, they were defeated by forces both outside and within. At the peak of the Roman Empire, it would have been hard to imagine that it would fall. At the height of the British Empire it would have been hard to imagine Britain's current state. Today's America is starting to show signs of the same future.

Excess energy has plummeted. Remember:

In 1919, 1 barrel of oil found you at least another 1,000.
In 1950, 1 barrel of oil found you another 100.
By 2010, 1 barrel found you another 5.

A theory of everyone reveals the true story of *Homo sapiens*. We have a genetic inheritance but also a cultural inheritance, and now a machine inheritance. Our intelligence is not just a product of the hardware of our big brain but also the socially acquired software that shapes it. To understand human intelligence we must understand social learning and cooperation, and how these laws work together to drive innovation and progress. And today, we have to understand how that psychology and sociality play out online in a world coinhabited by AI agents who may be smarter than we are.

The story of us that is emerging shows the fundamental drivers of our existence, how we organize ourselves on a micro level and a macro level. But there are some dark twists.

We have cooperated in the pursuit of great achievements – building an Internet and walking on the Moon.

We have cooperated toward great atrocities – slavery and genocides.

Innovation through our collective brains and enormous energy budgets has increased our capabilities for both creation and destruction.

Our story has always played out against the backdrop of the laws of life. New social and technological breakthroughs have been deployed against those who don't yet have the advantage. Over time, aggrieved groups learn, share, and grow, sometimes becoming aggressors themselves. The bright sparks that lead to progress are not specific people but all of us, sharing ideas that eventually meet in someone's head: our collective brains.

Those bright sparks are powered by energy.

The quest for energy is at the heart of the laws of life and the theory of everyone. It is the ultimate driver of all we do and the ultimate constraint on what we can achieve. Its scarcity, relative scarcity, or even perceived scarcity drive competition and conflict; its availability allows for compassion and cooperation. Each energy level gives us more power to create and more power to destroy. Energy is what powers our technologies, our economies, and everything we do. No matter how fancy and powerful our technologies may be, they're useless if we can't charge them. Other resources, such as water, may be scarce, but they are fundamentally different to energy. With abundant energy you can get water, but abundant water alone isn't enough to get energy.

As energy scientist and energy return on investment (EROI) pioneer Charles Hall often points out, the correlation between GDP growth in real terms and ability to consume oil is 0.7 (remember a maximum correlation is 1). The same correlation exists for the size of firms in a country and for energy per person. That means that 50% of economic growth, the size of companies, and your wealth can be attributed to the law of energy. The other 50% can be attributed to the laws of innovation and cooperation increasing efficiency. Given this close link between our economies and energy, it should be no surprise that before every recession in the past fifty years, including the 2008 financial crisis, the price of oil shot up. It should be no surprise that when the average capacity of US power plants fell in the 1970s, so too did economic growth.

The energy sector currently accounts for 5% of the economy, which means that 5% of our collective efforts are devoted to producing and managing energy. This leaves 95% of our resources free to be allocated toward other endeavors. However, as the EROI or energy abundance decreases, the 95% excess energy shrinks. This means that we have to dedicate more effort toward extracting energy than enjoying it. As I write, the early signs of this energy crisis are already becoming apparent.

At a societal level, it means an increase in the energy sector relative to the rest of the economy. It means higher energy prices.

At a household level this directly translates to more money spent on gas, heating, and electricity, and less money for luxuries, holidays, and fun. In other words, life becomes harder.

Everything runs on excess energy. The more excess energy we have, the better life becomes.

Watching world events sometimes leaves us with a sense of impotence. What can *I* do about all the world's ills? But there is reason for hope.

Our everyday experience tells us that what we see online, hear on the radio, or read in newspapers is not representative of most people. In real life, people don't really hate each other that much. On most topics, most people are ambivalent, apathetic to all but a handful of issues that directly affect their everyday existence, waiting instead for what is required to reinforce their group identity. Most people are kind and cooperative at a scale that would surprise our ancestors. Before those norms change, we must harness them to reunite humanity, develop models of governance for the twenty-first century, shatter the glass ceiling, trigger a creative explosion, improve the Internet, and become brighter. This will require a shift in our thinking and a willingness to challenge the status quo. It won't be easy, but it's a challenge worth taking on, because by setting ourselves on this brighter path we will have enough energy to overcome the challenges we face. With abundant energy, growth can be clean. Instead of destroying our planet, we can use that abundant energy to clean up the climate mess.

If there is a final message in this book, it is this. We don't have to accept the world as it is. Change is possible. The world was made by people no smarter than us, and thanks to rising IQ scores, probably less smart. A committed, well-connected, and well-resourced group powered by an easily understood idea is capable of moving us to a different equilibrium.

The Quakers, other Christian abolitionists, and the enslaved sparked the end of slavery – once commonplace throughout the world – which is now an unfathomable, unthinkable manifestation of unfairness and inequality. The Fabians and other socialists sparked a change that led to Britain transitioning from the world's largest empire to a smaller social-security welfare state. The Suffragettes sparked the expansion of women's rights and gender equality. Civil Rights activists challenged the legal inequality of Blacks and Whites in the United States. Rights have continued to expand to more and more areas of life in what might best be described as *runaway cultural evolution*: once a norm of equality is established, consistency and status for greater equality can drive attempts to expand the moral circle.

Norms, whether evidence-based or not, can be formalized in legal codes and constitutions. These can then be coordinated and called on for consistency. The belief that 'all are created equal' is one such norm, formalized in the American Constitution. In the United States, it has triggered cultural run-away toward ever-greater equality.

These were the words used in Abraham Lincoln's 1863 Gettysburg address to inspire a reunification in the midst of a civil war:

> Four score and seven years ago our fathers brought forth on this continent, a new nation, conceived in Liberty, and dedicated to the proposition that all men are created equal. Now we are engaged in a great civil war, testing whether that nation, or any nation so conceived and so dedicated, can long endure.

These were the words used by women's rights activists in the 1848 Declaration of Sentiments:

> We hold these truths to be self-evident; that all men and women are created equal.

These were the words used by Martin Luther King, Jr in 1963, when he called for America to live out its ideals:

> I have a dream that one day this nation will rise up and live out the true meaning of its creed: 'We hold these truths to be self-evident, that all men are created equal.'

Just as we ended mass slavery and lifted up women and minorities of all kinds in an ever-expanding circle of who matters, we can also tie our societies back together. We can shatter the glass ceiling of systematic, multigenerational inequality for a fairer world; we can evolve better governance structures; we can trigger a creative explosion; and we can become our best selves, a beacon for future generations.

I hope this book has provided tools for how to advocate and what to advocate for. Not proximate solutions that patch problems and polarize groups, creating more problems, but instead permanent systematic ultimate solutions. I hope I have helped you realize that our problems and their answers don't lie with any particular leader, any particular person, or any particular group. They require us to consider the rules of the system and what they inevitably lead to. Often, we cannot design the right rules, but we can *create conditions for the right rules to evolve*.

We have laws of life and a theory of everyone. We have a periodic table for people.

I hope you now know the answer to what Wallace's older fish asked. I hope you can now see the water. We have the power to shape our societies, to influence our systems, and to determine our future. We can crack the next energy revolution to create a world that is not just sustainable, but thriving; not just efficient, but just; not just innovative, but transformative.

The laws of life will go ever onwards. If we make the right decisions, so too will we.

Acknowledgments

This book is the product of my collective brain of family and friends, experts and colleagues, editors and publishers. Together, they contributed to and vastly improved the final product. I am eternally grateful. My agent, Eric Lupfer immediately understood the concepts I wanted to convey and helped me refine the message. From proposal to final form, Eric has been invaluable. Seth Stephens-Davidowitz: thank you for the introduction to Eric. In addition to Eric, I also had the support of several editors, including my wife Stephanie Salgado, who provided extensive feedback at every stage. I am grateful to Bob Prior at MIT Press and Nick Davies at John Murray Press for understanding the message of the book and believing in its importance. Jonathan Balcombe and Christine Fenner both provided detailed editorial support on early versions and Sarah Caro provided extensive comments on the final manuscript, really helping to polish the final product. Thanks also to Hilary Hammond and Caroline Westmore for their helpful copy-editing, to the design teams at MIT Press and John Murray Press for their gorgeous book covers, and to Veronika Plant and Abheeth Salgado who both offered feedback on drafts of the cover, even creating mock-ups and suggestions.

I am also incredibly grateful to Matthew Syed. I have long admired Matthew's seemingly effortless ability to communicate the most complex concepts so clearly. When I first described the research and shared a very early manuscript, Matthew's excitement provided much motivation. He generously sat down with me for hours, helping me craft better analogies, suggesting sections that needed to be better explained, and guiding me in writing a better introduction. Matthew is as generous as he is brilliant. Of my other author friends, David Bodanis offered a lot of early encouragement, feedback, and professional tips on the writing and publication process when this book was just an idea.

I am grateful to the many researchers and professionals who generously shared their expertise and feedback on relevant sections. Particular thanks goes to energy expert Charles Hall, who spent hours talking to me over Zoom and who offered detailed comments on the entire manuscript. Thank you also to Abdel Abdellaoui (genetics), Tim Besley (economics), Daniel Muthukrishna (physics — it was in a conversation with Dan that the title emerged), Rachel Spicer (biology), Emily Burdett (developmental psychology), Helen Elizabeth Davis, Will Gervais, and Ivan Kroupin (IQ and intelligence), Jamie Heywood (Uber and innovation), Cameron Murray (land taxes and Singapore's housing policies), Julian Ashwin (free speech), Nicolai Tangen (Norway's oil fund), David Yang (China's politics and meritocracy), Sandya Salgado (marketing), Kurtis Lockhart (start-up cities), Briitta van Staalduinen (data on how people find jobs around the world), Sarah Beck (for sharing her videos with me), Mark Nielsen (for information about his studies with Bushmen in South Africa), Philipp Koellinger (for permission to use the brain volume – IQ graph and confirming the correct interpretation), Ryutaro Uchiyama (kaizen and shuhari), Josh Tan (cryptocurrencies, programmable politics, and decentralized autonomous organizations), Marc Warner (social media and AI), and Adam D'Angelo (AI and energy requirements).

Thank you also to my students who patiently read the manuscript, offering fact checking and feedback. Robin Schimmelpfennig went through almost the entire manuscript with me, offering his invaluable unvarnished feedback. Zoé Vanhersecke read every chapter in detail, performed extensive fact checking, and helped with recreating graphs. Zoé was also the person who first suggested the acronym COMPASS. Theiss Bendixen offered helpful comments on overall framing during the early stages. Navdeep Kaur helped me with some final fact checking. Nicole George read the whole manuscript as a model reader, and gave extensive feedback on every chapter. Other early readers who offered feedback on various chapters include Chris Muthukrishna, Shanthi Muthukrishna, Penny Gray, Clinton Freeman, and David Ryan. Andrew McAfee read the whole book as a model academic reader, with helpful feedback on his wide range of expertise and thoughts on how others working on technology and innovation might react to the arguments.

More informal feedback came from conversations over the years with many people, in particular Joe Henrich, Mark Schaller, Steve Heine, Ara Norenzayan, Michael Doebeli, Patrick Francois, Nathan Nunn, Penny Sanderson, David Liu, Clinton Freeman, Ben Cole, Ben Cheung, Aiyana Willard, Dan Randles, Wanying Zhao, Charles Stafford, Natasha Griffiths, Roger Frosh, Matthew Hindhaugh, and Charlotte Rowan.

My family around the world opened their homes, helped with childcare, and offered me physical and mental space to focus and finish the book during my sabbatical. For three long months, Hiranya Dharmaratne helped look after our children in London. It was during this time that much of the book was finalized. Shanthi Muthukrishna and Ruvan Dharmaratne also helped with childcare at various stages. Manju, Minoli, and Peter Haththotuwa allowed us to stay for weeks in their beautiful home in the peaceful surrounds of Arlington, Virginia. Renata Dharmaratne graciously hosted our whole family and helped care for our children for two months in Brisbane, Australia. Damian Leggat also patiently helped with everything from childcare to meals. Swarnamali, Rajpal, Sudharma, and Kavisha Salgado hosted us and helped with childcare during an intense bout of final editing in Sri Lanka. Many other people offered us a peaceful place to stay, including Sandya and Priyath Salgado and Saro Ramakrishna. I am grateful to all for their hospitality and support.

Almost all the amazing illustrations in this book are thanks to the incredibly talented Veronika Plant. Veronika brilliantly and creatively brought these ideas to life in her characteristic style. I can't thank her enough for working so quickly – including on her anniversary – to get all the illustrations done on time.

Finally, I would like to thank my three children, Robert, Alexandra, and Gabriele, to whom this book is dedicated. Thank you for understanding why Daddy was working so much and sometimes seemed deeply lost in his own thoughts. Above all, I am grateful to my greatest critic, biggest supporter, first and last editor, and love of my life, Stephanie Salgado. She helped shaped the ideas in this book through our many debates and discussions, encouraged me and gave me the time and head space to work on it, and read more versions than anyone else. So much of this book is thanks to Steph.

Further Reading

'If all you have is a hammer, every problem looks like a nail.' Maslow's hammer reminds us that a diverse range of tools and approaches is essential for understanding and solving the complex problems of the world. Our collective brain has a vast repository of knowledge and expertise, but its strewn across the academic literature and popular sources. Within some of these books and papers are loose threads, partial answers, and pieces of the puzzle of who we are, how we got here, and where we're going. But once these are woven into a tapestry, that tapestry is almost obvious, its implications unmissable. Once seen, it cannot be unseen.

I have tried to show you the tapestry in this book, highlighting the threads that I think are most important, and yet by necessity each thread is missing fascinating details and important context that can often only be hinted at by a single paragraph or even single sentence. This is no doubt dissatisfying for readers with an undisciplined curiosity, courage to trample over traditional disciplines, and a willingness to deliberately and disrespectfully dismantle unquestioned assumptions that have been passed down to us by our culture. For those who want to delve deeper and explore all those fascinating details and important context, here are some recommended further readings.

Introduction

For those looking to further their knowledge in the fields of psychology, economics, evolutionary biology, anthropology, and data science, there is no shortage of excellent introductory material available.

For an introduction to psychology, I assign Peter Gray and David F. Bjorklund's popular textbook *Psychology*, 8th edition, Worth Publishers, 2018, alongside a popular book on cultural evolution, Joseph Henrich, *The Secret of Our Success*, Princeton University Press, 2015.

For an introduction to economics see Thomas Sowell, *Basic Economics*, Basic Books, 2014; and The Core Team, *The Economy: Economics for a Changing*

World, e-book, Core Economics Education, 2022, www.core-econ.org (available free). A deeper dive is Gregory Mankiw's excellent textbook *Principles of Economics*, Cengage Learning, 2020 (among others). For how economists think about mathematical models (with a balanced discussion on Milton Friedman's 1953 essay) see Dani Rodrik, *Economics Rules: Why Economics Works, When It Fails, and How to Tell the Difference*, Oxford University Press, 2015.

To contrast this with how evolutionary biologists and cultural evolution researchers think about models, see Hanna Kokko, *Modelling for Field Biologists and Other Interesting People*, Cambridge University Press, 2007. A further reading is Sarah P. Otto and Troy Day, *A Biologist's Guide to Mathematical Modeling in Ecology and Evolution*, Princeton University Press, 2011.

For a history of anthropology see Charles King, *Gods of the Upper Air: How a Circle of Renegade Anthropologists Reinvented Race, Sex, and Gender in the Twentieth Century*, Anchor, 2020.

For a slightly less mathematical but still rigorous approach to control systems, which also focuses on the context of humans, see Richard J. Jagacinski and John M. Flach, *Control Theory for Humans: Quantitative Approaches to Modeling Performance*, CRC Press, 2018.

There are few books that bridge these disciplines. For psychology meets economics see George A. Akerlof and Robert J. Shiller, *Animal Spirits: How Human Psychology Drives the Economy, and Why It Matters for Global Capitalism*, Princeton University Press, 2010; and R. H. Thaler, *Misbehaving: The Making of Behavioural Economics*, Allen Lane, 2016. For evolution meets economics there is Robert H. Frank, *The Darwin Economy*, Princeton University Press, 2012 and Lionel Page's *Optimally Irrational*, Cambridge University Press, 2022.

Data science is a rapidly changing space, but for a fundamental understanding of probability and statistics, you can't do much better than Richard McElreath's freely available online course and accompanying textbook, *Statistical Rethinking: A Bayesian Course with Examples in R and Stan*, Chapman & Hall/CRC, 2020.

There are many great books on the history of science and discoveries in astronomy. One older thought-provoking volume with a focus on astronomy that shaped me in my youth is Arthur Koestler's *The Sleepwalkers: A History of Man's Changing Vision of the Universe*, Penguin, 1959.

For the history and philosophy of science and the flow of ideas, in addition to classics such as Thomas Kuhn's *The Structure of Scientific Revolutions* (1962) and Karl Popper's *The Logic of Scientific Discovery* (1959), you might also enjoy the more recent Randall Collins, *The Sociology of Philosophies*, Harvard University Press, 2009; and David Wootton, *The Invention of Science: A New History of the Scientific Revolution*, Penguin, 2015.

For the role of theory in science, Bernard Forscher wrote a wonderful article expanding on Poincaré's quote about bricks versus houses: 'Chaos in the Brickyard', *Science* 142, no. 3590 (1963): 339; this article is beautifully illustrated in comic form by Matteo Farinella at https://massivesci.com/articles/chaos-in-the-brickyard-comic-matteo-farinella/

The full text of David Foster Wallace's commencement address was later published as a book, David Foster Wallace, *This Is Water: Some Thoughts, Delivered on a Significant Occasion, about Living a Compassionate Life*, Little, Brown, 2009.

Specific and specialist reading

Muthukrishna, Michael, and Joseph Henrich, 'A Problem in Theory', *Nature Human Behaviour* 3, no. 3 (2019): 221–9

Chapter 1: Laws of Life

Here are a few good introductions to the major themes in this chapter. Some of the enjoyable TOTTEE books I mention or hint at include: Jared M. Diamond, *Guns, Germs and Steel: A Short History of Everybody for the Last 13,000 Years*, Random House, 1998; Yuval Noah Harari, *Sapiens: A Brief History of Humankind*, Random House, 2014; Ian Morris, *Why the West Rules – for Now: The Patterns of History and What They Reveal about the Future*, Profile, 2010; and James A. Robinson and Daron Acemoglu, *Why Nations Fail: The Origins of Power, Prosperity and Poverty*, Profile, 2012.

For major transitions and key inventions in evolution there are two books by John Maynard Smith and Eörs Szathmáry, *The Origins of Life: From the Birth of Life to the Origin of Language* and *The Major Transitions in Evolution*, both published by Oxford University Press, respectively 2000 and 1997; and Nick Lane's *Life Ascending: The Ten Great Inventions of Evolution*, Profile, 2010.

Multilevel selection – cooperation and competition from cells to societies – is covered by Itai Yanai and Martin Lercher, *The Society of Genes*, Harvard University Press, 2016; Athena Aktipis, *The Cheating Cell*, Princeton University Press, 2020; and D. S. Wilson, *This View of Life: Completing the Darwinian Revolution*, Knopf Doubleday, 2019. For a deeper dive there is Samir Okasha, *Evolution and the Levels of Selection*, Clarendon Press, 2006; and Eva Jablonka and Marion J. Lamb, *Evolution in Four Dimensions: Genetic, Epigenetic, Behavioral, and Symbolic Variation in the History of Life*, MIT Press, 2014.

For more on ancient human migrations see David Reich, *Who We Are and How We Got Here: Ancient DNA and the New Science of the Human Past*, Oxford University Press, 2018.

The analogy of 'ideas having sex' was coined by Matt Ridley. Two relevant books of his are *The Rational Optimist: How Prosperity Evolves* and *How Innovation Works*, both published by Harper, in 2010 and 2020 respectively.

As a starting point for the role of energy in life, civilization, and economics, Ian Morris has a book with related ideas: *Foragers, Farmers, and Fossil Fuels*, Princeton University Press, 2015.

On the role of fire and farming in human evolution see Richard W. Wrangham, *Catching Fire: How Cooking Made Us Human*, Profile, 2010; and James C. Scott, *Against the Grain: A Deep History of the Earliest States*, Yale University Press, 2017. For more thorough and detailed discussion, I recommend two recent books by Vaclav Smil (for an additional endorsement, Smil is Bill Gates' favorite author): *Growth: From Microorganisms to Megacities* and *Energy and Civilization: A History*, both by MIT Press, published 2019 and 2018 respectively.

For the role of energy returns in the fall of civilizations see Joseph Tainter, *The Collapse of Complex Societies*, Cambridge University Press, 1988; and Thomas Homer-Dixon, *The Upside of Down: Catastrophe, Creativity, and the Renewal of Civilization*, Island Press, 2010.

On economic growth and productivity declines there is Tyler Cowen, *The Great Stagnation: How America Ate All the Low-hanging Fruit of Modern History, Got Sick, and Will (Eventually) Feel Better*, Penguin, 2011.

Specific and specialist reading

Andersson, G. E., Olof Karlberg, Björn Canbäck et al., 'On the Origin of Mitochondria: A Genomics Perspective', *Philosophical Transactions of the Royal Society of London. Series B: Biological Sciences* 358, no. 1429 (2003): 165–79

Asphaug, Erik, 'Impact Origin of the Moon?', *Annual Review of Earth and Planetary Sciences* 42 (2014): 551–78

De Visser, J., G. M. Arjan, and Santiago F. Elena, 'The Evolution of Sex: Empirical Insights into the Roles of Epistasis and Drift', *Nature Reviews Genetics* 8, no. 2 (2007): 139–49

Dyson, Freeman J., 'Search for Artificial Stellar Sources of Infrared Radiation', *Science* 131, no. 3414 (1960): 1667–8

Fix, Blair, 'Energy and Institution Size', *PLoS ONE* 12, no. 2 (2017): e0171823

Flannery, K., and J. Marcus, *The Creation of Inequality*, Harvard University Press, 2012

Gabaldón, Toni, 'Origin and Early Evolution of the Eukaryotic Cell', *Annual Review of Microbiology* 75 (2021): 631–47

Grosberg, R. K., and R. R. Strathmann, 'The Evolution of Multicellularity: A Minor Major Transition?', *Annual Review of Ecology Evolution and Systematics* 38, no. 1 (2007): 621–54

Hall, Charles A. S., *Energy Return on Investment: A Unifying Principle for Biology, Economics, and Sustainability*, Springer, 2017

Hall, Charles A. S., Cutler J. Cleveland, and Robert Kaufmann, 'Energy and Resource Quality: The Ecology of the Economic Process', Wiley-Blackwell, 1986

Hall, Charles A. S., and Kent Klitgaard, *Energy and the Wealth of Nations: An Introduction to Biophysical Economics*, Springer International, 2018

Hey, Jody, 'On the Number of New World Founders: A Population Genetic Portrait of the Peopling of the Americas', *PLoS Biology* 3, no. 6 (2005): e193

Higham, Tom, Katerina Douka, Rachel Wood et al., 'The Timing and Spatiotemporal Patterning of Neanderthal Disappearance', *Nature* 512, no. 7514 (2014): 306–9

Hohmann-Marriott, Martin F., and Robert E. Blankenship, 'Evolution of Photosynthesis', *Annual Review of Plant Biology* 62 (2011): 515–48

Jedwab, Remi, Noel D. Johnson, and Mark Koyama, 'The Economic Impact of the Black Death', *Journal of Economic Literature* 60, no. 1 (2022): 132–78

Knoll, Andrew H., 'The Multiple Origins of Complex Multicellularity', *Annual Review of Earth and Planetary Sciences* 39 (2011): 217–39

Koppelaar, Rembrandt H. E. M., 'Solar-PV Energy Payback and Net Energy: Meta-assessment of Study Quality, Reproducibility, and Results Harmonization', *Renewable and Sustainable Energy Reviews* 72 (2017): 1241–55

Koyama, Mark, and Jared Rubin, *How the World Became Rich: The Historical Origins of Economic Growth*, John Wiley, 2022

Langergraber, Kevin E., Kay Prüfer, Carolyn Rowney et al., 'Generation Times in Wild Chimpanzees and Gorillas Suggest Earlier Divergence Times in Great Ape and Human Evolution', *Proceedings of the National Academy of Sciences* 109, no. 39 (2012): 15716–21

Lyons, Timothy W., Christopher T. Reinhard, and Noah J. Planavsky, 'The Rise of Oxygen in Earth's Early Ocean and Atmosphere', *Nature* 506, no. 7488 (2014): 307–15

McDonald, Michael J., Daniel P. Rice, and Michael M. Desai, 'Sex Speeds Adaptation by Altering the Dynamics of Molecular Evolution', *Nature* 531, no. 7593 (2016): 233–6

Mankiw, N. Gregory, David Romer, and David N. Weil. 'A Contribution to the Empirics of Economic Growth', *Quarterly Journal of Economics* 107, no. 2 (1992): 407–37

Marciniak, Stephanie, Christina M. Bergey, Ana Maria Silva et al., 'An Integrative Skeletal and Paleogenomic Analysis of Stature Variation Suggests Relatively Reduced Health for Early European Farmers', *Proceedings of the National Academy of Sciences* 119, no. 15 (2022): e2106743119

Mignacca, Benito, and Giorgio Locatelli, 'Economics and Finance of Small Modular Reactors: A Systematic Review and Research Agenda', *Renewable and Sustainable Energy Reviews* 118 (2019): 109519

Mokyr, Joel, *The British Industrial Revolution: An Economic Perspective*, Routledge, 2018

Muthukrishna, Michael, Michael Doebeli, Maciej Chudek et al., 'The Cultural Brain Hypothesis: How Culture Drives Brain Expansion, Sociality, and Life History', *PLoS Computational Biology* 14, no. 11 (2018): e1006504

Navarrete, Ana, Carel P. van Schaik, and Karin Isler, 'Energetics and the Evolution of Human Brain Size', *Nature* 480, no. 7375 (2011): 91–3

Nielsen, Claus, Thibaut Brunet, and Detlev Arendt, 'Evolution of the Bilaterian Mouth and Anus', *Nature Ecology & Evolution* 2, no. 9 (2018): 1358–7

Orgel, Leslie E., 'The Origin of Life – A Review of Facts and Speculations', *Trends in Biochemical Sciences* 23, no. 12 (1998): 491–5

Pingali, Prabhu L., 'Green Revolution: Impacts, Limits, and the Path Ahead', *Proceedings of the National Academy of Sciences* 109, no. 31 (2012): 12302–8

Planck Collaboration, P. A. R. Ade, N. Aghanim et al., 'Planck 2015 Results: XIII: Cosmological Parameters', *Astronomy & Astrophysics* 594 (October 2016): A13

Romer, Paul M., 'The Origins of Endogenous Growth', *Journal of Economic Perspectives* 8, no. 1 (1994): 3–22

Shine, Rick, *Cane Toad Wars*, University of California Press, 2018

Tewksbury, J. J., and G. P. Nabhan, 'Directed Deterrence by Capsaicin in Chillies', *Nature* 412, no. 6845 (2001): 403–4

Weißbach, Daniel, G. Ruprecht, A. Huke et al., 'Energy Intensities, EROIs (Energy Returned on Invested), and Energy Payback Times of Electricity Generating Power Plants', *Energy* 52 (2013): 210–21

Wrigley, E. A., *Energy and the English Industrial Revolution*, Cambridge University Press, 2010

Zeder, Melinda A., 'The Origins of Agriculture in the Near East', *Current Anthropology* 52, no. S4 (2011): S221–35

Chapter 2: The Human Animal

An encompassing world history is Peter Frankopan, *The Silk Roads: A New History of the World*, Bloomsbury, 2015. For more on Franz Boas in the context of the history of anthropology see Charles King, *Gods of the Upper Air: How a Circle of Renegade Anthropologists Reinvented Race, Sex, and Gender in the Twentieth Century*, Anchor, 2020.

There are a few recent books on cultural evolution, dual inheritance theory, culture–gene co-evolution, and the new science of human evolution: Joseph Henrich, *The Secret of Our Success*, Princeton University Press, 2015, and *The WEIRDest People in the World: How the West became Psychologically Peculiar and Particularly Prosperous*, Penguin, 2020; Kevin N. Laland, *Darwin's Unfinished Symphony*, Princeton University Press, 2018; Robert Boyd, *A Different Kind of Animal*, Princeton University Press, 2017; and Lesley Newson and Peter Richerson, *A Story of Us: A New Look at Human Evolution*, Oxford University Press, 2021.

A complementary view from evolutionary psychology is provided by Steven Pinker, *The Blank Slate: The Modern Denial of Human Nature*, Viking, 2004; and David M. Buss, *Evolutionary Psychology: The New Science of the Mind*, Routledge, 2019.

Some of the language examples and the later Shakespeare example come from Mark Forsyth: see his *The Etymologicon: A Circular Stroll through the Hidden Connections of the English Language*, Penguin, 2012, and *The Elements of Eloquence: How to Turn the Perfect English Phrase*, Icon Books, 2013.

For the illusion of explanatory depth, how everything is obvious once you know the answer, and one of my favorite essays on the 'reductive seduction of other people's problems', see Steven Sloman and Philip Fernbach, *The Knowledge Illusion: Why We Never Think Alone*, Penguin, 2018; Duncan J. Watts, *Everything Is Obvious: Why Common Sense Is Nonsense*, Atlantic Books, 2011; and Courtney Martin, 'The Reductive Seduction of Other People's Problems', *Bright Magazine* online, 11 January 2016.

Specific and specialist reading

Al-Andalusi, Said, Semaan I. Salem, and Alok Kumar, *Science in the Medieval World*, vol. 5, University of Texas Press, 1996

Australian Institute of Health and Welfare, *Skin Cancer in Australia*, AIHW, 2016

Barsh, Gregory S., 'What Controls Variation in Human Skin Color?', *PLoS Biology* 1, no. 1 (2003): e27

Beckwith, Christopher I., *Empires of the Silk Road*, Princeton University Press, 2009

Boyd, Robert, and Peter J. Richerson, *Culture and the Evolutionary Process*, University of Chicago Press, 1985

Burns, Thomas S., *Rome and the Barbarians, 100 BC–AD 400*, Johns Hopkins University Press, 2003

Cashman, Kevin D., Kirsten G. Dowling, Zuzana Škrabáková et al., 'Vitamin D Deficiency in Europe: Pandemic?', *American Journal of Clinical Nutrition* 103, no. 4 (2016): 1033–44

Cavalli-Sforza, Luigi Luca, and Marcus W. Feldman, *Cultural Transmission and Evolution: A Quantitative Approach*, Princeton University Press, 1981

Chua, Amy, *Battle Hymn of the Tiger Mother*, Bloomsbury, 2011

Chudek, Maciej, Michael Muthukrishna, and Joe Henrich, 'Cultural Evolution', in *The Handbook of Evolutionary Psychology*, pp. 1–21, Wiley, 2015

Cosmides, Leda, 'The Logic of Social Exchange: Has Natural Selection Shaped How Humans Reason? Studies with the Wason Selection Task', *Cognition* 31, no. 3 (1989): 187–276

Dunbar, Robin Ian MacDonald, *Grooming, Gossip, and the Evolution of Language*, Harvard University Press, 1998

Garland, Cedric F., Frank C. Garland, Edward D. Gorham et al., 'The Role of Vitamin D in Cancer Prevention', *American Journal of Public Health* 96, no. 2 (2006): 252–61

Grogger, Jeffrey, Andreas Steinmayr, and Joachim Winter, 'The Wage Penalty of Regional Accents', working paper 26719, National Bureau of Economic Research, 2020

Henrich, Joseph, 'The Evolution of Costly Displays, Cooperation and Religion: Credibility Enhancing Displays and Their Implications for Cultural Evolution', *Evolution and Human Behavior* 30, no. 4 (2009): 244–60

Henrich, Joseph, and Francisco J. Gil-White, 'The Evolution of Prestige: Freely Conferred Deference as a Mechanism for Enhancing the Benefits of Cultural Transmission', *Evolution and Human Behavior* 22, no. 3 (2001): 165–96

Henrich, Joseph, and Natalie Henrich, 'The Evolution of Cultural Adaptations: Fijian Food Taboos Protect against Dangerous Marine Toxins', *Proceedings of the Royal Society B: Biological Sciences* 277, no. 1701 (2010): 3715–24

Herrmann, Esther, Josep Call, María Victoria Hernández-Lloreda et al., 'Humans Have Evolved Specialized Skills of Social Cognition: The Cultural Intelligence Hypothesis', *Science* 317, no. 5843 (2007): 1360–6

Horner, Victoria, and Andrew Whiten, 'Causal Knowledge and Imitation/ Emulation Switching in Chimpanzees (Pan Troglodytes) and Children (Homo Sapiens)', *Animal Cognition* 8 (2005): 164–81

Jablonski, Nina G., and George Chaplin, 'Human Skin Pigmentation as an Adaptation to UV radiation', *Proceedings of the National Academy of Sciences* 107, supplement no. 2 (2010): 8962–8

——, 'The Colours of Humanity: The Evolution of Pigmentation in the Human Lineage', *Philosophical Transactions of the Royal Society B: Biological Sciences* 372, no. 1724 (2017): 20160349

Kahneman, Daniel, and Amos Tversky, 'Prospect Theory: An Analysis of Decision under Risk', *Econometrica* 47, no. 2 (1979): 263–92

Keil, Frank C., 'Folkscience: Coarse Interpretations of a Complex Reality', *Trends in Cognitive Sciences* 7, no. 8 (2003): 368–73

——, 'Explanation and Understanding', *Annual Review of Psychology* 57, no. 1 (2006): 227–54

Kendal, Rachel L., Neeltje J. Boogert, Luke Rendell et al., 'Social Learning Strategies: Bridge-building between Fields', *Trends in Cognitive Sciences* 22, no. 7 (2018): 651–65

Kinzler, Katherine D., Kristin Shutts, Jasmine DeJesus et al., 'Accent Trumps Race in Guiding Children's Social Preferences', *Social Cognition* 27, no. 4 (2009): 623–34

Kipling, Rudyard, 'The White Man's Burden', *The Times*, 4 February 1899

Knight, Phil, *Shoe Dog: A Memoir by the Creator of Nike*, Simon & Schuster, 2016

Kurzban, Robert, John Tooby, and Leda Cosmides, 'Can Race Be Erased? Coalitional Computation and Social Categorization', *Proceedings of the National Academy of Sciences* 98, no. 26 (2001): 15387–92

Lanman, Jonathan A., and Michael D. Buhrmester, 'Religious Actions Speak Louder than Words: Exposure to Credibility-enhancing Displays Predicts Theism', *Religion, Brain & Behavior* 7, no. 1 (2017): 3–16

Lewis, Bernard, *The Muslim Discovery of Europe*, W. W. Norton, 2001

Luria, Alexander Romanovich, *Cognitive Development: Its Cultural and Social Foundations*, Harvard University Press, 1976

Norenzayan, Ara, Azim F. Shariff, Will M. Gervais et al., 'The Cultural Evolution of Prosocial Religions', *Behavioral and Brain Sciences* 39 (2016): e1

Parasher-Sen, Aloka, ed., *Subordinate and Marginal Groups in Early India*, Oxford University Press, 2007

Persky, Joseph, 'Retrospectives: The Ethology of Homo Economicus', *Journal of Economic Perspectives* 9, no. 2 (1995): 221–31

Phelps, Michael, and Alan Abrahamson, *No Limits: The Will to Succeed*, Simon & Schuster, 2008

Rozenblit, Leonid, and Frank Keil, 'The Misunderstood Limits of Folk Science: An Illusion of Explanatory Depth', *Cognitive Science* 26, no. 5 (2002): 521–62

Saussy, Haun, *The Making of Barbarians: Chinese Literature and Multilingual Asia*, vol. 48, Princeton University Press, 2022

Simon, Herbert A., '*Models of Man: Social and Rational*', Wiley, 1957

Spiro, A., and J. L. Buttriss, 'Vitamin D: An Overview of Vitamin D Status and Intake in Europe', *Nutrition Bulletin* 39, no. 4 (2014): 322–50

Thaler, Richard H, and Cass R. Sunstein, *Nudge: Improving Decisions about Health, Wealth, and Happiness*, Yale University Press, 2008

Tillman, B., 'Are We to Spread the Christian Religion with the Bayonet Point as Mahomet Spread Islam with a Scimitar?', Address to the U.S. Senate, 7 February 1899

Todd, Peter M., and Gerd Ed Gigerenzer, *Ecological Rationality: Intelligence in the World*, Oxford University Press, 2012

Uchiyama, Ryutaro, Rachel Spicer, and Michael Muthukrishna, 'Cultural Evolution of Genetic Heritability', *Behavioral and Brain Sciences* 45 (2022): e152

Chapter 3: Human Intelligence

A good history of eugenics is Adam Rutherford, *Control: The Dark History and Troubling Present of Eugenics*, Orion, 2023. For a summary of the history and research on intelligence from those who argue for the importance of culture see Richard E. Nisbett, *Intelligence and How to Get It: Why Schools and Cultures Count*, W. W. Norton, 2009; and for two from those who argue for the importance of genes see Robert Plomin, *Blueprint, with a New Afterword: How DNA Makes Us Who We Are*, MIT Press, 2019; and Kathryn Paige Harden, *The Genetic Lottery: Why DNA Matters for Social Equality*, Princeton University Press, 2021.

How language and learning change our brains is described by Guy Deutscher, *Through the Language Glass: Why the World Looks Different in Other Languages*, Metropolitan Books, 2010; and by Stanislas Dehaene in three books: *How We Learn: The New Science of Education and the Brain*, Penguin, 2020; *Reading in the Brain: The New Science of How We Read*, Penguin, 2010; and *The Number Sense: How the Mind Creates Mathematics*, Oxford University Press, 2011.

A longer case for how popular culture has become more complex and made us cleverer is presented by Steven Johnson in *Everything Bad Is Good for You: How Today's Popular Culture Is Actually Making Us Smarter*, Penguin, 2006.

For how to think about race from a genetic perspective, dispelling many myths, see Adam Rutherford, *How to Argue with a Racist: History, Science, Race and Reality*, Weidenfeld & Nicolson, 2020; David Reich, *Who We Are and How We Got Here: Ancient DNA and the New Science of the Human Past*, Pantheon Books, 2018; and Steven J. Heine, *DNA Is Not Destiny: The Remarkable, Completely Misunderstood Relationship between You and Your Genes*, W. W. Norton, 2017.

Specific and specialist reading

Antman, Francisca, and Brian Duncan, 'Incentives to Identify: Racial Identity in the Age of Affirmative Action', *Review of Economics and Statistics* 97, no. 3 (2015): 710–13

Archer, John, 'The Reality and Evolutionary Significance of Human Psychological Sex Differences', *Biological Reviews* 94, no. 4 (2019): 1381–415

Auton, Adam, Adi Fledel-Alon, Susanne Pfeifer et al., 'A Fine-scale Chimpanzee Genetic Map from Population Sequencing', *Science* 336, no. 6078 (2012): 193–8

Beck, Sarah R., Ian A. Apperly, Jackie Chappell et al., 'Making Tools Isn't Child's Play', *Cognition* 119, no. 2 (2011): 301–6

Bender, Andrea, and Sieghard Beller, 'Fingers as a Tool for Counting – Naturally Fixed or Culturally Flexible?', *Frontiers in Psychology* 2 (2011): 256

Binet, Alfred, and Theodore Simon, 'New Methods for the Diagnosis of the Intellectual Level of Subnormals, *L'Année Psych.*, 1905, pp. 191–244

Botvinick, Matthew M., Jonathan D. Cohen, and Cameron S. Carter, 'Conflict Monitoring and Anterior Cingulate Cortex: An Update', *Trends in Cognitive Sciences* 8, no. 12 (2004): 539–46

Bowden, Rory, Tammie S. MacFie, Simon Myers et al., 'Genomic Tools for Evolution and Conservation in the Chimpanzee: Pan Troglodytes Ellioti Is a Genetically Distinct Population', *PLoS Genetics* 8, no. 3 (2012): e1002504

Brinch, Christian N., and Taryn Ann Galloway, 'Schooling in Adolescence Raises IQ Scores', *Proceedings of the National Academy of Sciences* 109, no. 2 (2012): 425–30

Burgoyne, Alexander P., David Z. Hambrick, and Brooke N. Macnamara,

'How Firm Are the Foundations of Mind-Set Theory? The Claims Appear Stronger Than the Evidence', *Psychological Science* 31, no. 3 (2020): 258–67

Campbell, Michael C., and Sarah A. Tishkoff, 'African Genetic Diversity: Implications for Human Demographic History, Modern Human Origins, and Complex Disease Mapping', *Annual Review of Genomics and Human Genetics* 9 (2008): 403–33

Card, David, Stefano DellaVigna, Patricia Funk et al., 'Gender Gaps at the Academies', *Proceedings of the National Academy of Sciences* 120, no. 4 (24 January 2023): e2212421120

Ceci, Stephen J., and Wendy M. Williams, 'Understanding Current Causes of Women's Underrepresentation in Science', *Proceedings of the National Academy of Sciences* 108, no. 8 (2011): 3157–62

Dev, Pritha, Blessing U. Mberu, and Roland Pongou, 'Ethnic Inequality: Theory and Evidence from Formal Education in Nigeria', *Economic Development and Cultural Change* 64, no. 4 (2016): 603–60

Dickens, William T., and James R. Flynn, 'Black Americans Reduce the Racial IQ Gap: Evidence from Standardization Samples', *Psychological Science* 17, no. 10 (2006): 913–20

Durvasula, Arun, and Sriram Sankararaman, 'Recovering Signals of Ghost Archaic Introgression in African Populations', *Science Advances* 6, no. 7 (2020): eaax5097

Fan, Shaohua, Matthew E. B. Hansen, Yancy Lo et al., 'Going Global by Adapting Local: A Review of Recent Human Adaptation', *Science* 354, no. 6308 (2016): 54–9

Flynn, James R., *What Is Intelligence? Beyond the Flynn Effect*, Cambridge University Press, 2007

Galton, Francis, 'Hereditary Talent and Character', *Macmillan's Magazine* 12, nos 157–166 (1865): 318–27

——, *Hereditary Genius*, Macmillan & Co., 1869

Giangrande, Evan J., Christopher R. Beam, Sarah Carroll et al., 'Multivariate Analysis of the Scarr–Rowe Interaction across Middle Childhood and Early Adolescence', *Intelligence* 77, no. 1 (2019): 101400

Halpern, Diane F., and Mary L. LaMay, 'The Smarter Sex: A Critical Review of Sex Differences in Intelligence', *Educational Psychology Review* 12, no. 2 (2000): 229–46

Hanscombe, Ken B., Maciej Trzaskowski, Claire M. A. Haworth et al., 'Socioeconomic Status (SES) and Children's Intelligence (IQ): In a UK-Representative Sample SES Moderates the Environmental, Not Genetic, Effect on IQ', *PLoS ONE* 7, no. 2 (2012): e30320

Herrnstein, Richard J., and Charles Murray, *The Bell Curve: Intelligence and Class Structure in American Life*, Free Press, 1994

Hsin, Amy, and Yu Xie, 'Explaining Asian Americans' Academic Advantage over Whites', *Proceedings of the National Academy of Sciences* 111, no. 23 (2014): 8416–21

Hulance, Jo, Mark Kowalski, and Robert Fairhurst, *Long-term Strategies to Reduce Lead Exposure from Drinking Water*, report number DWI14372.2, Drinking Water Inspectorate (UK), 26 January 2021

Hyde, Janet S., and Janet E. Mertz, 'Gender, Culture, and Mathematics Performance', *Proceedings of the National Academy of Sciences* 106, no. 22 (2 June 2009): 8801–7

Khalid, Muhammed Abdul, and Li Yang, 'Income Inequality and Ethnic Cleavages in Malaysia: Evidence from Distributional National Accounts (1984–2014)', *Journal of Asian Economics* 72 (2021): 101252

Kiesow, Hannah, Robin I. M. Dunbar, Joseph W. Kable et al., '10,000 Social Brains: Sex Differentiation in Human Brain Anatomy', *Science Advances* 6, no. 12 (March 2020): eaaz1170

König, Peter, and Sabine U. König, 'Learning a New Sense by Sensory Augmentation', in *2016 4th International Winter Conference on Brain-Computer Interface (BCI)*, pp. 1–3, Institute of Electrical and Electronics Engineers, 2016

Laland, Kevin N., John Odling-Smee, and Sean Myles, 'How Culture Shaped the Human Genome: Bringing Genetics and the Human Sciences Together', *Nature Reviews Genetics* 11, no. 2 (2010): 137–44

Lassek, William D., and Steven J. C. Gaulin, 'Costs and Benefits of Fat-free Muscle Mass in Men: Relationship to Mating Success, Dietary Requirements, and Native Immunity', *Evolution and Human Behavior* 30, no. 5 (2009): 322–8.

Levinson, Stephen C., 'Language and Cognition: The Cognitive Consequences of Spatial Description in Guugu Yimithirr', *Journal of Linguistic Anthropology* 7, no. 1 (1997): 98–131

Loftus, Elizabeth F., 'Eyewitness Testimony', *Applied Cognitive Psychology* 33, no. 4 (2019): 498–503

Luria, Alexander Romanovich, *Cognitive Development: Its Cultural and Social Foundations*, Harvard University Press, 1976

Machin, Stephen, and Tuomas Pekkarinen, 'Global Sex Differences in Test Score Variability', *Science* 322, no. 5906 (2008): 1331–2

Muthukrishna, Michael, and Joseph Henrich, 'Innovation in the Collective Brain', *Philosophical Transactions of the Royal Society B: Biological Sciences* 371, no. 1690 (2016): 20150192

Nave, Gideon, Wi Hoon Jung, Richard Karlsson Linnér et al., 'Are Bigger Brains Smarter? Evidence from a Large-scale Preregistered Study', *Psychological Science* 30, no. 1 (2019): 43–54

Nielsen, Mark, Keyan Tomaselli, Ilana Mushin et al., 'Exploring Tool Innovation: A Comparison of Western and Bushman Children', *Journal of Experimental Child Psychology* 126 (2014): 384–94

Nisbett, Richard E., Joshua Aronson, Clancy Blair et al., 'Intelligence: New Findings and Theoretical Developments', *American Psychologist* 67, no. 2 (2012): 130

OECD, *PISA 2018 Results (Volume II): Where All Students Can Succeed*, OECD Publishing, 2019

Pemberton, Trevor J., Michael DeGiorgio, and Noah A. Rosenberg, 'Population Structure in a Comprehensive Genomic Data Set on Human Microsatellite Variation', *G3: Genes, Genomes, Genetics* 3, no. 5 (2013): 891–907

Pickrell, Joseph K., and David Reich, 'Toward a New History and Geography of Human Genes Informed by Ancient DNA', *Trends in Genetics* 30, no. 9 (2014): 377–89

Pietschnig, J., and M. Voracek, 'One Century of Global IQ Gains: A Formal Meta-analysis of the Flynn Effect (1909–2013)', *Perspectives on Psychological Science* 10, no. 3 (2015): 282–306

Platt, Jonathan M., Katherine M. Keyes, Katie A. McLaughlin et al., 'The Flynn Effect for Fluid IQ May Not Generalize to All Ages or Ability Levels: A Population-based Study of 10,000 US Adolescents', *Intelligence* 77 (2019): 101385

Plomin, Robert, *Blueprint, with a New Afterword: How DNA Makes Us Who We Are*, MIT Press, 2019

Puamau, Priscilla Qolisaya, 'Understanding Fijian Under-achievement: An Integrated Perspective', *Directions: Journal of Educational Studies* 21, no. 2 (1999): 100–12

Ramachandran, Sohini, Omkar Deshpande, Charles C. Roseman et al., 'Support from the Relationship of Genetic and Geographic Distance in Human Populations for a Serial Founder Effect Originating in Africa', *Proceedings of the National Academy of Sciences* 102, no. 44 (2005): 15942–7

Ritchie, Stuart, *Intelligence: All That Matters*, John Murray, 2015

Ritchie, Stuart, and Elliot M. Tucker-Drob, 'How Much Does Education Improve Intelligence? A Meta-analysis', *Psychological Science* 29, no. 8 (2018): 1358–69

Rosenberg, Noah A., Jonathan K. Pritchard, James L. Weber et al.,

'Genetic Structure of Human Populations', *Science* 298, no. 5602 (2002): 2381–5

Roser, Max, and Esteban Ortiz-Ospina, 'Literacy', Our World in Data, 2016, https://ourworldindata.org/literacy

St Clair, James J. H., Barbara C. Klump, Shoko Sugasawa et al., 'Hook Innovation Boosts Foraging Efficiency in Tool-using Crows', *Nature Ecology & Evolution* 2, no. 3 (2018): 441–4

Samuelsson, Stefan, Brian Byrne, Richard K. Olson et al., 'Response to Early Literacy Instruction in the United States, Australia, and Scandinavia: A Behavioral–Genetic Analysis', *Learning and Individual Differences* 18, no. 3 (2008): 289–95

Sowell, Thomas, *Affirmative Action around the World: An Empirical Study*, Yale University Press, 2004

Teferra, Damtew, and Philip G. Altbach, *African Higher Education: An International Reference Handbook*, Indiana University Press, 2003

Thöni, Christian, and Stefan Volk, 'Converging Evidence for Greater Male Variability in Time, Risk, and Social Preferences', *Proceedings of the National Academy of Sciences* 118, no. 23 (2021): e2026112118

Thöni, Christian, Stefan Volk, and Jose M. Cortina, 'Greater Male Variability in Cooperation: Meta-analytic Evidence for an Evolutionary Perspective', *Psychological Science* 32, no. 1 (2021): 50–63

Trahan, Lisa, Karla K. Stuebing, Merril K. Hiscock et al., 'The Flynn Effect: A Meta-analysis', *Psychological Bulletin* 140, no. 5 (2014): 1332–60.

Tucker-Drob, Elliot M., and Timothy C. Bates, 'Large Cross-national Differences in Gene × Socioeconomic Status Interaction on Intelligence', *Psychological Science* 27, no. 2 (2016): 138–49

Turkheimer, Eric, Andreana Haley, Mary Waldron et al., 'Socio-economic Status Modifies Heritability of IQ in Young Children', *Psychological Science* 14, no. 6 (2003): 623–8

Uchiyama, Ryutaro, Rachel Spicer, and Michael Muthukrishna, 'Cultural Evolution of Genetic Heritability', *Behavioral and Brain Sciences* 45 (2022): e152

US Census Bureau, 'S0201: Selected Population Profile in the United States', 2018, https://api.census.gov/data/2018/acs/acs1/spp/groups/S0201.html

Wickramasinghe, Nira, *Sri Lanka in the Modern Age: A History*, Oxford University Press, 2014

Yeager, David S., Paul Hanselman, Gregory M. Walton et al., 'A National Experiment Reveals Where a Growth Mindset Improves Achievement', *Nature* 573, no. 7774 (2019): 364–9

Yuan, Kai, Xumin Ni, Chang Liu et al., 'Refining Models of Archaic Admixture in Eurasia with ArchaicSeeker 2.0', *Nature Communications* 12, no. 1 (2021): 6232

Chapter 4: Innovation in the Collective Brain

For examples of the path dependence that has led to modern US regional cultures see Colin Woodard, *American Nations: A History of the Eleven Rival Regional Cultures of North America*, Penguin, 2012; and David Hackett Fischer, *Albion's Seed: Four British Folkways in America*, Oxford University Press, 1989. Eric Hobsbawm and Terence Ranger, eds, *The Invention of Tradition*, Cambridge University Press, 2012, provides examples of path dependence in traditions we think are more ancient than they are.

On tolerance for diversity and innovation see Michele Gelfand, *Rule Makers, Rule Breakers: Tight and Loose Cultures and the Secret Signals That Direct Our Lives*, Scribner, 2019.

Two books that reveal the complexity of even very ordinary aspects of the world are Lewis Dartnell, *The Knowledge: How to Rebuild Our World from Scratch*, Random House, 2014; and Ryan North, *How to Invent Everything: Rebuild All of Civilization (with 96% Fewer Catastrophes This Time)*, Random House, 2018.

Discussions of the conditions that led to the Enlightenment and Industrial Revolution are offered by Joel Mokyr in *The Enlightened Economy: An Economic History of Britain, 1700–1850*, Yale University Press, 2009; and *A Culture of Growth*, Princeton University Press, 2016.

Specific and specialist reading

Becker, Sascha O., Erik Hornung, and Ludger Woessmann, 'Education and Catch-up in the Industrial Revolution', *American Economic Journal: Macroeconomics* 3, no. 3 (2011): 92–126

Bernard, Diane, *How a Miracle Drug Changed the Fight against Infection during World War II*, Washington Post, 2020

Bradley, David, 'Impossibly Amorphous Material Synthesized', *Materials Today* 9, no. 16 (2013): 304

De Sousa, Telma, Miguel Ribeiro, and Carolina Sabença, 'The 10,000-Year Success Story of Wheat!', *Foods* 10, no. 9 (2021): 2124

Dehaene, Stanislas, *The Number Sense: How the Mind Creates Mathematics*, Oxford University Press, 2011

Edvinsson, Rodney, and Johan Söderberg, 'Prices and the Growth of the

European Knowledge Economy, 1200–2007', in *Swedish Economic History Meeting, Uppsala, 5–7 March 2009*, pp. 1–21

Forsyth, Mark, *The Elements of Eloquence: How to Turn the Perfect English Phrase*, Icon Books, 2013

Gilbert, Will, Tuomo Tanttu, Wee Han Lim et al., 'On-demand Electrical Control of Spin Qubits', *Nature Nanotechnology* (12 January 2023): 1–6. https://doi.org/10.1038/s41565-022-01280-4

Henrich, Joseph, 'Demography and Cultural Evolution: How Adaptive Cultural Processes Can Produce Maladaptive Losses – the Tasmanian Case', *American Antiquity* 69, no. 2 (2004): 197–214

Kauffman, Stuart A., *Investigations*, Oxford University Press, 2000

Kebric, Robert B., *Roman People*, McGraw-Hill, 2005

Kline, Michelle A., and Robert Boyd, 'Population Size Predicts Technological Complexity in Oceania', *Proceedings of the Royal Society B: Biological Sciences* 277, no. 1693 (2010): 2559–64

Lewis, Michael, *The Undoing Project: A Friendship That Changed the World*, Penguin, 2016

Liu, Yi-Ping, Gui-Sheng Wu, Yong-Gang Yao et al., 'Multiple Maternal Origins of Chickens: Out of the Asian Jungles', *Molecular Phylogenetics and Evolution* 38, no. 1 (2006): 12–19

Love, John F., *McDonald's: Behind the Arches*, Random House, 1995

McTavish, Emily Jane, Jared E. Decker, Robert D. Schnabel et al., 'New World Cattle Show Ancestry from Multiple Independent Domestication Events', *Proceedings of the National Academy of Sciences* 110, no. 15 (2013): e1398–406

March, Richard, *Charles Goodyear and the Strange Story of Rubber*, Reader's Digest, 1958

Muthukrishna, Michael, and Joseph Henrich, 'Innovation in the Collective Brain', *Philosophical Transactions of the Royal Society B: Biological Sciences* 371, no. 1690 (2016): 20150192

Muthukrishna, Michael, Ben W. Shulman, Vlad Vasilescu et al., 'Sociality Influences Cultural Complexity', *Proceedings of the Royal Society B: Biological Sciences* 281, no. 1774 (2014): 20132511

National Center for Education Statistics, 'Education Expenditures by Country', in *Condition of Education*, US Department of Education, Institute of Education Sciences, 2022, https://nces.ed.gov/programs/coe/indicator/cmd

Parthasarathy, N., 'Origin of Noble Sugar-Canes (Saccharum Officinarum.)', *Nature* 161, no. 4094 (1948): 608

Schimmelpfennig, Robin, Layla Razek, Eric Schnell et al., 'Paradox of

Diversity in the Collective Brain', *Philosophical Transactions of the Royal Society B* 377, no. 1843 (2022): 20200316

Vance, Ashlee, *Elon Musk: How the Billionaire CEO of SpaceX and Tesla is Shaping Our Future*, Virgin Books, 2016

Chapter 5: Created by Culture

Some of the most accessible introductions to cultural evolution, dual inher itance theory, and culture-gene co-evolution are Joseph Henrich's *The Secret of Our Success*, Princeton University Press, 2015, and *The WEIRDest People in the World: How the West became Psychologically Peculiar and Particularly Prosperous*, Penguin, 2020; Kevin N. Laland, *Darwin's Unfinished Symphony*, Princeton University Press, 2018; Robert Boyd, *A Different Kind of Animal*, Princeton University Press, 2017; Leslie Newson and Peter Richerson, *A Story of Us: A New Look at Human Evolution*, Oxford University Press, 2021; and Alex Mesoudi, *Cultural Evolution*, University of Chicago Press, 2011. The self-domestication hypothesis is nicely described by Brian Hare and Vanessa Woods in *Survival of the Friendliest: Understanding Our Origins and Rediscovering Our Common Humanity*, Random House, 2021.

The psychology of childhood is examined by Alison Gopnik in *The Gardener and the Carpenter: What the New Science of Child Development Tells Us about the Relationship between Parents and Children*, Macmillan, 2016; and the history of menopause is charted by Susan Mattern in *The Slow Moon Climbs: The Science, History, and Meaning of Menopause*, Princeton University Press, 2019.

Specific and specialist reading

Aiello, Leslie C., and Peter Wheeler, 'The Expensive-Tissue Hypothesis: The Brain and the Digestive System in Human and Primate Evolution', *Current Anthropology* 36, no. 2 (1995): 199–221

Alesina, Alberto, Paola Giuliano, and Nathan Nunn, 'On the Origins of Gender Roles: Women and the Plough', *Quarterly Journal of Economics* 128, no. 2 (2013): 469–530

Becker, Anke, 'On the Economic Origins of Restricting Women's Promiscuity', *Review of Economic Studies* (forthcoming)

Betran, Ana Pilar, Jiangfeng Ye, Ann-Beth Moller et al., 'Trends and Projections of Caesarean Section Rates: Global and Regional Estimates', *BMJ Global Health* 6, no. 6 (2021): e005671

Christiansen, Morten H., and Nick Chater, *Creating Language: Integrating Evolution, Acquisition, and Processing*, MIT Press, 2016

Chudek, Maciej, Michael Muthukrishna, and Joe Henrich, 'Cultural Evolution', in *The Handbook of Evolutionary Psychology*, pp. 1–21, Wiley, 2015

Dominguez-Bello, Maria G., Kassandra M. De Jesus-Laboy, Nan Shen et al., 'Partial Restoration of the Microbiota of Cesarean-born Infants via Vaginal Microbial Transfer', *Nature Medicine* 22, no. 3 (2016): 250–3

Dunbar, Robin I. M., 'The Social Brain Hypothesis', *Evolutionary Anthropology* 6, no. 5 (1998): 178–90

Enard, Wolfgang, Molly Przeworski, Simon E. Fisher et al., 'Molecular Evolution of FOXP2, a Gene Involved in Speech and Language', *Nature* 418, no. 6900 (2002): 869–72

Fox, K. C., M. Muthukrishna, and S. Shultz, 'The Social and Cultural Roots of Whale and Dolphin Brains', *Nature Ecology & Evolution* 1, no. 11 (2017): 1699–705

Gneezy, Uri, Kenneth L. Leonard, and John A. List, 'Gender Differences in Competition: Evidence from a Matrilineal and a Patriarchal Society', *Econometrica* 77, no. 5 (2009): 1637–64

Haeusler, Martin, Nicole D. S. Grunstra, Robert D. Martin et al., 'The Obstetrical Dilemma Hypothesis: There's Life in the Old Dog Yet', *Biological Reviews* 96, no. 5 (2021): 2031–57

Hare, Brian, 'Survival of the Friendliest: Homo Sapiens Evolved Via Selection for Prosociality', *Annual Review of Psychology* 68 (2017): 155–86

Hare, Brian, and Vanessa Woods, *Survival of the Friendliest: Understanding Our Origins and Rediscovering Our Common Humanity*, Random House, 2021

Hawkes, Kristen, James F. O'Connell, N. G. Blurton Jones et al., 'Grandmothering, Menopause, and the Evolution of Human Life Histories', *Proceedings of the National Academy of Sciences* 95, no. 3 (1998): 1336–9

Henrich, Joseph, Robert Boyd, and Peter J. Richerson, 'The Puzzle of Monogamous Marriage', *Philosophical Transactions of the Royal Society B: Biological Sciences* 367, no. 1589 (2012): 657–69

Henrich, Joseph, and Richard McElreath, 'The Evolution of Cultural Evolution', *Evolutionary Anthropology* 12, no. 3 (2003): 123–35

Henrich, Joseph, and Michael Muthukrishna, 'The Origins and Psychology of Human Cooperation', *Annual Review of Psychology* 72 (2021): 207–40

Hrdy, Sarah Blaffer, 'Evolutionary Context of Human Development: The Cooperative Breeding Model', in *Family Relationships: An Evolutionary Perspective*, ed. Catherine A. Salmon and Todd K. Shackelford, pp. 39–68, Oxford University Press, 2006

——, *Mothers and Others: The Evolutionary Origins of Mutual Understanding*, Harvard University Press, 2009

Kramer, Karen L., 'Cooperative Breeding and Its Significance to the Demographic Success of Humans', *Annual Review of Anthropology* 39 (2010): 417–36

Lassek, William D., and Steven J. C. Gaulin, 'Costs and Benefits of Fat-free Muscle Mass in Men: Relationship to Mating Success, Dietary Requirements, and Native Immunity', *Evolution and Human Behavior* 30, no. 5 (2009): 322–8

Lind, Johan, and Patrik Lindenfors, 'The Number of Cultural Traits Is Correlated with Female Group Size but Not with Male Group Size in Chimpanzee Communities', *PloS ONE* 5, no. 3 (2010): e9241

Lipschuetz, Michal, Sarah M. Cohen, Eliana Ein-Mor et al., 'A Large Head Circumference Is more Strongly Associated with Unplanned Cesarean or Instrumental Delivery and Neonatal Complications than High Birthweight', *American Journal of Obstetrics and Gynecology* 213, no. 6 (2015): 833.e1–833.e12

Lonsdorf, Elizabeth V., 'What Is the Role of Mothers in the Acquisition of Termite-Fishing Behaviors in Wild Chimpanzees (Pan Troglodytes Schweinfurthii)?', *Animal Cognition* 9 (2006): 36–46

Mattison, Siobhán M., Brooke Scelza, and Tami Blumenfield, 'Paternal Investment and the Positive Effects of Fathers among the Matrilineal Mosuo of Southwest China', *American Anthropologist* 116, no. 3 (2014): 591–610

Mesoudi, Alex, Andrew Whiten, and Kevin N. Laland, 'Towards a Unified Science of Cultural Evolution', *Behavioral and Brain Sciences* 29, no. 4 (2006): 329–47

Muthukrishna, Michael, Michael Doebeli, Maciej Chudek et al., 'The Cultural Brain Hypothesis: How Culture Drives Brain Expansion, Sociality, and Life History', *PLoS Computational Biology* 14, no. 11 (2018): e1006504

Navarrete, Ana, Carel P. Van Schaik, and Karin Isler, 'Energetics and the Evolution of Human Brain Size', *Nature* 480, no. 7375 (2011): 91–3

Pruetz, Jill D., and Paco Bertolani, 'Savanna Chimpanzees, Pan Troglodytes Verus, Hunt with Tools', *Current Biology* 17, no. 5 (2007): 412–17

Ruff, C. B., E. Trinkaus, and T. W. Holliday, 'Body Mass and Encephalization in Pleistocene Homo', *Nature*, 387 (1997): 173–6

Smith, J., F. Plaat, and Nicholas M. Fisk, 'The Natural Caesarean: A Woman-centred Technique', *BJOG: An International Journal of Obstetrics & Gynaecology* 115, no. 8 (2008): 1037–42

Somers, Ali, *The Intergenerational Programme at Nightingale House: A Study into the Impact on the Well-being of Elderly Residents*, Nightingale Hammerson, 2019

Wrangham, Richard, *Catching Fire: How Cooking Made Us Human*, Basic Books, 2009

Chapter 6: Cooperation

Some accessible books on cooperation are Joseph Henrich, *The WEIRDest People in the World: How the West became Psychologically Peculiar and Particularly Prosperous*, Penguin, 2020; Nichola Raihani, *The Social Instinct: How Cooperation Shaped the World*, Random House, 2021; Martin A. Nowak and Roger Highfield, *Supercooperators*, Canongate, 2011; Sarah Blaffer Hrdy, *Mothers and Others: The Evolutionary Origins of Mutual Understanding*, Harvard University Press, 2009; David Sloan Wilson, *This View of Life: Completing the Darwinian Revolution*, Vintage, 2020; and Oren Harman, *The Price of Altruism: George Price and the Search for the Origins of Kindness*, W. W. Norton, 2011.

For the psychology and evolution of morality and religion see Ara Norenzayan, *Big Gods: How Religion Transformed Cooperation and Conflict*, Princeton University Press, 2013; Henrich's *The WEIRDest People in the World*; and Joshua Greene, *Moral Tribes: Emotion, Reason, and the Gap between Us and Them*, Penguin, 2014.

There are two books by Steven Pinker on the long peace: *The Better Angels of Our Nature: The Decline of Violence in History and Its Causes*, Penguin, 2011, and *Enlightenment Now: The Case for Reason, Science, Humanism, and Progress*, Penguin, 2018.

For a history of microchip manufacturing see Chris Miller, *Chip War: The Fight for the World's Most Critical Technology*, Simon & Schuster, 2022.

Specific and specialist reading

Acemoglu, Daron, David Autor, David Dorn et al., 'Return of the Solow Paradox? IT, Productivity, and Employment in US Manufacturing', *American Economic Review* 104, no. 5 (2014): 394–9

Alexander, Richard D., 'The Biology of Moral Systems', Routledge, 1987

Baldwin, Richard, *The Great Convergence: Information Technology and the New Globalization*, Harvard University Press, 2016

Biden, Joseph, 'Remarks by President Biden on Afghanistan', speech, The White House, 16 August 2021, https://www.whitehouse.gov/briefing-room/speeches-remarks/2021/08/16/remarks-by-president-biden-on-afghanistan/

Cowen, Tyler, *The Great Stagnation: How America Ate All the Low-hanging Fruit of Modern History, Got Sick, and Will (Eventually) Feel Better: A Penguin eSpecial from Dutton*, Penguin, 2011

Dawkins, Richard, *The Selfish Gene*, Oxford University Press, 1976

Eurostat, 'Employees by Sex, Age, Educational Attainment Level, Work Experience While Studying and Method Used for Finding Current Job', LFSO_16FINDMET, Eurostat, 2022

Fehr, Ernst, and Simon Gächter, 'Cooperation and Punishment in Public Goods Experiments', *American Economic Review* 90, no. 4 (2000): 980–94

Finke, Roger, and Rodney Stark, *The Churching of America, 1776–2005: Winners and Losers in Our Religious Economy*, Rutgers University Press, 2005

Fisher, Ronald A., *The Genetical Theory of Natural Selection*, Clarendon Press, 1930

Haji, Nafisa, *The Sweetness of Tears*, William Morrow, 2011

Hamilton, William D., 'The Genetical Theory of Social Behaviour. I, II', *Journal of Theoretical Biology* 7 (1964): 1–52

Hardin, Garrett, 'The Tragedy of the Commons', *Science* 162, no. 3859 (1968): 1243–8

Henrich, Joseph, 'Does Culture Matter in Economic Behavior? Ultimatum Game Bargaining among the Machiguenga of the Peruvian Amazon', *American Economic Review* 90, no. 4 (2000): 973–9

——, 'Cultural Group Selection, Coevolutionary Processes and Large-scale Cooperation', *Journal of Economic Behavior & Organization* 53, no. 1 (2004): 3–35

Henrich, Joseph, Robert Boyd, Samuel Bowles et al., 'In Search of Homo Economicus: Behavioral Experiments in 15 Small-scale Societies', *American Economic Review* 91, no. 2 (2001): 73–8

Henrich, Joseph, Robert Boyd, Ernst Fehr et al., eds, *Foundations of Human Sociality: Economic Experiments and Ethnographic Evidence from Fifteen Small-scale Societies*, Oxford University Press on Demand, 2004

Henrich, Joseph, and Michael Muthukrishna, 'The Origins and Psychology of Human Cooperation', *Annual Review of Psychology* 72 (2021): 207–40

Ito, Koichi, and Michael Doebeli, 'The Joint Evolution of Cooperation and Competition', *Journal of Theoretical Biology* 480 (2019): 1–12

Kant, Immanuel, *The 'Metaphysics of Ethics'*, T. & T. Clark, 1796

——, *Physical Geography*, T. & T. Clark, 1802

Klitgaard, Robert E., Ronald MacLean Abaroa, and H. Lindsey Parris, *Corrupt Cities: A Practical Guide to Cure and Prevention*, World Bank Publications, 2000

Lugo, Luis, Alan Cooperman, James Bell et al., 'The World's Muslims: Religion, Politics and Society', Pew Research Center, 2013

Morris, Ian, *Why the West Rules – For Now: The Patterns of History and What They Reveal about the Future*, Profile, 2010

Muthukrishna, Michael, Patrick Francois, Shayan Pourahmadi et al., 'Corrupting Cooperation and How Anti-corruption Strategies May Backfire', *Nature Human Behaviour* 1, no. 7 (2017): 0138

Muthukrishna, Michael, Joseph Henrich, and Edward Slingerland, 'Psychology as a Historical Science', *Annual Review of Psychology* 72 (2021): 717–49

Norenzayan, Ara, Azim F. Shariff, Will M. Gervais et al., 'The Cultural Evolution of Prosocial Religions', *Behavioral and Brain Sciences* 39 (2016): e1

Nowak, Martin A., 'Five Rules for the Evolution of Cooperation', *Science* 314, no. 5805 (2006): 1560–3

Nowak, Martin A., and K. Sigmund, 'The Dynamics of Indirect Reciprocity', *Journal of Theoretical Biology* 194 (1998): 561–74

——, 'Evolution of Indirect Reciprocity', *Nature* 437, no. 7063 (2005): 1291–8

Okasha, Samir, *Evolution and the Levels of Selection*, Clarendon Press, 2006

Ostrom, Elinor, *Governing the Commons: The Evolution of Institutions for Collective Action*, Cambridge University Press, 1990

Pacheco, Jorge M., Francisco C. Santos, Max O. Souza et al., 'Evolutionary Dynamics of Collective Action in N-Person Stag Hunt Dilemmas', *Proceedings of the Royal Society B: Biological Sciences* 276, no. 1655 (2009): 315–21

Panchanathan, Karthik, and Robert Boyd, 'Indirect Reciprocity Can Stabilize Cooperation without the Second-order Free Rider Problem', *Nature* 432, no. 7016 (2004): 499–502

Pennisi, Elizabeth, 'How Did Cooperative Behavior Evolve?', *Science* 309, no. 5731 (2005): 93

Pomeranz, Kenneth, *The Great Divergence: China, Europe, and the Making of the Modern World Economy*, Princeton University Press, 2021

Price, George R., 'Selection and Covariance', *Nature* 227 (1970): 520–1

——, 'Extension of Covariance Selection Mathematics', *Annals of Human Genetics* 35, no. 4 (1972): 485–90

Richerson, Peter, Ryan Baldini, Adrian V. Bell et al., 'Cultural Group Selection Plays an Essential Role in Explaining Human Cooperation: A Sketch of the Evidence', *Behavioral and Brain Sciences* 39 (2016): e30

Saify, Khyber, and Mostafa Saadat, 'Consanguineous Marriages in Afghanistan', *Journal of Biosocial Science* 44, no. 1 (2012): 73–81

Schnell, Eric, Robin Schimmelpfennig, and Michael Muthukrishna, 'The Size of the Stag Determines the Level of Cooperation', *bioRxiv* (2021): 2021–2

Skyrms, Brian, *The Stag Hunt and the Evolution of Social Structure*, Cambridge University Press, 2004

Smith, J. Maynard, 'Group Selection and Kin Selection', *Nature* 201, no. 4924 (1964): 1145–7

Solow, Robert, 'We'd Better Watch Out', *New York Times Book Review* 36 (1987): 36

Toje, A., and N. V. Steen, *The Causes of Peace: What We Know Now*, Nobel Peace Prize Research & Information, 2019

Treisman, Daniel, 'What Have We Learned about the Causes of Corruption from Ten Years of Cross-national Empirical Research?', *Annual Review of Political Science* 10, no. 1 (June 2007): 211–44

Trivers, Robert L., 'The Evolution of Reciprocal Altruism', *Quarterly Review of Biology* 46, no. 1 (1971): 35–57

Trompenaars, Fons, and Charles Hampden-Turner, *Riding the Waves of Culture: Understanding Diversity in Global Business*, Nicholas Brealey International, 2011

Wilson, David Sloan, 'A Theory of Group Selection', *Proceedings of the National Academy of Sciences* 72, no. 1 (1975): 143–6

Part II: Where We're Going

The opening quote from Edward Wilson comes from the second paragraph of Chapter 1 of *The Social Conquest of Earth*, W. W. Norton, 2012.

Specific and specialist reading

'Does the Current Migrant Crisis in Europe Make You More or Less Likely to Vote to Leave the EU?', What UK Thinks: EU opinion poll, 2015

Ash, Konstantin, and Nick Obradovich, 'Climatic Stress, Internal Migration, and Syrian Civil War Onset', *Journal of Conflict Resolution* 64, no. 1 (2020): 3–31

Burke, Marshall B., Edward Miguel, Shanker Satyanath et al., 'Warming Increases the Risk of Civil War in Africa', *Proceedings of the National Academy of Sciences* 106, no. 49 (2009): 20670–4

Golec de Zavala, Agnieszka, Rita Guerra, and Cláudia Simão, 'The Relationship between the Brexit Vote and Individual Predictors of

Prejudice: Collective Narcissism, Right Wing Authoritarianism, Social Dominance Orientation', *Frontiers in Psychology* 8 (2017): 2023

Hall, Charles A. S., and Kent Klitgaard, *Energy and the Wealth of Nations: An Introduction to Biophysical Economics*, 2nd edn, Springer, 2018

Hsiang, Solomon M., Marshall Burke, and Edward Miguel, 'Quantifying the Influence of Climate on Human Conflict', *Science* 341, no. 6151 (2013): 1235367

Kelley, Colin P., Shahrzad Mohtadi, Mark A. Cane et al., 'Climate Change in the Fertile Crescent and Implications of the Recent Syrian Drought', *Proceedings of the National Academy of Sciences* 112, no. 11 (2015): 3241–6

Koubi, Vally, 'Climate Change and Conflict', *Annual Review of Political Science* 22 (2019): 343–60

Schleussner, Carl-Friedrich, Jonathan F. Donges, Reik V. Donner et al., 'Armed-Conflict Risks Enhanced by Climate-related Disasters in Ethnically Fractionalized Countries', *Proceedings of the National Academy of Sciences* 113, no. 33 (2016): 9216–21

Chapter 7: Reuniting Humanity

On polarization in America see Jonathan Haidt, *The Righteous Mind: Why Good People are Divided by Politics and Religion*, Vintage, 2012; Chris Bail, *Breaking the Social Media Prism: How to Make Our Platforms Less Polarizing*, Princeton University Press, 2022; Sinan Aral, *The Hype Machine: How Social Media Disrupts Our Elections, Our Economy, and Our Health—and How We Must Adapt*, Currency, 2021; Lilliana Mason, *Uncivil Agreement: How Politics Became Our Identity*, University of Chicago Press, 2018; and James E. Campbell, *Polarized: Making Sense of a Divided America*, Princeton University Press, 2018.

For the effects of the private school pipeline and similar issues of inequality in the UK, US, and elsewhere see Simon Kuper, *Chums: How a Tiny Caste of Oxford Tories Took Over the UK*, Profile, 2022; Richard Reeves, *Dream Hoarders: How the American Upper Middle Class Is Leaving Everyone Else in the Dust, Why That Is a Problem, and What to Do About It*, Brookings Institution Press, 2017; and Anand Giridharadas, *Winners Take All: The Elite Charade of Changing the World*, Knopf, 2018.

The role East–West versus North–South geography plays in innovation is explored by Jared M. Diamond in *Guns, Germs and Steel: A Short History of Everybody for the Last 13,000 Years*, Random House, 1998. The role of coder culture in creating the Internet is described by Clive Thompson in

Coders: The Making of a New Tribe and the Remaking of the World, Penguin, 2019; and corruption in Australia is examined in Cameron K. Murray and Paul Frijters, *Rigged: How Networks of Powerful Mates Rip Off Everyday Australians*, Allen & Unwin, 2022.

On immigration see the following: Bryan Caplan, *Open Borders: The Science and Ethics of Immigration*, First Second, 2019; Garett Jones, *The Culture Transplant: How Migrants Make the Economies They Move to a Lot Like the Ones They Left*, Stanford University Press, 2022; and Ran Abramitzky and Leah Boustan, *Streets of Gold: America's Untold Story of Immigrant Success*, Hachette, 2022.

Specific and specialist reading

Alesina, Alberto, Paola Giuliano, and Nathan Nunn, 'On the Origins of Gender Roles: Women and the Plough', *Quarterly Journal of Economics* 128, no. 2 (2013): 469–530

Algan, Yann, Christian Dustmann, Albrecht Glitz et al., 'The Economic Situation of First and Second-Generation Immigrants in France, Germany and the United Kingdom', *Economic Journal* 120 (2010): F4–30

Andersen, S. N., and T. Kornstad, 'Time since Immigration and Crime amongst Adult Immigrants in Norway', Government report, Statistics Norway, 2017, https://www.ssb.no/sosiale-forhold-og-kriminalitet/artikler-og-publikasjoner/_attachment/332400?_ts=1603007dd70

Andersen, S. N., B. Holtsmark, and S. B. Mohn, 'Crime amongst Immigrants, Children of Immigrants and the Remaining Population: An Analysis of Register Data 1992–2015 Reports 2017/36', Government report, Statistics Norway, 2017, https://www.ssb.no/sosiale-forhold-ogkriminalitet/artikler-og-publikasjoner/_attachment/332143?_ts=16035 d6f0d8

Australian Government, Department of Home Affairs, 'Australian Cultural Orientation (AUSCO) Program', Immigration and Citizenship online overview, 2019, https://immi.homeaffairs.gov.au/settling-in-australia/ausco

Aydemir, Abdurrahman, and George J. Borjas, 'Cross-country Variation in the Impact of International Migration: Canada, Mexico, and the United States', *Journal of the European Economic Association* 5, no. 4 (2007): 663–708

Beaulier, Scott A., 'Explaining Botswana's Success: The Critical Role of Post-colonial Policy', *Cato Journal* 23, no. 2 (2003): 227–40

Bertrand, Marianne, and Sendhil Mullainathan, 'Are Emily and Greg More Employable than Lakisha and Jamal? A Field Experiment on Labor

Market Discrimination', *American Economic Review* 94, no. 4 (2004): 991–1013

Beveridge, William, 'Social Insurance and Allied Services', command paper 6404 (The Beveridge Report), His Majesty's Stationery Office, 1942

Bloemraad, Irene, Anna Korteweg, and Gökçe Yurdakul, 'Citizenship and Immigration: Multiculturalism, Assimilation, and Challenges to the Nation-state', *Annual Review of Sociology* 34 (2008): 153–79

Brubaker, Rogers, *Citizenship and Nationhood in France and Germany*, Harvard University Press, 2009

Chetty, Raj, John N. Friedman, and Emmanuel Saez, 'Using Differences in Knowledge across Neighborhoods to Uncover the Impacts of the EITC on Earnings', *American Economic Review* 103, no. 7 (2013): 2683–721

Christophers, Brett, *Rentier Capitalism: Who Owns the Economy, and Who Pays for It?*, Verso, 2022

Cohen, Roger, 'Can-Do Lee Kuan Yew', *New York Times*, 24 March 2015

Day, Richard J. F., *Multiculturalism and the History of Canadian Diversity*, University of Toronto Press, 2000

Dustmann, Christian, and Tommaso Frattini, 'The Fiscal Effects of Immigration to the UK', *Economic Journal* 124, no. 580 (2014): F593–643

Dustmann, Christian, and Ian P. Preston, 'Free Movement, Open Borders, and the Global Gains from Labor Mobility', *Annual Review of Economics* 11 (2019): 783–808

Fasting, Mathilde, and Øystein Sørensen, *The Norwegian Exception? Norway's Liberal Democracy since 1814*, Hurst, 2021

Fischer, David Hackett, *Albion's Seed: Four British Folkways in America*, Oxford University Press, 1989

Fisher, Matthew C., Sarah J. Gurr, Christina A. Cuomo et al., 'Threats Posed by the Fungal Kingdom to Humans, Wildlife, and Agriculture', *mBio* 11, no. 3 (2020): e00449–20

Frijters, Paul, and Cameron Murray, *Rigged: How Networks of Powerful Mates Rip Off Everyday Australians*, Allen & Unwin, 2022

'Germany's Vice Chancellor Says Merkel Underestimated Migrant Challenge', *Reuters*, 27 August 2016

Gerring, John, Michael Hoffman, and Dominic Zarecki, 'The Diverse Effects of Diversity on Democracy', *British Journal of Political Science* 48, no. 2 (2018): 283–314

Giuliano, Paola, and Nathan Nunn, 'The Transmission of Democracy: From the Village to the Nation-State', *American Economic Review* 103, no. 3 (2013): 86–92

———, 'Understanding Cultural Persistence and Change', *Review of Economic Studies* 88, no. 4 (2021): 1541–81

Gleason, Philip, 'The Melting Pot: Symbol of Fusion or Confusion?', *American Quarterly* 16, no. 1 (1964): 20–46

Gudelunas, David, 'There's an App for That: The Uses and Gratifications of Online Social Networks for Gay Men', *Sexuality & Culture* 16 (2012): 347–65

Heine, Steven J., *Cultural Psychology: Fourth International Student Edition*, W. W. Norton, 2020

Holden, Steinar, 'Avoiding the Resource Curse the Case Norway', *Energy Policy* 63 (2013): 870–6

Huntington, Samuel P., and Steve R. Dunn, *Who Are We? The Challenges to America's National Identity*, Simon & Schuster, 2004

Jack, Rachael E., Oliver G. B. Garrod, Hui Yu et al., 'Facial Expressions of Emotion Are Not Culturally Universal', *Proceedings of the National Academy of Sciences* 109, no. 19 (2012): 7241–4

Khoshnood, Ardavan, Henrik Ohlsson, Jan Sundquist et al., 'Swedish Rape Offenders – a Latent Class Analysis', *Forensic Sciences Research* 6, no. 2 (2021): 124–32

Kriminalität im Kontext von Zuwanderung: Bundeslagebild (*Crime in the Context of Immigration: National Situation Report*), Bundeskriminalamt, 2017

Landgrave, Michelangelo, and Alex Nowrasteh, *Illegal Immigrant Incarceration Rates, 2010–2018: Demographics and Policy Implications*, Cato Institute, 2020

Lochmann, Alexia, Hillel Rapoport, and Biagio Speciale, 'The Effect of Language Training on Immigrants' Economic Integration: Empirical Evidence from France', *European Economic Review* 113 (2019): 265–96

Luttmer, Erzo F. P., and Monica Singhal, 'Culture, Context, and the Taste for Redistribution', *American Economic Journal: Economic Policy* 3, no. 1 (2011): 157–79

Meng, Xin, and Robert G. Gregory, 'Intermarriage and the Economic Assimilation of Immigrants', *Journal of Labor Economics* 23, no. 1 (2005): 135–74

Michalopoulos, Stelios, and Elias Papaioannou, 'The Long-run Effects of the Scramble for Africa', *American Economic Review* 106, no. 7 (2016): 1802–48

Miller, Paul W., 'Immigration Policy and Immigrant Quality: The Australian Points System', *American Economic Review* 89, no. 2 (1999): 192–7

Moser, Petra, and Shmuel San, 'Immigration, Science, and Invention:
Lessons from the Quota Acts', Social Science Research Network, 2020
Muthukrishna, Michael, Adrian V. Bell, Joseph Henrich et al.,
'Beyond Western, Educated, Industrial, Rich, and Democratic (WEIRD)
Psychology: Measuring and Mapping Scales of Cultural and
Psychological Distance', *Psychological Science* 31, no. 6 (2020): 678–701
Nisbett, Richard E., and Dov Cohen, *Culture of Honor: The Psychology of
Violence in the South*, Routledge, 2018
Nowrasteh, Alex, Andrew C. Forrester, and Michelangelo Landgrave,
'Illegal Immigration and Crime in Texas', working paper no. 60, Cato
Institute, 2020
Oreopoulos, Philip, 'Why Do Skilled Immigrants Struggle in the Labor
Market? A Field Experiment with Thirteen Thousand Resumes',
American Economic Journal: Economic Policy 3, no. 4 (2011): 148–71
Pirsig, Robert M., *Zen and the Art of Motorcycle Maintenance: An Inquiry into
Values*, Random House, 1999
Poole, Michael, and Rachel Bell, 'Middle Classes – Their Rise and Sprawl',
BBC, 2001.
Pratchett, Terry, *Raising Steam* (Discworld novel 40), Random House, 2013
Read, Michael, 'Australians Are the World's Richest People', *Financial
Review*, 20 September 2022
Rosenberg, Steve, 'Why Russian Workers Are Being Taught How to Smile',
BBC News, 9 June 2018
Ross, Michael L., 'What Have We Learned about the Resource Curse?',
Annual Review of Political Science 18 (2015): 239–59
Rychlowska, Magdalena, Yuri Miyamoto, David Matsumoto et al.,
'Heterogeneity of Long-History Migration Explains Cultural Differences
in Reports of Emotional Expressivity and the Functions of Smiles',
Proceedings of the National Academy of Sciences 112, no. 19 (12 May 2015):
E2429–36.
Schimmelpfennig, Robin, Layla Razek, Eric Schnell et al., 'Paradox of
Diversity in the Collective Brain', *Philosophical Transactions of the Royal
Society B* 377, no. 1843 (2022): 20200316
Schulz, Jonathan F., Duman Bahrami-Rad, Jonathan P. Beauchamp et al.,
'The Church, Intensive Kinship, and Global Psychological Variation',
Science 366, no. 6466 (2019): eaau5141
Sequeira, Sandra, Nathan Nunn, and Nancy Qian, 'Immigrants and the
Making of America', *Review of Economic Studies* 87, no. 1 (2020):
382–419
Silver, Laura, 'Populists in Europe – Especially Those on the Right – Have

Increased Their Vote Shares in Recent Elections', Pew Research Center, 2022

Simon, Rita J., and Keri W. Sikich, 'Public Attitudes toward Immigrants and Immigration Policies across Seven Nations', *International Migration Review* 41, no. 4 (2007): 956–62

Skardhamar, Torbjørn, Mikko Aaltonen, and Martti Lehti, 'Immigrant Crime in Norway and Finland', *Journal of Scandinavian Studies in Criminology and Crime Prevention* 15, no. 2 (2014): 107–27

Steinmayr, Andreas, 'Contact versus Exposure: Refugee Presence and Voting for the Far Right', *Review of Economics and Statistics* 103, no. 2 (2021): 310–27

Strafurteilsstatistik 2020: Nationalitäten der verurteilten Personen (*Criminal Conviction Statistics 2020: Nationalities of the Convicted Persons*), Swiss Federal Office for Statistics, 2020

Tabellini, Marco, 'Gifts of the Immigrants, Woes of the Natives: Lessons from the Age of Mass Migration', *Review of Economic Studies* 87, no. 1 (2020): 454–86

Van der Ploeg, Frederick, 'Natural Resources: Curse or Blessing?', *Journal of Economic Literature* 49, no. 2 (2011): 366–420

Watson, James L., ed., *Golden Arches East: McDonald's in East Asia*, Stanford University Press, 2006

Wood, Adrienne, Magdalena Rychlowska, and Paula M. Niedenthal, 'Heterogeneity of Long-History Migration Predicts Emotion Recognition Accuracy', *Emotion* 16, no. 4 (2016): 413–20

Woodard, Colin, *American Nations: A History of the Eleven Rival Regional Cultures of North America*, Penguin, 2012

Chapter 8: Governance in the Twenty-first Century

For a history of democracy, cooperation, and political order there are two books by Francis Fukuyama: *The Origins of Political Order: From Prehuman Times to the French Revolution*, Farrar, Straus & Giroux, 2011; and *Political Order and Political Decay: From the Industrial Revolution to the Globalization of Democracy*, Macmillan, 2014. For a similar vision for start-up cities see the Charter Cities Institute's continually updated reading list, https://chartercitiesinstitute.org/reading/

For a concept close to programmable politics see Balaji Srinivasan, *The Network State: How to Start a New Country*, 2022. Commentary from Vitalik Buterin (inventor of Ethereum) is at https://vitalik.ca/general/2022/07/13/networkstates.html, and Buterin has also co-authored (with Nathan Schneider) a book on blockchain and Ethereum, *Proof of Stake: The Making of Ethereum and the Philosophy of Blockchains*, Seven Stories Press, 2022.

Some of the radical ideas being proposed that could be tested within start-up cities or programmed polities can be found in Eric A. Posner and E. Glen Weyl's *Radical Markets*, Princeton University Press, 2018.

Specific and specialist reading

Berlin, Leslie, *The Man Behind the Microchip: Robert Noyce and the Invention of Silicon Valley*, Oxford University Press, 2005

Chesterton, Gilbert Keith, *Orthodoxy*, Bodley Head, 1908

Etzkowitz, Henry, and Chunyan Zhou, *The Triple Helix: University–Industry–Government Innovation and Entrepreneurship*, Routledge, 2017

Finke, Roger, and Rodney Stark, *The Churching of America, 1776–2005: Winners and Losers in Our Religious Economy*, Rutgers University Press, 2005

Fowler, Anthony, 'Electoral and Policy Consequences of Voter Turnout: Evidence from Compulsory Voting in Australia', *Quarterly Journal of Political Science* 8, no. 2 (2013): 159–82

Fukuyama, Francis, *The End of History and the Last Man*, Simon & Schuster, 2006

——, *Identity: Contemporary Identity Politics and the Struggle for Recognition*, Profile, 2018

Gilson, Ronald J., 'The Legal Infrastructure of High-technology Industrial Districts: Silicon Valley, Route 128, and Covenants Not to Compete', *New York University Law Review* 74 (1999): 575

Glaeser, Edward, *Triumph of the City: How Our Greatest Invention Makes Us Richer, Smarter, Greener, Healthier, and Happier*, Penguin, 2012

Huang, Cary, 'Tocqueville's Advice on French Revolution Captures Chinese Leaders' Attention', *South China Morning Post*, 22 January 2013

'International Migrant Stock', United Nations, Department of Economic and Social Affairs, Population Division, 2019, www.unmigration.org

Jacques, Martin, *When China Rules the World: The End of the Western World and the Birth of a New Global Order*, Penguin, 2009

Kline, Patrick, and Enrico Moretti, 'People, Places, and Public Policy: Some Simple Welfare Economics of Local Economic Development Programs', *Annual Review of Economics* 6, no. 1 (2014): 629–62

Li, Hongbin, and Li-An Zhou, 'Political Turnover and Economic Performance: The Incentive Role of Personnel Control in China', *Journal of Public Economics* 89, nos 9–10 (2005): 1743–62

Liker, Jeffrey K., *Toyota Way: 14 Management Principles from the World's Greatest Manufacturer*, McGraw-Hill Education, 2021

McGregor, Richard, *The Party: The Secret World of China's Communist Rulers*, Penguin, 2010

Mackerras, Malcolm, and Ian McAllister, 'Compulsory Voting, Party Stability and Electoral Advantage in Australia', *Electoral Studies* 18, no. 2 (1999): 217–33

Martine, George, and Population Fund, *Unleashing the Potential of Urban Growth*, United Nations Population Fund report, UNFPA, 2007

Muthukrishna, Michael, and Joseph Henrich, 'Innovation in the Collective Brain', *Philosophical Transactions of the Royal Society B: Biological Sciences* 371, no. 1690 (2016): 20150192

Nakamoto, Satoshi, 'Bitcoin: A Peer-to-Peer Electronic Cash System', Whitepaper, 2008

New State Ice Co. v. Liebmann, 285 U.S. 262, 52 S. Ct. 371, 76 L. Ed. 747 (1932)

O'Mara, Margaret, *The Code: Silicon Valley and the Remaking of America*, Penguin, 2020

Reynolds, Andrew, Ben Reilly, and Andrew Ellis, *Electoral System Design: The New International IDEA Handbook*, International Institute for Democracy and Electoral Assistance, 2008

Sassen, Saskia, *Cities in a World Economy*, Sage, 2018

Saxenian, AnnaLee, *Regional Advantage: Culture and Competition in Silicon Valley and Route 128*, Harvard University Press, 1994

Scheidel, Walter, *Escape from Rome: The Failure of Empire and the Road to Prosperity*, Princeton University Press, 2019

Schimmelpfennig, Robin, Layla Razek, Eric Schnell et al., 'Paradox of Diversity in the Collective Brain', *Philosophical Transactions of the Royal Society B* 377, no. 1843 (2022): 20200316

Shum, Desmond, *Red Roulette: An Insider's Story of Wealth, Power, Corruption and Vengeance in Today's China*, Simon & Schuster, 2021

Special Economic Zones: Progress, Emerging Challenges, and Future Directions, World Bank Publications, 2011

Tauberer, Joshua, 'How I Changed the Law with a GitHub Pull Request', *Arstechnica*, 25 November 2018

Tsang, Steve, *A Modern History of Hong Kong: 1841–1997*, Bloomsbury Academic, 2004

Twain, Mark, *Following the Equator: A Journey around the World*, Dover, 1897

Vogel, Ezra F., *Deng Xiaoping and the Transformation of China*, Belknap Press of Harvard University Press, 2011

Wang, Jin, 'The Economic Impact of Special Economic Zones: Evidence from Chinese Municipalities', *Journal of Development Economics* 101 (2013): 133–47

Wang, Shaoda, and David Y. Yang, *Policy Experimentation in China: The Political Economy of Policy Learning*, working paper no. 29402, National Bureau of Economic Research, 2021

Wang, Shuai, Wenwen Ding, Juanjuan Li et al., 'Decentralized Autonomous Organizations: Concept, Model, and Applications', *IEEE Transactions on Computational Social Systems* 6, no. 5 (2019): 870–8

Chapter 9: Shattering the Glass Ceiling

On inequality see the following: Joseph E. Stiglitz, *The Price of Inequality: How Today's Divided Society Endangers Our Future*, W. W. Norton, 2012; Angus Deaton, *The Great Escape*, Princeton University Press, 2013; Richard Baldwin, *The Great Convergence*, Harvard University Press, 2018; and Walter Scheidel, *The Great Leveler*, Princeton University Press, 2017.

Case studies of affirmative action around the world may be found in Thomas Sowell, *Affirmative Action around the World: An Empirical Study*, Yale University Press, 2004.

The history of money is presented by Niall Ferguson in *The Ascent of Money: A Financial History of the World*, Penguin, 2008; and by Jeffrey E. Garten in *Three Days at Camp David: How a Secret Meeting in 1971 Transformed the Global Economy*, Amberley, 2021.

For the history of key economic ideas see Lawrence H. White, *The Clash of Economic Ideas: The Great Policy Debates and Experiments of the Last Hundred Years*, Cambridge University Press, 2012; and two books by Ben Bernanke, *21st Century Monetary Policy: The Federal Reserve from the Great Inflation to COVID-19* and *The Courage to Act: A Memoir of a Crisis and Its Aftermath*, both W. W. Norton, 2022 and 2015 respectively.

For land value taxes see Lars A. Doucet, *Land Is a Big Deal*, Shack Simple Press, 2022; Josh Ryan-Collins, Toby Lloyd, and Laurie Macfarlane, *Rethinking the Economics of Land and Housing*, Bloomsbury, 2017; and Henry George's 1879 work *Progress and Poverty* (available free of charge online). On the challenges of tax more generally (all deeper dives) see Chuck Collins, *The Wealth Hoarders: How Billionaires Pay Millions to Hide Trillions*, John Wiley, 2021; and Kenneth Scheve and David Stasavage, *Taxing the Rich*, Princeton University Press, 2016.

How wealth, politics, and corruption interact in wealthier countries such as the United States and Australia see Jane Mayer, *Dark Money: The Hidden History of the Billionaires Behind the Rise of the Radical Right*, Anchor, 2017;

Nancy MacLean, *Democracy in Chains: The Deep History of the Radical Right's Stealth Plan for America*, Penguin, 2018; and Cameron K. Murray and Paul Frijters, *Rigged: How Networks of Powerful Mates Rip Off Everyday Australians*, Allen & Unwin, 2022.

Two insightful documentary films directed by Jamie Johnson (heir to the Johnson & Johnson fortune) offer a rare glimpse into wealth through interviews with young heirs (including a young Ivanka Trump, twenty-two years old at the time): *Born Rich*, Shout! Factory, 2003, and *The One Percent*, Wise and Good Films, 2008.

Specific and specialist reading

Arnott, Richard J., and Joseph E. Stiglitz, 'Aggregate Land Rents, Expenditure on Public Goods, and Optimal City Size', *Quarterly Journal of Economics* 93, no. 4 (1979): 471–500

Bell, Alex, Raj Chetty, Xavier Jaravel et al., 'Who Becomes an Inventor in America? The Importance of Exposure to Innovation', *Quarterly Journal of Economics* 134, no. 2 (2019): 647–713

Bernanke, Ben S., Mark Gertler, Mark Watson et al., 'Systematic Monetary Policy and the Effects of Oil Price Shocks', *Brookings Papers on Economic Activity* no. 1 (1997): 91–157

Buchholz, Katharina, 'The Top 10 Percent Own 70 Percent of U.S. Wealth', Statista, 31 August 2021, https://www.statista.com/chart/19635/wealth-distribution-percentiles-in-the-us

Campbell, Cameron, and James Z. Lee, 'Kinship and the Long-term Persistence of Inequality in Liaoning, China, 1749–2005', *Chinese Sociological Review* 44, no. 1 (2011): 71–103

Cannadine, David, *The Decline and Fall of the British Aristocracy*, Knopf Doubleday, 1999

Clark, Gregory, *The Son Also Rises*, Princeton University Press, 2014

Clark, Gregory, and Neil Cummins, 'Intergenerational Wealth Mobility in England, 1858–2012: Surnames and Social Mobility', *Economic Journal* 125, no. 582 (2015): 61–85

Clark, Gregory, Neil Cummins, Yu Hao et al., 'Surnames: A New Source for the History of Social Mobility', *Explorations in Economic History* 55 (2015): 3–24

Cleveland, Cutler J., Robert Costanza, Charles A. S. Hall et al., 'Energy and the US Economy: A Biophysical Perspective', *Science* 225, no. 4665 (1984): 890–7

Evans, Judith, and Richard Milne, 'Duke of Westminster Dies', *Financial Times*, 10 August 2016

Ferris, Nick, 'Weekly Data: China's Nuclear Pipeline as Big as the Rest of the World's Combined', Energy Monitor, 20 December 2021, https://www.energymonitor.ai/sectors/power/weekly-data-chinas-nuclear-pipeline-as-big-as-the-rest-of-the-worlds-combined/

Gould, Stephen Jay, *The Panda's Thumb: More Reflections in Natural History*, W. W. Norton, 1992

Haan, Marco A., Pim Heijnen, Lambert Schoonbeek et al., 'Sound Taxation? On the Use of Self-declared Value', *European Economic Review* 56, no. 2 (2012): 205–15

Hall, Charles A. S., and Kent Klitgaard, *Energy and the Wealth of Nations: An Introduction to Biophysical Economics*, Springer International, 2018

Hall, Charles A. S., Jessica G. Lambert, and Stephen B. Balogh, 'EROI of Different Fuels and the Implications for Society', *Energy Policy* 64 (2014): 141–52

Hao, Yu, 'Social Mobility in China, 1645–2012: A Surname Study', *China Economic Quarterly International* 1, no. 3 (2021): 233–43

Jiménez-Rodríguez, Rebeca, and Marcelo Sánchez, 'Oil Price Shocks and Real GDP Growth: Empirical Evidence for Some OECD Countries', *Applied Economics* 37, no. 2 (2005): 201–28

Joulfaian, David, *The Federal Estate Tax: History, Law, and Economics*, MIT Press, 2019

King, Mervyn A., *The End of Alchemy: Money, Banking, and the Future of the Global Economy*, W. W. Norton, 2017

Lindert, P. H., and J. G. Williamson, *Unequal Gains: American Growth and Inequality since 1700*, Princeton University Press, 2016

Meltzer, Allan H., *A History of the Federal Reserve*, vols 1 and 2, University of Chicago Press, 2010

Michels, Robert, *Political Parties: A Sociological Study of the Oligarchical Tendencies of Modern Democracy*, Hearst's International Library Company, 1915

Mill, John Stuart, *Principles of Political Economy*, D. Appleton, 1884

Mirrlees, James, ed., *Tax by Design: The Mirrlees Review*, Oxford University Press, 2011

Murphy, David J., and Charles A. S. Hall, 'Energy Return on Investment, Peak Oil, and the End of Economic Growth', *Annals of the New York Academy of Sciences* 1219, no. 1 (2011): 52–72

Park, Jungwook, and Ronald A. Ratti, 'Oil Price Shocks and Stock Markets in the US and 13 European Countries', *Energy Economics* 30, no. 5 (2008): 2587–608

Piketty, Thomas, *Capital in the Twenty-first Century*, Harvard University Press, 2014

Pilon, Mary, *The Monopolists: Obsession, Fury, and the Scandal Behind the World's Favorite Board Game*, Bloomsbury, 2015

Rawls, John, *A Theory of Justice*, Harvard University Press, 1971

Ryan-Collins, Josh, Tony Greenham, Richard Werner et al., *Where Does Money Come From? A Guide to the UK Monetary and Banking System*, New Economics Foundation, 2012

Shrubsole, Guy, *Who Owns England? How We Lost Our Green and Pleasant Land, and How to Take It Back*, HarperCollins, 2019

Steil, Benn, *The Battle of Bretton Woods*, Princeton University Press, 2013

Young, H. Peyton, *Equity: In Theory and Practice*, Princeton University Press, 1995

Chapter 10: Triggering a Creative Explosion

The history of the shipping container has been documented by Marc Levinson: see *The Box*, Princeton University Press, 2016.

On economic growth and productivity declines see Tyler Cowen, *The Great Stagnation: How America Ate All the Low-hanging Fruit of Modern History, Got Sick, and Will (Eventually) Feel Better*, Penguin, 2011. On social mobility there is Gregory Clark's *The Son Also Rises*, Princeton University Press, 2014.

The way in which the Industrial Revolution was disruptive and how AI and other technologies will be too is described in two books: Carl Benedikt Frey's *The Technology Trap*, Princeton University Press, 2019; and Erik Brynjolfsson and Andrew McAfee's *The Second Machine Age: Work, Progress, and Prosperity in a Time of Brilliant Technologies*, W. W. Norton, 2014.

For some examples of Silicon Valley's unique culture, see Andrew McAfee's *The Geek Way: The Radical Mindset that Drives Extraordinary Results*, Little, Brown, 2023.

Case studies of how the process of science is really conducted by scientists are Harry M. Collins and Trevor Pinch, *The Golem: What You Should Know about Science*, Cambridge University Press, 1998; and Stuart Firestein, *Failure: Why Science Is So Successful*, Oxford University Press, 2015.

Some books on the importance of diversity are Caroline Criado Perez, *Invisible Women: Data Bias in a World Designed for Men*, Abrams, 2019; Matthew Syed, *Rebel Ideas: The Power of Diverse Thinking*, John Murray, 2019; and Scott Page, *The Diversity Bonus*, Princeton University Press, 2017.

For examples of structured diversity, innovation, and changes to collective brains see Walter Scheidel, *Escape from Rome: The Failure of Empire and the Road to Prosperity*, Princeton University Press, 2019; Gen. Stanley McChrystal,

Tantum Collins, David Silverman et al., *Team of Teams: New Rules of Engagement for a Complex World*, Penguin, 2015; and Satya Nadella, Greg Shaw, Jill Tracie Nichols et al., *Hit Refresh: The Quest to Rediscover Microsoft's Soul and Imagine a Better Future for Everyone*, Harper Business, 2017.

Specific and specialist reading

AlShebli, Bedoor, Kinga Makovi, and Talal Rahwan, Retracted articles: 'The Association between Early Career Informal Mentorship in Academic Collaborations and Junior Author Performance', *Nature Communications* 11, no. 1 (2020): 1–8

AlShebli, Bedoor K., Talal Rahwan, and Wei Lee Woon, 'The Preeminence of Ethnic Diversity in Scientific Collaboration', *Nature Communications* 9, no. 1 (2018): 5163

Berkes, Enrico, and Peter Nencka, 'Knowledge Access: The Effects of Carnegie Libraries on Innovation', 22 December 2021, Social Science Research Network, ssrn.3629299

Bertrand, Marianne, and Esther Duflo, 'Field Experiments on Discrimination', *Handbook of Economic Field Experiments* 1 (2017): 309–93

Chen, Yixing, Vikas Mittal, and Shrihari Sridhar, 'Investigating the Academic Performance and Disciplinary Consequences of School District Internet Access Spending', *Journal of Marketing Research* 58, no. 1 (2021): 141–62

Dawkins, Richard, *The Selfish Gene*, Oxford University Press, 1976

Draghi, Jeremy, and Günter P. Wagner, 'Evolution of Evolvability in a Developmental Model', *Evolution* 62, no. 2 (2008): 301–15

Duhigg, Charles, 'What Google Learned from Its Quest to Build the Perfect Team', *New York Times Magazine*, 25 February 2016

Elman, Benjamin A., *Civil Examinations and Meritocracy in Late Imperial China*, Harvard University Press, 2013

Forscher, Patrick S., Calvin K. Lai, Jordan R. Axt et al., 'A Meta-analysis of Procedures to Change Implicit Measures', *Journal of Personality and Social Psychology* 117, no. 3 (2019): 522

Gopnik, Alison, 'Childhood as a Solution to Explore–Exploit Tensions', *Philosophical Transactions of the Royal Society B: Biological Sciences* 375, no. 1803 (20 July 2020): 20190502

Healy, Thomas, *The Great Dissent: How Oliver Wendell Holmes Changed His Mind – and Changed the History of Free Speech in America*, Metropolitan Books, 2013

Hirschman, Daniel, 'Controlling for What? Movements, Measures, and Meanings in the US Gender Wage Gap Debate', *History of Political Economy* 54, no. S1 (2022): 221–57

Jackson, Joshua Conrad, Michele Gelfand, Soham De et al., 'The Loosening of American Culture over 200 Years Is Associated with a Creativity–Order Trade-Off', *Nature Human Behaviour* 3, no. 3 (2019): 244–50

Kleven, Henrik, Camille Landais, and Jakob Egholt Søgaard, 'Children and Gender Inequality: Evidence from Denmark', *American Economic Journal: Applied Economics* 11, no. 4 (2019): 181–209

Kleven, Henrik, Camille Landais, Johanna Posch et al., 'Child Penalties across Countries: Evidence and Explanations', *AEA Papers and Proceedings* 109 (2019): 122–6

Lane, Melissa, Henry Lee, and Henry Desmond Pritchard, *The Republic*, Penguin, 2007

McLean, Bethany, and Peter Elkind, *The Smartest Guys in the Room: The Amazing Rise and Scandalous Fall of Enron*, Penguin, 2013

Mill, John Stuart, *On Liberty*, J. W. Parker & Son, 1859

Moser, Petra, 'Patents and Innovation: Evidence from Economic History', *Journal of Economic Perspectives* 27, no. 1 (2013): 23–44

Muthukrishna, Michael, and Joseph Henrich, 'Innovation in the Collective Brain', *Philosophical Transactions of the Royal Society B: Biological Sciences* 371, no. 1690 (2016): 20150192

Nix, Emily, and Martin Eckhoff Andresen, *What Causes the Child Penalty? Evidence from Same Sex Couples and Policy Reforms*, discussion paper no. 902, Statistics Norway, Research Department, 2019

Paluck, Elizabeth Levy, and Donald P. Green, 'Prejudice Reduction: What Works? A Review and Assessment of Research and Practice', *Annual Review of Psychology* 60 (2009): 339–67

Paluck, Elizabeth Levy, Roni Porat, Chelsey S. Clark et al., 'Prejudice Reduction: Progress and Challenges', *Annual Review of Psychology* 72 (2021): 533–60

Payne, Joshua L., and Andreas Wagner, 'The Causes of Evolvability and Their Evolution', *Nature Reviews Genetics* 20, no. 1 (January 2019): 24–38

Pigliucci, Massimo, 'Is Evolvability Evolvable?' *Nature Reviews Genetics* 9, no. 1 (2008): 75–82

Pronin, Emily, Daniel Y. Lin, and Lee Ross, 'The Bias Blind Spot: Perceptions of Bias in Self versus Others', *Personality and Social Psychology Bulletin* 28, no. 3 (2002): 369–81

Qian, Yi, 'Do National Patent Laws Stimulate Domestic Innovation in a Global Patenting Environment? A Cross-country Analysis of Pharmaceutical Patent Protection, 1978–2002', *Review of Economics and Statistics* 89, no. 3 (2007): 436–53

Scopelliti, Irene, Carey K. Morewedge, Erin McCormick et al., 'Bias Blind
Spot: Structure, Measurement, and Consequences', *Management
Science* 61, no. 10 (2015): 2468–86
Sobel, Robert, *Car Wars: The Untold Story*, Dutton, 1984
Weatherford, Jack, *Genghis Khan and the Making of the Modern World*,
Crown, 2005
Williams, Heidi L., 'Intellectual Property Rights and Innovation: Evidence
from the Human Genome', *Journal of Political Economy* 121, no. 1 (2013):
1–27

Chapter 11: Improving the Internet

A history of human rights is Lynn Hunt, *Inventing Human Rights: A History*,
W. W. Norton, 2007.
On training to spot misinformation see Carl T. Bergstrom and Jevin D.
West, *Calling Bullshit: The Art of Skepticism in a Data-driven World*, Random
House, 2021; Stuart Ritchie, *Science Fictions: Exposing Fraud, Bias, Negligence
and Hype in Science*, Random House, 2020; and Nassim Nicholas Taleb, *Incerto
5-Book Bundle: Fooled by Randomness, The Black Swan, The Bed of Procrustes,
Antifragile, Skin in the Game*, Random House, 2021.
On the use of game theory to perform incentive audits see E. Yoeli and
M. Hoffman, *Hidden Games: The Surprising Power of Game Theory to Explain
Irrational Human Behavior*, Basic Books, 2022.

Specific and specialist reading

Allcott, Hunt, Luca Braghieri, Sarah Eichmeyer et al., 'The Welfare
Effects of Social Media', *American Economic Review* 110, no. 3 (2020):
629–76
Banerjee, Abhijit, Arun G. Chandrasekhar, Esther Duflo et al., 'The
Diffusion of Microfinance', *Science* 341, no. 6144 (2013): 1236498
——, 'Using Gossips to Spread Information: Theory and Evidence from
Two Randomized Controlled Trials', *Review of Economic Studies* 86, no. 6
(2019): 2453–90
Brashier, Nadia M., and Daniel L. Schacter, 'Aging in an Era of Fake
News', *Current Directions in Psychological Science* 29, no. 3 (2020): 316–23
Brynjolfsson, Erik, Xiang Hui, and Meng Liu, 'Does Machine Translation
Affect International Trade? Evidence from a Large Digital Platform',
Management Science 65, no. 12 (2019): 5449–60

Carron-Arthur, Bradley, John A. Cunningham, and Kathleen M. Griffiths, 'Describing the Distribution of Engagement in an Internet Support Group by Post Frequency: A Comparison of the 90-9-1 Principle and Zipf's Law', *Internet Interventions* 1, no. 4 (2014): 165–8

Cheng, Joey T., Jessica L. Tracy, Tom Foulsham et al., 'Two Ways to the Top: Evidence That Dominance and Prestige Are Distinct Yet Viable Avenues to Social Rank and Influence', *Journal of Personality and Social Psychology* 104, no. 1 (2013): 103

Chudek, Maciej, Michael Muthukrishna, and Joe Henrich, 'Cultural Evolution', in *The Handbook of Evolutionary Psychology*, pp. 1–21, Wiley, 2015

Gergely, György, Harold Bekkering, and Ildikó Király, 'Rational Imitation in Preverbal Infants', *Nature* 415, no. 6873 (2002): 755

Henrich, Joseph, *The Secret of Our Success*, Princeton University Press, 2015

Henrich, Joseph, Maciej Chudek, and Robert Boyd, 'The Big Man Mechanism: How Prestige Fosters Cooperation and Creates Prosocial Leaders', *Philosophical Transactions of the Royal Society B: Biological Sciences* 370, no. 1683 (2015): 2015001

Henrich, Joseph, and Francisco J. Gil-White, 'The Evolution of Prestige: Freely Conferred Deference as a Mechanism for Enhancing the Benefits of Cultural Transmission', *Evolution and Human Behavior* 22, no. 3 (2001): 165–96

Hilmert, Clayton J., James A. Kulik, and Nicholas J. S. Christenfeld, 'Positive and Negative Opinion Modeling: The Influence of Another's Similarity and Dissimilarity', *Journal of Personality and Social Psychology* 90, no. 3 (2006): 440

Hoffman, Moshe, and Erez Yoeli, *Hidden Games: The Surprising Power of Game Theory to Explain Irrational Human Behaviour*, Basic Books UK, 2022

Johnson, Jamie (dir.), *Born Rich*, Shout! Factory, 2003

Norenzayan, Ara, Azim F. Shariff, Will M. Gervais et al., 'The Cultural Evolution of Prosocial Religions', *Behavioral and Brain Sciences* 39 (2016): e1

Van Mierlo, Trevor, 'The 1% Rule in Four Digital Health Social Networks: An Observational Study', *Journal of Medical Internet Research* 16, no. 2 (2014): e2966

Chapter 12: Becoming Brighter

For the history of Estonia's Tiger Leap program see *Tiger Leap: 1997–2007*, Tiger Leap Foundation, 2007. On some of the educational differences and outcomes around the world see Amanda Ripley, *The Smartest Kids in the World: And How They Got That Way*, Simon & Schuster, 2013. Conrad Wolfram's manifesto on math education is *The Math(s) Fix: An Education Blueprint for the AI Age*, Wolfram Media, Incorporated, 2020.

On inequalities, prejudice, and bias see Linda Scott, *The Double X Economy: The Epic Potential of Empowering Women*, Faber & Faber, 2020; Caroline Criado Perez, *Invisible Women: Data Bias in a World Designed for Men*, Abrams, 2019; and Jennifer L. Eberhardt, *Biased: Uncovering the Hidden Prejudice That Shapes What We See, Think, and Do*, Penguin, 2020.

Specific and specialist reading

Algan, Yann, Christian Dustmann, Albrecht Glitz et al., 'The Economic Situation of First and Second-Generation Immigrants in France, Germany and the United Kingdom', *Economic Journal* (2010): F4–30

Arcidiacono, Peter, and Michael Lovenheim, 'Affirmative Action and the Quality–Fit Trade-off', *Journal of Economic Literature* 54, no. 1 (2016): 3–51

Arcidiacono, Peter, Esteban M. Aucejo, and V. Joseph Hotz, 'University Differences in the Graduation of Minorities in STEM Fields: Evidence from California', *American Economic Review* 106, no. 3 (2016): 525–62

Arcidiacono, Peter, Esteban M. Aucejo, and Ken Spenner, 'What Happens after Enrollment? An Analysis of the Time Path of Racial Differences in GPA and Major Choice', *Journal of Labor Economics* 1 (2012): 1–24

Beraja, Martin, Andrew Kao, David Y. Yang et al., 'Exporting the Surveillance State via Trade in AI', *Brookings Institute Report*, 2023

Bergman, Peter, Raj Chetty, Stefanie DeLuca et al., *Creating Moves to Opportunity: Experimental Evidence on Barriers to Neighborhood Choice*, working paper no. 26164, National Bureau of Economic Research, 2019

Bertrand, Marianne, and Sendhil Mullainathan, 'Are Emily and Greg More Employable Than Lakisha and Jamal? A Field Experiment on labor Market Discrimination', *American Economic Review* 94, no. 4 (2004): 991–1013

Chetty, Raj, 'Improving Equality of Opportunity: New Insights from Big Data', *Contemporary Economic Policy* 39, no. 1 (2021): 7–41

Chetty, Raj, John N. Friedman, Nathaniel Hendren et al., *The Opportunity Atlas: Mapping the Childhood Roots of Social Mobility*, working paper no. 25147, National Bureau of Economic Research, 2018

Chetty, Raj, David Grusky, Maximilian Hell et al., 'The Fading American Dream: Trends in Absolute Income Mobility since 1940', *Science* 356, no. 6336 (2017): 398–406

Chetty, Raj, Matthew O. Jackson, Theresa Kuchler et al., 'Social Capital I: Measurement and Associations with Economic Mobility', *Nature* 608, no. 7921 (2022): 108–21

——, 'Social Capital II: Determinants of Economic Connectedness', *Nature* 608, no. 7921 (2022): 122–34

Christian, Brian, and Tom Griffiths, *Algorithms to Live By: The Computer Science of Human Decisions*, Macmillan, 2016

Durlauf, Steven N., Andros Kourtellos, and Chih Ming Tan, 'The Great Gatsby Curve', *Annual Review of Economics* 14 (2022): 571–605

Dustmann, Christian, and Tommaso Frattini, 'The Fiscal Effects of Immigration to the UK', *Economic Journal* 124, no. 580 (2014): F593–643

Flynn, James R., *What Is Intelligence? Beyond the Flynn Effect*, Cambridge University Press, 2007

Green, John, 'Samurai, Daimyo, Matthew Perry, and Nationalism: Crash Course World History #34', CrashCourse, YouTube, 14 September 2012, https://www.youtube.com/watch?v=Nosq94oCl_M

Hinton, Geoffrey E., and Ruslan R. Salakhutdinov, 'Reducing the Dimensionality of Data with Neural Networks', *Science* 313, no. 5786 (2006): 504–7

Hulance, Jo, Mark Kowalksi, and Robert Fairhurst, *Long-term Strategies to Reduce Lead Exposure from Drinking Water*, report no. 14372.2, UK Government Drinking Water Inspectorate, 2021

Jobs Are Changing, So Should Education, Royal Society, 2019

Johnson, George, 'To Test a Powerful Computer, Play an Ancient Game', *New York Times*, 29 July 1997

Kline, Michelle Ann, 'How to Learn about Teaching: An Evolutionary Framework for the Study of Teaching Behavior in Humans and Other Animals', *Behavioral and Brain Sciences* 38 (2015): e31

LeCun, Yann, Yoshua Bengio, and Geoffrey Hinton, 'Deep Learning', *Nature* 521, no. 7553 (2015): 436–44

Lu, Jackson G., Richard E. Nisbett, and Michael W. Morris, 'Why East Asians but Not South Asians Are Underrepresented in Leadership Positions in the United States', *Proceedings of the National Academy of Sciences* 117, no. 9 (2020): 4590–600

McFarland, Michael J., Matt E. Hauer, and Aaron Reuben, 'Half of US Population Exposed to Adverse Lead Levels in Early Childhood', *Proceedings of the National Academy of Sciences* 119, no. 11 (2022): e2118631119

Metz, Cade, *Genius Makers: The Mavericks Who Brought AI to Google, Facebook, and the World*, Penguin, 2022

Muthukrishna, Michael, and Joseph Henrich, 'Innovation in the Collective Brain', *Philosophical Transactions of the Royal Society B: Biological Sciences* 371, no. 1690 (2016): 20150192

Muthukrishna, Michael, Michael Doebeli, Maciej Chudek et al., 'The Cultural Brain Hypothesis: How Culture Drives Brain Expansion, Sociality, and Life History', *PLoS Computational Biology* 14, no. 11 (2018): e1006504

Nisbett, Richard E., *Mindware: Tools for Smart Thinking*, Farrar, Straus & Giroux, 2015

Oreopoulos, Philip, 'Why Do Skilled Immigrants Struggle in the Labor Market? A Field Experiment with Thirteen Thousand Resumes', *American Economic Journal: Economic Policy* 3, no. 4 (2011): 148–71

Pietschnig, J., and M. Voracek, 'One Century of Global IQ Gains: A Formal Meta-analysis of the Flynn Effect (1909–2013)', *Perspectives on Psychological Science* 10, no. 3 (2015): 282–306

Reeves, Richard V., and Dimitrios Halikias, *Race Gaps in SAT Scores Highlight Inequality and Hinder Upward Mobility*, Brookings Institute, 2017

Trahan, Lisa, Karla K. Stuebing, Merril K Hiscock et al., 'The Flynn Effect: A Meta-analysis', *Psychological Bulletin* 140, no. 5 (2014): 1332–60

Vision for Science and Mathematics Education, Royal Society, 2014

'Why Brahmins Lead Western Firms but Rarely Indian Ones', *The Economist*, 1 January 2022

Conclusion

For speculation on some of the challenges that face us, see Vaclav Smil, *Global Catastrophes and Trends: The Next 50 Years*, MIT Press, 2008; and Toby Ord, *The Precipice: Existential Risk and the Future of Humanity*, Hachette, 2020.

Specific and specialist reading

Carey, Brycchan, *From Peace to Freedom: Quaker Rhetoric and the Birth of American Antislavery, 1657–1761*, Yale University Press, 2012

Gore, Al, *An Inconvenient Truth: The Planetary Emergency of Global Warming and What We Can Do About It*, Rodale, 2006

Hall, Charles A. S., and Kent Klitgaard, *Energy and the Wealth of Nations: An Introduction to Biophysical Economics*, Springer International, 2018

Hall, Charles A. S., Jessica G. Lambert, and Stephen B. Balogh, 'EROI of Different Fuels and the Implications for Society', *Energy Policy* 64 (2014): 141–52

Hochschild, Adam, *Bury the Chains: Prophets and Rebels in the Fight to Free an Empire's Slaves*, Houghton Mifflin Harcourt, 2006

Pankhurst, E. Sylvia, *The Suffragette – The History of The Women's Militant Suffrage Movement – 1905–1910*, Read Books, 2009

Renwick, Chris, *Bread for All: The Origins of the Welfare State*, Penguin, 2017

Williams, Juan, *Eyes on the Prize: Civil Rights Reader*, Penguin, 1991

Index

Aaviksoo, Jaak, 347
ability, rewarding, 329–30
abiogenesis, 30–1
accent, discrimination, 94
accumulated wisdom, 113–19
ad hominem, 82–3
Adams, John Quincy, 341
adaptability, 332–3
adaptive evolutionary systems, 86–8
adjacent possibilities (A), COMPASS,
    155–6, 315
adolescence, 169–70
affirmative action, 360–1
Afghanistan, lessons from, 191–5
Africa
    climate change and, 215
    diversity in, 253–6
    genetic diversity in, 132–3
    ghost population, 137
    low resources in, 253–6
    Scramble for Africa, 253
Africa, studying intelligence in, 109–13
Agricultural Revolution, 164
agriculture, 48–50
Alexander the Great, 346
algorithms, 87, 89, 117, 258–9, 337,
    343, 347, 349, 368
Alphabet, 287, 363
AlphaGo, 369
AlShebli, Bedoor, 321
Amazon, 287, 298
Amazon Web Services (AWS), 58, 287
American Dream, 358
American Psychological Association,
    102, 104
Andalusi, Said al-, 66
Angola, 109–13

Ankh-Morpork *see* melting pot model
Antony and Cleopatra, 152
apes, studying, 71–5
apple pie, making, 142
Araud, Gérard, 241
arbitrage, 151
aristocracy *see* entrepreneurs
ARPANET, 38
arriéré, 99
Arthur Andersen, 330
artificial intelligence (AI), 366–71
Asian Tigers, 315
'Association between Early Career
    Informal Mentorship in Academic
    Collaborations and Junior Author
    Performance', 321
ATP, 31
Australia, 149, 235
    cane toads in, 22–3
    democratic polices in, 264
    heritability in, 121–2
    lithium reserves in, 216
    Sandline Affair, 3–4
    solar panels in, 47
    sustainably managed migration in,
        249–52
    wealth in, 284
Australian Cultural Orientation
    (AUSCO) Program, 249–50
Austria, 174
Azerbaijan, 236

Baby Boomers, 311
Back to the Future (film), 318
bags, 35–6
Bahrami-Rad, Duman, 255
Baldwin, James, 176–8